Principles, Materials, and Applications in Facade Engineering

Arkar Htet
Mullion Facade Engineering School, Myanmar

Theingi Aung
Mullion Facade Engineering School, Myanmar

IGI Global
Scientific Publishing
Publishing Tomorrow's Research Today

Published in the United States of America by
IGI Global Scientific Publishing
701 East Chocolate Avenue
Hershey, PA, 17033, USA
Tel: 717-533-8845 | Fax: 717-533-7115
Website: https://www.igi-global.com E-mail: cust@igi-global.com

Library of Congress Cataloging-in-Publication Data

Names: Htet, Arkar, 1986- editor | Aung, Theingi, 1985- editor
Title: Principles, materials, and applications in facade engineering / edited by Arkar Htet, Theingi Aung.
Description: Hershey, PA : IGI Global Scientific Publishing, [2026] | Includes bibliographical references and index. | Summary: "This book provides a comprehensive reference on the core principles, modern materials, emerging technologies, and design strategies in façade engineering"-- Provided by publisher.
Identifiers: LCCN 2026002172 (print) | LCCN 2026002173 (ebook) | ISBN 9798337360232 hardcover | ISBN 9798337360249 paperback | ISBN 9798337360256 ebook
Subjects: LCSH: Facades--Design and construction | Facades--Materials
Classification: LCC TH2235 .P75 2026 (print) | LCC TH2235 (ebook)
LC record available at https://lccn.loc.gov/2026002172
LC ebook record available at https://lccn.loc.gov/2026002173

British Cataloguing in Publication Data
A Cataloguing in Publication record for this book is available from the British Library.

Table of Contents

Detailed Table of Contents

Smart materials have revolutionized the realm of architectural design, particularly in the construction and optimization of building façades. These materials—capable of altering their properties in response to external stimuli such as temperature, light, moisture, or electric fields—offer dynamic solutions to long-standing challenges in energy efficiency, sustainability, aesthetics, and occupant comfort. In the context of façade design, smart materials are redefining the way buildings interact with their environment, marking a paradigm shift from static, passive exteriors to adaptive, responsive skins that enhance performance while embodying cutting-edge design principles. At the core of the importance of smart materials in façade design is their potential to significantly improve energy efficiency and thermal performance. Traditional façades are often limited by fixed thermal resistance and light transmission values, which can lead to excessive reliance on mechanical heating, ventilation, and air conditioning (HVAC) systems.

This chapter reviews high-performance cladding materials and facade systems for greener, climate-responsive urban buildings. It covers bio-based composites, smart and responsive materials, building-integrated photovoltaics, and vegetated facades, evaluating thermal performance, durability, lifecycle impact, carbon reduction, and urban environmental quality. It also examines the growing role of digital tools, including AI-assisted analysis, BIM, and predictive monitoring, in material selection,

geometries. Further, connection strategies are evaluated for resisting wind loads and supporting cladding systems. Beam modeling and FEA help assess stresses and optimize cross-sections for stiffness and strength, strengthening the role of built-up beams in façade engineering for durable and impactful structural integration.

This chapter develops a durability-informed lifecycle assessment perspective for aluminium composite panel (ACP) façades in tropical and temperate climates. It argues that ACP sustainability cannot be judged by initial embodied carbon alone, because climatic exposure influences coating degradation, corrosion risk, sealant ageing, moisture ingress, maintenance frequency, service life, and end-of-life recovery. Tropical climates tend to intensify humidity-, rainfall-, UV-, and corrosion-related deterioration, while temperate climates present challenges associated with thermal movement, freeze-thaw exposure, sealant fatigue, and seasonal moisture management. The chapter integrates lifecycle assessment, façade durability science, and green building certification implications to show that ACP façade sustainability is a dynamic whole-life performance outcome. It concludes that more reliable assessment requires climate-sensitive durability assumptions, preventive maintenance planning, realistic recycling pathways, and transparent environmental documentation.

This chapter presents a comprehensive treatment of energy performance modelling and optimization for nanofluid-based cooling systems used in renewable energy applications. It covers the physical principles of nanofluids, effective-property models, convective heat transfer correlations, numerical and reduced-order modelling approaches, experimental considerations, and optimization strategies (design of experiments, response surface methodology, evolutionary algorithms, and multi-

operation. The framework extends conventional FMEA through façade-specific severity, occurrence, and detection rubrics for unitized curtain walls and ventilated rainscreen systems; a façade-weighted priority index used alongside the traditional Risk Priority Number; dependency-based treatment of coupled failures; and full-matrix application across three façade scenarios with sensitivity testing under alternative weighting schemes. The chapter also positions F-REF against Fault Tree Analysis, Event Tree Analysis, Bow-Tie analysis, and qualitative checklist methods. Although further expert calibration is needed, the framework provides a more rigorous and actionable basis for façade risk screening, inspection planning, and lifecycle decision support.

Preface

INTRODUCTION

Façade engineering is rapidly evolving as an interdisciplinary field that bridges architecture, structural engineering, environmental science, and advanced construction technologies. With the increasing demand for energy-efficient, sustainable, and aesthetically compelling building envelopes, the façade has become far more than an exterior skin; it now operates as a dynamic system capable of responding to climatic conditions, energy performance requirements, and complex functional demands. This volume emerges from a recognition of that transformation and seeks to capture the expanding scope and significance of façade design in contemporary practice.

Principles, Materials, and Applications in Façade Engineering brings together contributions that examine the theoretical foundations, material innovations, design methodologies, performance simulations, and real-world applications of façade systems across diverse building typologies and climatic contexts. As editors, our objective has been to assemble a body of work that not only reflects the current state of knowledge but also anticipates future directions in this rapidly advancing field. The chapters presented herein highlight how façades are increasingly conceived as intelligent, adaptive, and high-performance systems integral to the overall behavior of buildings.

This book aims to serve as both a comprehensive reference and a forward-looking compendium on façade engineering. It addresses core principles alongside emerging technologies and evolving design strategies. The contributions collectively:

- Present theoretical frameworks for understanding façade performance and structural behavior.
- Examine advances in material science, including composites, smart glass, and adaptive components.

- Analyze the application of simulation tools such as CFD, FEA, and BIM in evaluating environmental and structural performance.
- Highlight regional and contextual design approaches while maintaining global relevance.
- Offer industry-based case studies that demonstrate successful façade integration in practice.

By bringing together multidisciplinary perspectives, this volume contributes to ongoing discourse in academic research, professional practice, and education. It underscores the importance of collaboration across disciplines to address the increasingly complex challenges associated with building envelopes in a changing environmental and technological landscape.

The book is intended for a broad audience, including researchers, academics, and postgraduate students in architecture, civil and structural engineering, environmental design, and building science. It also serves professional engineers, architects, and façade consultants seeking to expand their knowledge and engage with current innovations. In addition, urban planners, sustainability experts, and construction managers will find value in its insights on building performance and integration. Educators and institutions may use this work to support curricula in façade technologies and sustainable construction, while policymakers and technical authorities can draw upon its discussions in shaping green building regulations and urban climate strategies.

As editors, we hope this volume will not only inform but also inspire continued exploration and innovation in façade engineering, encouraging readers to rethink the building envelope as a critical interface between built form, environment, and human experience.

ORGANIZATION OF THE BOOK

The chapters assembled in this volume collectively reflect the breadth and depth of contemporary façade engineering, moving from material innovation and structural analysis to digital intelligence, lifecycle assessment, and strategic leadership. Each contribution advances a distinct yet interconnected dimension of façade performance, underscoring the interdisciplinary nature of the field.

Chapter 1 introduces the transformative role of smart materials and nanomaterials in façade applications, emphasizing their capacity to respond dynamically to environmental stimuli. The discussion highlights how such materials shift façades from static enclosures to adaptive systems that enhance energy efficiency, thermal regulation, and occupant comfort. By addressing the limitations of conventional building envelopes, this chapter establishes a foundation for understanding responsive façade technologies as a cornerstone of future design.

Chapter 2 expands this material focus by examining innovative cladding systems for high-performance, climate-responsive urban façades. It explores bio-based composites, responsive materials, photovoltaic integration, and vegetated systems, while also considering lifecycle impacts and urban environmental benefits. The integration of digital tools—including AI and predictive modeling—demonstrates how material selection and façade performance can be optimized, even as practical challenges such as cost, regulation, and maintenance remain critical considerations.

Chapter 3 investigates the emerging application of technical textiles in façade systems, presenting them as a convergence of structural efficiency and architectural expression. Through analysis of materials such as ETFE and PTFE, the chapter evaluates their mechanical strength, environmental adaptability, and design flexibility in comparison to traditional cladding. It also addresses the practical constraints of durability and scalability, offering a balanced perspective on their potential within innovative façade design.

Chapter 4 shifts attention to the structural dimension of façade engineering through an in-depth study of built-up beam systems. By analyzing parameters such as shear force, bending moment, and stress distribution, the chapter demonstrates how optimized beam configurations contribute to load transfer, material efficiency, and adaptability in complex façade geometries. The integration of finite element analysis further reinforces the importance of precision engineering in achieving safe and high-performing façade structures.

Chapter 5 presents a nuanced lifecycle assessment of aluminium composite panel façades, emphasizing the influence of climatic conditions on durability and sustainability. By comparing tropical and temperate environments, the chapter reveals how factors such as moisture, UV exposure, and thermal variation affect degradation, maintenance, and service life. It argues for a climate-sensitive, durability-informed approach to sustainability, moving beyond simplistic evaluations based solely on embodied carbon.

Chapter 6 broadens the discussion of performance optimization through an exploration of nanofluid cooling systems in renewable energy applications. While not limited to façades, the chapter provides valuable insights into thermal management strategies that can inform building envelope design. Through detailed modeling techniques and optimization methods, it demonstrates how advanced cooling systems can enhance the efficiency of energy-integrated façade technologies.

Chapter 7 examines the growing role of artificial intelligence and machine learning in façade performance prediction. It highlights how data-driven approaches enable more accurate forecasting, adaptive control, and energy optimization in increasingly complex building systems. By moving beyond traditional rule-based models, this chapter positions AI and ML as essential tools for managing the dynamic behavior of modern façades.

Chapter 8 introduces a managerial and strategic perspective by exploring the role of entrepreneurial leadership in advancing façade performance and sustainability. It emphasizes how innovation in materials, technologies, and business models is often driven by leadership that can navigate technical uncertainty and market demands. The chapter underscores the importance of interdisciplinary collaboration and strategic vision in shaping the future of façade engineering.

Chapter 9 addresses the critical issue of façade failures, offering a systematic examination of common deterioration mechanisms such as thermal bridging, sealant degradation, structural fatigue, and water ingress. It presents diagnostic methodologies and targeted retrofitting strategies aimed at extending service life while minimizing waste and disruption. The chapter advocates for evidence-based, lifecycle-oriented approaches to façade maintenance and renewal.

Finally, Chapter 10 proposes a structured framework for façade risk evaluation based on Failure Mode and Effects Analysis. By adapting and extending established risk assessment methodologies to façade-specific conditions, the chapter provides a practical tool for identifying, prioritizing, and managing risks across the building lifecycle. Its emphasis on systematic analysis and decision support highlights the growing importance of risk-informed design and management in façade engineering.

Together, these chapters offer a comprehensive exploration of the principles, materials, technologies, and strategies shaping contemporary façade engineering. They reflect a field that is not only technically sophisticated but also increasingly responsive to environmental, economic, and societal imperatives.

CONCLUSION

As editors, we recognize façade engineering as a discipline in continual transformation—one that sits at the intersection of technological innovation, environmental responsibility, and architectural expression. The contributions in this volume collectively demonstrate that façades are no longer passive building envelopes, but active, intelligent systems that mediate complex interactions between structure, climate, energy, and human experience.

Across the chapters, a consistent theme emerges: the future of façade engineering will be defined by integration. Integration of advanced materials with adaptive performance, of structural logic with environmental responsiveness, and of digital intelligence with design and construction practice. From smart and nanostructured materials to lifecycle-based sustainability assessment, from AI-driven performance prediction to risk-informed design frameworks, this book highlights the expanding toolkit available to researchers and practitioners. Equally important, it underscores the necessity of aligning these innovations with real-world constraints such as durability, cost, regulation, and long-term maintainability.

We also observe a clear shift toward systems thinking in façade design. Performance can no longer be understood in isolation; it must be evaluated across the entire lifecycle of the building envelope and within the broader context of urban resilience and environmental impact. This perspective calls for closer collaboration between architects, engineers, material scientists, data analysts, and industry leaders, reinforcing the inherently interdisciplinary nature of the field.

While significant progress has been made, the chapters also reveal areas where further inquiry is essential. These include improving the scalability of emerging materials, refining predictive models for complex environmental conditions, strengthening retrofit strategies for aging building stock, and developing more robust frameworks for risk and lifecycle assessment. Addressing these challenges will require sustained research, innovation, and knowledge exchange between academia and industry.

Ultimately, this volume is intended not as a final statement, but as a reference point within an evolving conversation. It is our hope that it will support educators in shaping curricula, guide practitioners in advancing design and construction practices, and inspire researchers to explore new frontiers in façade engineering.

We conclude by affirming that the façade, as both a technical system and an architectural interface, will remain central to the pursuit of sustainable, high-performance, and responsive built environments. The continued advancement of this field will depend on our collective ability to integrate knowledge, challenge conventions, and embrace innovation with both rigor and responsibility.

Chapter 1
Smart Materials and Nanomaterials in Façade Applications

P. Selvakumar
http://orcid.org/0000-0002-3650-4548
Department of Science and Humanities, Nehru Institute of Technology, Coimbatore, India

T. C. Manjunath
http://orcid.org/0000-0003-2545-9160
Rajarajeswari College of Engineering, Bengaluru, India

Santosh Kumar Nathsharma
http://orcid.org/0000-0003-3935-2938
Stewart Science College, India

Imran zahoor Khan
http://orcid.org/0009-0005-9847-3125
Datta Meghe Institute of Management Studies, India

Mukesh Gulani
Datta Meghe Institute of Management Studies, India

ABSTRACT

Smart materials have revolutionized the realm of architectural design, particularly in the construction and optimization of building façades. These materials—capable of altering their properties in response to external stimuli such as temperature, light, moisture, or electric fields—offer dynamic solutions to long-standing challenges in energy efficiency, sustainability, aesthetics, and occupant comfort. In the context of

DOI: 10.4018/979-8-3373-6023-2.ch001

façade design, smart materials are redefining the way buildings interact with their environment, marking a paradigm shift from static, passive exteriors to adaptive, responsive skins that enhance performance while embodying cutting-edge design principles. At the core of the importance of smart materials in façade design is their potential to significantly improve energy efficiency and thermal performance. Traditional façades are often limited by fixed thermal resistance and light transmission values, which can lead to excessive reliance on mechanical heating, ventilation, and air conditioning (HVAC) systems.

1. INTRODUCTION

Contemporary façade engineering increasingly treats the building envelope as an environmental interface rather than a fixed external shell. Responsive materials, adaptive façades, and sensor-aware envelope systems are therefore receiving growing attention because they allow buildings to negotiate solar gain, daylight, temperature variation, and user comfort more dynamically than conventional passive skins. (Jamilu et al., 2024; Gonçalves et al., 2024; Lee et al., 2021).

This chapter reviews key smart and nano-enabled material strategies for façades, including adaptive glazing, latent-heat storage, motion-capable components, multifunctional coatings, and durability-enhancing nano-additives. The introduction frames the topic briefly, while the later sections retain the original material-specific discussion so that the chapter remains full-length without repeating the same argument twice. (Grillo et al., 2022; Kuda & Yadav, 2022; Brzezicki, 2021; Faragalla & Asadi, 2022).

2. THE ROLE OF SMART MATERIALS IN FAÇADE DESIGN: IMPORTANCE AND IMPACT

Smart materials have revolutionized the realm of architectural design, particularly in the construction and optimization of building façades. These materials-capable of altering their properties in response to external stimuli such as temperature, light, moisture, or electric fields-offer dynamic solutions to long-standing challenges in energy efficiency, sustainability, aesthetics, and occupant comfort. In the context of façade design, smart materials are redefining the way buildings interact with their environment, marking a paradigm shift from static, passive exteriors to adaptive, responsive skins that enhance performance while embodying cutting-edge design principles. At the core of the importance of smart materials in façade design is their potential to significantly improve energy efficiency and thermal performance. Traditional façades are often limited by fixed thermal resistance and light transmission values, which can lead to excessive reliance on mechanical heating, ventilation, and air conditioning (HVAC) systems. Smart materials such as thermochromic and photochromic glazing adapt their transparency or reflectivity based on ambient temperature or light intensity, thereby optimizing the amount of solar heat gain and natural light entering the building. For instance, thermochromic materials darken when exposed to high temperatures, reducing heat ingress during hot days, while remaining transparent during cooler periods. This dynamic control of solar radiation reduces the building's energy consumption, cuts utility costs, and supports net-zero energy goals. In addition to regulating temperature and light, smart materials also contribute to enhanced occupant comfort and well-being. Materials such as electrochromic glass allow occupants to manually or automatically adjust the transparency of windows, providing a personalized visual and thermal environment. This adaptability not only improves indoor comfort by mitigating glare and controlling temperature fluctuations but also positively impacts occupant health and productivity, especially in workplaces and educational settings. Moreover, the use of phase-change materials (PCMs) in façades can buffer internal spaces against extreme outdoor temperature variations by absorbing and releasing latent heat, further stabilizing indoor conditions. Environmental sustainability is another key domain where smart materials offer significant advantages. By reducing energy use and enhancing resource efficiency, these materials contribute to lowering the carbon footprint of buildings throughout their lifecycle. Some smart materials, such as photovoltaic-integrated glass, serve a dual function-generating renewable energy while also acting as a shading or cladding element. Such multifunctional components support the principles of green building design and align with global environmental targets such as the United Nations Sustainable Development Goals (SDGs), particularly SDG 7 (Affordable and Clean Energy) and SDG 11 (Sustainable Cities and Communities) (Jamilu et al., 2024; Gonçalves et al., 2024; Lee et al., 2021).

Smart façades also play a crucial role in resilience and climate adaptability. With climate change leading to increased frequency and intensity of extreme weather events, buildings must be designed to withstand fluctuating environmental conditions. Smart materials can help façades respond to rapid changes in temperature, humidity, and solar radiation, thereby protecting the building envelope and ensuring long-term durability. For instance, hydrophobic and breathable materials can repel water during heavy rains while allowing vapor diffusion, preventing mold growth and structural damage. Similarly, materials with adaptive insulation properties can maintain building performance across seasonal extremes, improving resilience and lowering operational costs. Beyond individual performance metrics, smart materials facilitate data integration and intelligent building systems. When embedded with sensors and connected to a building management system (BMS), smart façades can monitor environmental conditions in real time and respond accordingly. For example, a façade integrated with smart glass, temperature sensors, and automated actuators can dynamically adjust window tinting or open shading devices based on weather forecasts or user preferences. This integration of materials with digital technologies exemplifies the convergence of architecture and the Internet of Things (IoT), enabling smarter, more responsive urban environments. In conclusion, smart materials are playing an increasingly vital role in transforming façade design from a passive shell into an active, intelligent interface between a building and its environment. By enhancing energy efficiency, occupant comfort, aesthetic value, environmental sustainability, and resilience, smart façades are helping shape the future of architecture and urban development. While challenges related to cost, complexity, and awareness remain, the trajectory of innovation, combined with growing environmental and social imperatives, points toward a future where smart materials are central to sustainable, high-performance building design. Embracing these technologies is not merely a matter of architectural advancement but a critical step toward achieving adaptive, human-centric, and ecologically responsible built environments (Matin & Eydgahi, 2019; Brzezicki, 2021; Faragalla & Asadi, 2022).

3. NANOMATERIALS: ENHANCING FAÇADE PERFORMANCE

The integration of nanomaterials into façade design represents a transformative advancement in architecture, offering unprecedented improvements in performance, sustainability, durability, and aesthetics. Nanomaterials-engineered at the scale of 1 to 100 nanometers-possess unique physical, chemical, optical, and mechanical properties that differ significantly from their bulk counterparts. Their incorporation into façade systems has enabled architects and engineers to create building envelopes that are not only more efficient and environmentally responsive but also more durable, multifunctional, and visually compelling . As the demands for high-performance, sustainable buildings continue to rise in response to climate change, urbanization, and energy concerns, the strategic use of nanomaterials in façades emerges as a critical pathway toward achieving future-ready architectural solutions. Nanomaterials also contribute to light management and optical control, crucial for optimizing daylighting and visual comfort in buildings. Nanostructured coatings, such as nano-titania or nano-silica, can be engineered to reduce glare, scatter light, and enhance indoor illumination levels. These coatings improve the transmission of diffuse light while controlling direct solar penetration, thereby reducing the need for artificial lighting and enhancing the overall indoor environmental quality . The manipulation of light at the nanoscale opens doors to advanced façade technologies like smart windows and dynamic glazing, where nanomaterials respond to external stimuli (e.g., light, heat, or voltage) to adjust transparency and tint, supporting energy-saving and user-centric design goals. Another crucial advantage of nanomaterials is their role in self-cleaning and low-maintenance façades. One of the most widely used nanomaterials in this context is titanium dioxide (TiO_2), which exhibits photocatalytic properties. When exposed to sunlight, TiO_2 breaks down organic dirt and pollutants on the façade surface, while its hydrophilic nature allows rainwater to wash away the residues, maintaining a clean appearance without the use of harsh chemicals or frequent manual cleaning. This self-cleaning mechanism not only lowers maintenance costs but also supports environmental sustainability by reducing water consumption and the use of detergents . In polluted urban environments, such self-cleaning nanocoatings can also contribute to air purification, as TiO_2 can decompose airborne pollutants such as nitrogen oxides (NOx), improving local air quality. Durability and protection are further enhanced by nanomaterials, which provide superior resistance to environmental degradation. Nanocoatings can impart hydrophobic, oleophobic, anti-corrosive, and UV-resistant properties to façade surfaces, thereby extending the lifespan of materials like concrete, metal, and glass. For example, silane- or fluoropolymer-based nanocoatings can prevent water ingress, chloride penetration, and freeze-thaw damage in concrete façades, preserving structural integrity and appearance over time. Similarly, nano-aluminum oxide and zinc oxide particles can shield metal surfaces from oxidation and corrosion, particularly in coastal or industrial environments. By protecting the façade against weathering and aging, nanomaterials help reduce long-term maintenance costs and the environmental impact associated with repair and replacement (Kuda & Yadav, 2022; Hussein, 2023; Ahmed et al., 2022).

Nanomaterials also enable lightweight and multifunctional façade solutions, which are increasingly important in high-rise and modular construction. By integrating carbon nanotubes (CNTs), graphene, and nanoclay composites into traditional materials, manufacturers can produce façade panels that are lighter, stronger, and more versatile. These advanced composites exhibit high tensile strength, flexibility, and impact resistance while retaining excellent thermal and acoustic insulation properties. Such characteristics are vital in designing façades that can withstand extreme wind loads, seismic activity, and mechanical stress without compromising occupant comfort or energy performance. In addition, lightweight nanomaterial-enhanced panels contribute to reduced structural loads and transportation energy, supporting sustainable construction practices . The integration of nanophotonic and nanoplasmonic materials in façades is unlocking innovative functionalities, such as energy harvesting and data sensing. For example, quantum dot-based solar concentrators and nano-enhanced transparent photovoltaics can be embedded within glazing systems to convert sunlight into electricity without compromising visibility. This enables façades to serve as active energy-generating surfaces, contributing to building self-sufficiency and renewable energy goals. Moreover, nanomaterials can be used in sensor coatings that detect environmental parameters such as humidity, temperature, pollution, and structural strain. When connected to a building management system, these sensors facilitate real-time monitoring and adaptive control of building systems, turning façades into intelligent, responsive interfaces. In conclusion, nanomaterials are revolutionizing façade design by introducing a new dimension of performance, adaptability, and innovation . Their ability to enhance energy efficiency, durability, aesthetics, and functionality positions them as indispensable tools in the pursuit of high-performance, sustainable architecture. While challenges related to cost, safety, and knowledge dissemination must be addressed, the potential benefits of nanomaterials far outweigh these obstacles, especially as the construction industry moves toward smarter, greener, and more resilient building solutions. By embracing nanotechnology, architects and engineers can reimagine façades not as static boundaries but as intelligent, multifunctional systems that actively contribute to the well-being of people and the planet (Aziz & Abdelall, 2021; Grillo et al., 2022; Malik et al., 2025).

4. THERMOCHROMIC MATERIALS FOR DYNAMIC FAÇADES: IMPORTANCE AND APPLICATION

Thermochromic materials are a class of smart materials that exhibit reversible changes in color or transparency in response to temperature variations. Their application in dynamic façade systems marks a significant advancement in sustainable and responsive architectural design. As buildings consume a substantial portion of global energy, especially for heating, cooling, and lighting, the need for adaptive, energy-efficient envelope systems has become more urgent than ever. Thermochromic materials address this need by enabling façades to self-regulate solar heat gain and light transmission, thereby enhancing indoor comfort, reducing energy consumption, and contributing to the creation of climate-responsive buildings . Their integration into façades not only advances environmental performance but also supports aesthetic innovation, occupant well-being, and adaptive architecture that aligns with future urban sustainability goals. At the heart of the importance of thermochromic materials is their ability to dynamically manage solar radiation. In conventional building envelopes, windows and curtain walls have static optical properties that may not be suitable throughout the day or across seasons. Thermochromic glazing or coatings overcome this limitation by adjusting their light and heat transmittance according to ambient temperature. During hot conditions, these materials become darker or more opaque, reducing solar heat gain and mitigating indoor overheating. In contrast, during cooler periods, the material remains transparent, allowing maximum solar penetration for passive heating. This passive modulation of thermal loads reduces the building's dependence on mechanical heating and cooling systems, contributing significantly to energy conservation and cost savings . In terms of occupant well-being and comfort, thermochromic façades offer several advantages. By regulating the amount of sunlight and glare entering indoor spaces, they contribute to visual comfort, which is critical in workplaces, educational facilities, and residences. Excessive glare can lead to eye strain and discomfort, while insufficient daylight can affect mood, circadian rhythms, and productivity. Thermochromic glazing helps maintain optimal lighting levels without requiring blinds or external shading devices, promoting a clean, unobstructed view and greater connection to the outdoors. Additionally, by stabilizing indoor temperatures naturally, thermochromic materials help create thermally comfortable environments that improve occupant satisfaction and health. Thermochromic materials also align with the principles of passive design and smart building systems . Unlike electrochromic or photochromic technologies that require electrical input or constant exposure to light to change state, thermochromic materials function autonomously and passively, requiring no external power source. This makes them ideal for applications in regions with limited energy infrastructure or in retrofitting older buildings where integrating complex control systems may not be feasible. Their low-maintenance and energy-independent nature enhances system reliability and reduces operational complexity, supporting cost-effective and resilient façade solutions (Arnesano et al., 2021; Perez et al., 2021; Rashidzadeh & Matin, 2023).

Material composition and technological development play a critical role in the performance and viability of thermochromic façades. Most commercial thermochromic systems are based on either leuco dyes or vanadium dioxide (VO_2). Leuco dyes are organic compounds that undergo color change in response to temperature and are widely used in decorative or novelty applications. In architectural settings, more durable and functional thermochromic materials such as vanadium dioxide are preferred. VO_2 undergoes a phase transition near 68°C, switching from a transparent insulator to a reflective metal, thereby controlling infrared radiation transmission . Through doping and material engineering, the transition temperature of VO_2 can be lowered to make it suitable for building applications in moderate climates. Ongoing research is focused on optimizing these materials for clarity, switching speed, durability, and environmental responsiveness. Despite their promising benefits, the widespread adoption of thermochromic materials in façade design faces several challenges. One of the key limitations is the relatively high transition temperature of traditional thermochromic compounds, which may not activate effectively in some building environments. Advances in material science are gradually overcoming this barrier through temperature tuning and nanostructuring techniques. Optical clarity and color neutrality are additional concerns, as some thermochromic glazings may introduce color tints or reduce transparency, which can affect visual quality and user acceptance . Moreover, longevity and degradation under prolonged UV exposure or repeated phase changes must be addressed to ensure the long-term performance and aesthetic consistency of the façade. In conclusion, thermochromic materials represent a powerful tool for realizing dynamic, sustainable, and occupant-responsive façade systems. Their ability to passively adapt to temperature changes allows for improved energy efficiency, indoor comfort, visual appeal, and environmental integration. While technical and economic challenges persist, ongoing research and market development continue to expand their capabilities and applications. By integrating thermochromic materials into building envelopes, architects and engineers can create smart façades that are not only energy-efficient but also expressive, future-proof, and aligned with the evolving demands of urban living and environmental stewardship. In building practice, the evaluation of thermochromic façade systems often depends on switching temperature, visible transmittance, solar modulation capability, durability under repeated cycling, and the extent to which façade operation can reduce cooling demand without undermining daylight quality (Perez et al., 2021; Arnesano et al., 2021; Gonçalves et al., 2024).

5. PHASE CHANGE MATERIALS FOR ENERGY EFFICIENCY: IMPORTANCE AND BENEFITS

Phase change materials (PCMs) have emerged as a vital innovation in building science, offering a powerful solution for improving energy efficiency and thermal comfort in the built environment. As global energy consumption rises-driven largely by buildings and their heating, cooling, and lighting demands-there is an urgent need to implement materials and technologies that passively regulate indoor temperatures while reducing reliance on mechanical systems. PCMs provide a unique approach to this challenge through their ability to store and release large amounts of thermal energy during phase transitions, such as melting and solidification . By incorporating PCMs into building façades and envelope systems, architects and engineers can design thermally adaptive structures that buffer temperature fluctuations, reduce peak energy loads, and contribute to sustainable, low-carbon buildings. The significance of PCMs in energy-efficient design lies not only in their thermal properties but also in their potential to support occupant well-being, operational cost reduction, and resilience in the face of climate variability. From an environmental and sustainability perspective, the use of PCMs supports broader climate and energy goals. By reducing building energy consumption, PCMs help lower greenhouse gas emissions and contribute to achieving net-zero energy targets. When combined with renewable energy sources, PCM systems can store surplus solar heat during the day and use it at night, enhancing the efficiency of photovoltaic and solar thermal systems. Furthermore, the use of bio-based and recyclable PCMs offers a sustainable alternative to petroleum-derived options, aligning with circular economy principles and green building certification systems such as LEED, BREEAM, and WELL. Despite their many advantages, the widespread use of PCMs in façade applications is limited by several technical and economic challenges. Material selection and design are critical to ensuring that the PCM performs optimally within the specific thermal and architectural context of a building . The phase change temperature must be carefully chosen to align with the building's operating temperature range, and the thermal conductivity of the PCM may need to be enhanced with additives or encapsulation to ensure rapid heat transfer. Improper integration or thermal mismatch can lead to underperformance or even material degradation over time. Durability and long-term performance also remain concerns. PCMs must withstand repeated thermal cycles without significant changes in phase change temperature, latent heat capacity, or structural integrity. Inorganic PCMs, such as salt hydrates, are prone to phase segregation and supercooling, while organic PCMs like paraffins may suffer from flammability, low thermal conductivity, or leakage issues. To address these concerns, researchers have developed microencapsulation and macroencapsulation techniques, wherein PCMs are enclosed in polymer shells or containers to prevent leakage, enhance safety, and improve thermal exchange. Nanocomposite PCMs, which incorporate nanoparticles such as graphene or metal oxides, are also being explored for their superior thermal conductivity and stability. Cost and market adoption represent additional hurdles. Although PCM prices have decreased in recent years due to advancements in manufacturing, they remain more expensive than traditional thermal insulation or construction materials . The lack of standardization, testing protocols, and integration guidelines also limits adoption, especially in markets unfamiliar with advanced materials. Education, awareness-building, and demonstration projects are needed to showcase the real-world benefits and cost-effectiveness of PCM-integrated façades. Government incentives, green building regulations, and energy performance mandates can further stimulate demand and scale deployment. In conclusion, phase change materials represent a transformative approach to enhancing building energy efficiency through passive thermal energy storage . Their ability to absorb and release heat during phase transitions allows façades to respond intelligently to environmental changes, improving thermal comfort, reducing energy use, and supporting climate-responsive design. While challenges related to material properties, integration, and cost remain, the ongoing evolution of PCM technologies-coupled with increasing global demand for sustainable and resilient buildings-positions them as a key component in the future of façade engineering. By embracing PCMs, architects, engineers, and developers can contribute to smarter, healthier, and more energy-efficient built environments that meet the demands of the 21st century (Al-Yasiri & Szabó, 2021; Farid et al., 2004; Rathore et al., 2022; Wang et al., 2022; Suresh et al., 2022).

6. SHAPE MEMORY ALLOYS FOR ADAPTIVE FAÇADES: IMPORTANCE AND APPLICATIONS

Shape Memory Alloys (SMAs) represent a groundbreaking class of smart materials with the remarkable ability to undergo deformation and then return to their original shape when subjected to specific thermal or mechanical stimuli. Their unique mechanical behavior, rooted in solid-state phase transformations, has positioned them as critical enablers of adaptive façades-dynamic building skins that respond intelligently to changing environmental conditions. In an era of increasing energy demands, climate variability, and the urgent need for resilient and sustainable architecture, SMAs offer a compelling solution for the design of responsive building envelopes that actively regulate heat, light, airflow, and aesthetics . The importance of SMAs in façade design lies in their multifunctional capability to replace complex mechanical systems, enhance building performance, support energy conservation, and contribute to a new generation of biomimetic, intelligent architecture. The core principle behind SMAs is the thermoelastic martensitic transformation, where the material can exist in two different crystallographic phases: martensite (low-temperature phase) and austenite (high-temperature phase) . When deformed in the martensitic phase, the alloy "remembers" its original shape and returns to it upon heating to the transformation temperature, demonstrating the shape memory effect. Certain SMAs also exhibit superelasticity, enabling them to recover large strains without permanent deformation. These properties allow SMAs to function as self-actuating components in façades, eliminating the need for motors, springs, or sensors to drive kinetic or movable systems. This passive actuation makes SMA-integrated façades inherently energy-efficient, silent, and low-maintenance, ideal for sustainable building designs. SMAs also promote simplified system design and low maintenance requirements, which are essential in large-scale and high-rise buildings . Because the material itself acts as both sensor and actuator, there is no need for complex control systems, cabling, or software integration. This results in lower installation costs, reduced maintenance burdens, and greater architectural freedom. SMA-based solutions are also well-suited for retrofitting existing buildings, providing adaptive functionality without requiring invasive structural modifications or heavy equipment (Bagheri, 2025; Naeem et al., 2024; Tabrizikahou et al., 2021).

However, despite their numerous advantages, there are still several technical, economic, and practical challenges to the widespread adoption of SMAs in adaptive façades. One of the primary limitations is the cost of high-performance

SMAs-particularly NiTi alloys-which can be significantly higher than conventional actuation components. Although prices are gradually decreasing due to advancements in manufacturing and materials science, cost remains a barrier for large-scale applications. Additionally, SMAs have a relatively narrow operating temperature window, which may not be suitable for all climate zones or building types . Accurate tuning of transformation temperatures during alloy processing is required to ensure reliable actuation under real-world conditions. Fatigue and material degradation under repeated thermal and mechanical cycling are another concern. While SMAs are durable, their performance can degrade over time, especially under high-strain conditions or suboptimal environmental exposure. Advances in alloy composition, encapsulation, and protective coatings are addressing these issues, but lifecycle assessment and long-term monitoring remain essential. Moreover, the lack of standardization and limited industry familiarity with SMA technology can deter adoption by architects, engineers, and contractors. Education, demonstration projects, and interdisciplinary collaboration are necessary to build trust and competence in SMA-based systems within the construction sector. In conclusion, Shape Memory Alloys hold immense potential for transforming static building envelopes into adaptive, intelligent, and climate-responsive façades. Their ability to passively actuate without external energy inputs, combined with mechanical simplicity, durability, and aesthetic flexibility, makes them an ideal material for 21st-century architecture . By enabling dynamic control over light, heat, and air, SMAs contribute to improved energy efficiency, occupant comfort, and architectural creativity. While challenges related to cost, standardization, and material longevity must be addressed, the benefits of SMA technology-especially in the context of sustainability, resilience, and smart cities-underscore its critical role in shaping the future of adaptive building systems. Embracing SMAs in façade design is not only a technical advancement but also a visionary step toward buildings that can think, move, and breathe with their environment (Naeem et al., 2024; Bagheri, 2025; Faragalla & Asadi, 2022).

7. NANOCOATINGS FOR SELF-CLEANING AND WATER REPELLENCY: IMPORTANCE AND BENEFITS

Nanocoatings have emerged as a revolutionary advancement in materials science and surface engineering, offering powerful functionalities that significantly enhance the performance, durability, and sustainability of building façades. Among their most impactful properties are self-cleaning and water repellency, which are critical in maintaining façade aesthetics, reducing maintenance costs, and preserving building materials over time . These ultra-thin surface treatments, typically engineered at the nanometer scale, form protective barriers that alter the surface energy and structure of materials, enabling them to repel water, resist contaminants, and interact with the environment in a highly efficient, passive manner. In an age where sustainable design, durability, and low-maintenance solutions are in high demand, nanocoatings represent a transformative technology that enables smarter, cleaner, and more resilient architectural envelopes. The primary benefit of nanocoatings in façades is their ability to enable self-cleaning surfaces, thereby maintaining the aesthetic quality of buildings without the need for frequent manual cleaning. This is achieved through two main mechanisms: superhydrophobicity and photocatalysis . Superhydrophobic nanocoatings, inspired by the lotus leaf effect, create micro- and nanoscale roughness on the surface that causes water droplets to bead up and roll off effortlessly, collecting dirt, dust, and pollutants along the way. Photocatalytic coatings, on the other hand, typically based on titanium dioxide (TiO_2) nanoparticles, break down organic dirt and biological contaminants when exposed to ultraviolet (UV) light. The decomposition of grime, combined with a hydrophilic surface that allows water to spread and wash away debris, ensures that façades remain clean with minimal rainwater or moisture . This dual functionality enables a passive cleaning mechanism that is both energy-free and chemical-free, supporting long-term cleanliness in diverse urban environments. From an economic and maintenance standpoint, nanocoatings offer substantial long-term savings by reducing the frequency and cost of façade cleaning. In urban and industrial settings, façades are exposed to pollutants, dust, soot, bird droppings, and other contaminants that typically require regular cleaning with high-pressure washers, chemicals, and labor-intensive processes. Nanocoated surfaces, however, shed dirt and water naturally, reducing the need for detergents and cleaning equipment, lowering labor costs, and decreasing building downtime. This is particularly beneficial for high-rise buildings, where façade cleaning is not only costly but also risky. Additionally, by reducing water usage and eliminating chemical runoff, nanocoatings contribute to environmentally responsible building maintenance practices, aligning with green building certifications and corporate sustainability goals (Dahmani, 2024; Garg et al., 2024; Fu & Gray, 2021).

Figure 1. Nanofabrication of heat management materials for energy-efficient building facades

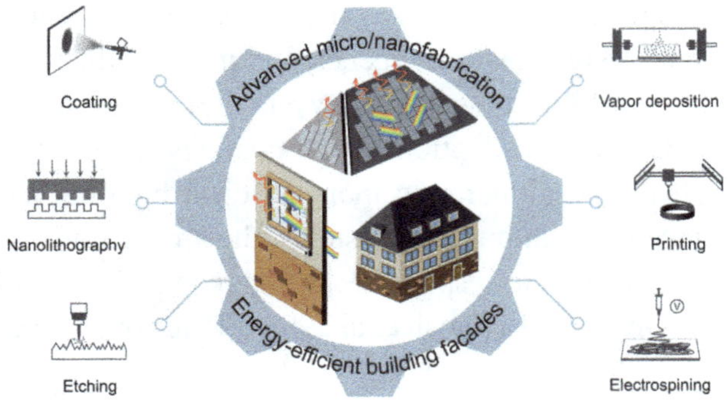

Nanocoatings also contribute to energy savings in buildings by improving the optical and thermal performance of façades. When applied to glass, nanostructured coatings can reduce glare, improve light transmission, and block infrared radiation, helping to control indoor temperatures and reduce the cooling load. In cold climates, hydrophobic coatings prevent frost formation and ice buildup, maintaining façade performance and visibility in winter months. When combined with photocatalytic air-purifying functions, such as those offered by TiO_2-based nanocoatings, building façades can also help decompose airborne pollutants like nitrogen oxides (NO_X) and volatile organic compounds (VOCs), contributing to cleaner urban air and healthier cities. Nanocoatings further enhance façade performance by improving resistance to ultraviolet (UV) radiation, abrasion, and chemical exposure . UV radiation is a leading cause of material fading, cracking, and surface degradation, especially in transparent or colored façade elements. Nanocoatings with UV-stabilizers and anti-aging additives prevent photodegradation, preserving surface integrity and color vibrancy over time. For buildings in coastal, industrial, or polluted environments, nanocoatings provide a protective shield against salt, acid rain, and urban grime, preventing corrosion and extending the life expectancy of external elements such as steel, aluminum cladding, and anodized finishes. In terms of environmental sustainability, nanocoatings play a significant role in reducing the ecological footprint of building operations . By enabling passive cleaning and water conservation, they reduce the need for harsh cleaning agents, thereby minimizing chemical pollution and resource consumption. Their role in air purification and energy conservation also supports climate action strategies and green infrastructure initiatives. Moreover, advances in eco-friendly nanocoatings, such as those based on bio-derived silica or water-based dispersions, further enhance their environmental compatibility by reducing VOC emissions during application and disposal. In conclusion, nanocoatings that offer self-cleaning and water-repellent properties are revolutionizing façade performance by delivering enhanced durability, reduced maintenance, aesthetic preservation, and environmental sustainability. Their passive functionality, combined with long-term economic and ecological benefits, makes them a valuable tool for architects and developers aiming to design smarter, greener, and more resilient buildings. While challenges related to durability, cost, and safety must be addressed through continued innovation and regulation, the adoption of nanocoatings is poised to grow as cities worldwide seek cleaner, healthier, and more efficient building technologies. By incorporating nanocoatings into façade systems, the architecture of the future can be more self-sustaining, climate-responsive, and aligned with the evolving demands of urban life (Fu & Gray, 2021; Ghosh et al., 2025; Perez et al., 2021).

8. NANOPARTICLES FOR ENHANCED THERMAL INSULATION: IMPORTANCE AND BENEFITS

Nanoparticles have emerged as a game-changing innovation in the field of thermal insulation, offering remarkable enhancements to the energy efficiency, performance, and sustainability of modern building envelopes. As climate change intensifies and urban populations grow, the architectural and construction industries face increasing pressure to reduce energy consumption, minimize heat transfer, and develop smarter, more adaptive building systems. Traditional insulation materials, while effective to a degree, often suffer from limitations such as bulkiness, degradation over time, or limited resistance to moisture and fire . The integration of nanoparticles into insulation systems offers a revolutionary leap forward, delivering superior thermal resistance, reduced material thickness, multifunctionality, and compatibility with diverse façade applications. In essence, nanoparticles enable façades to become high-performance, thermally efficient barriers that significantly reduce heating and cooling loads while supporting aesthetic flexibility and sustainability goals. At the heart of nanoparticle-enhanced insulation is the manipulation of material properties at the nanoscale (1-100 nm), where matter behaves differently than in bulk form. Nanoparticles can dramatically improve a material's thermal resistance by reducing heat conduction, convection, and radiation-the three primary modes of heat transfer. In conventional materials, heat typically travels through solid conduction pathways, air gaps, or via radiative transfer through pores and surfaces. By embedding nanoparticles-such as aerogels, silica, titania, alumina, or carbon-based nanomaterials like graphene and carbon nanotubes (CNTs)-into insulation matrices, these pathways can be disrupted at the molecular level. Nanoparticles scatter phonons (vibrational heat energy) and reflect infrared radiation, which effectively impedes thermal transmission through the material . This makes nanoparticle-enhanced insulation more effective per unit thickness compared to conventional options like fiberglass, mineral wool, or expanded polystyrene. One of the most prominent applications of nanoparticles in thermal insulation is in the development of aerogels-ultra-lightweight, highly porous solids that offer exceptional thermal resistance. Silica-based aerogels, which often contain nanoparticles to enhance their structural and thermal properties, are known as some of the best insulating materials available, with thermal conductivities as low as 0.012 W/m·K. Their low density, fire resistance, and translucency make them particularly suitable for façade glazing systems, curtain walls, skylights, and translucent panels, where insulation must be balanced with daylighting and aesthetic goals . Nanoparticle-infused aerogel blankets or composite boards can be easily integrated into façade cavities, rainscreens, or modular panels, improving insulation without increasing bulk or compromising design. The importance of nanoparticles in energy-efficient building design cannot be overstated. Buildings are responsible for more than 30% of global energy use, with heating and cooling accounting for a significant share. By improving the thermal insulation of façades through nanotechnology, buildings can dramatically reduce U-values (thermal transmittance), achieving energy savings of up to 50% in certain climates. This not only lowers operational energy costs but also supports compliance with strict energy codes and sustainability standards such as LEED, BREEAM, Passive House, and Net-Zero Energy Building (NZEB) criteria. Moreover, by reducing peak energy loads, nanoparticle-enhanced façades help stabilize grid demand and support the integration of renewable energy sources, which are often variable in supply (Abdelrady et al., 2021; Saleh et al., 2021; Toader et al., 2024).

Nanoparticles also contribute to indoor environmental quality and occupant comfort, two essential dimensions of sustainable architecture. By reducing unwanted heat gain in the summer and minimizing heat loss in winter, nanoparticle-enhanced façades maintain consistent indoor temperatures, improving thermal comfort without over-reliance on HVAC systems. This also helps reduce air stratification, drafts, and humidity imbalances. In addition, nanoparticle-infused insulation materials often possess sound-dampening properties, contributing to acoustic insulation in noisy urban environments. Some nanoparticles offer antibacterial, fire-retardant, and moisture-resistant properties, further improving the safety, hygiene, and resilience of building envelopes. From a design perspective, nanoparticle-enhanced insulation supports thinner wall profiles and greater design flexibility, especially in high-density urban environments where space is at a premium. Unlike bulky insulation materials, nanoparticle composites can deliver high thermal resistance in compact layers, freeing up valuable interior space and allowing for more slender or expressive architectural forms . This is particularly beneficial in high-rise construction, modular prefabricated façades, and envelope retrofits, where weight and thickness constraints are critical. In conclusion, nanoparticles are redefining the capabilities of thermal insulation in façade systems, offering a powerful combination of energy efficiency, material compactness, and multifunctionality. Their ability to reduce heat transfer at the molecular level, improve insulation performance without bulk, and support sustainable design makes them a vital technology in the development of next-generation building envelopes. While economic, environmental, and technical challenges remain, the strategic use of nanoparticles-supported by responsible innovation and performance validation-has the potential to revolutionize how we design, construct, and inhabit thermally efficient buildings. Embracing nanoparticles for enhanced thermal insulation is a critical step toward achieving low-carbon, high-performance architecture for the 21st century and beyond (Bueno et al., 2021; Zhang et al., 2022; Arowoiya et al., 2024).

9. NANOMATERIALS FOR IMPROVED DURABILITY AND MAINTENANCE: IMPORTANCE AND BENEFITS

Nanomaterials have become a transformative force in the construction and architectural sectors, particularly in enhancing the durability, longevity, and ease of maintenance of building envelope systems. As modern buildings are increasingly expected to withstand extreme weather, pollution, mechanical stress, biological degradation, and aging over decades of operation, the need for advanced materials that offer extended performance with minimal upkeep is greater than ever . Traditional materials like concrete, steel, stone, and glass are vulnerable to environmental degradation over time, resulting in high maintenance costs, frequent repairs, aesthetic decay, and safety concerns. Nanomaterials, through their unique nanoscale structure and surface interactions, offer tailored properties that significantly enhance resistance to wear, moisture, chemicals, UV radiation, biological agents, and environmental pollutants-thus playing a crucial role in improving the durability and maintainability of façade systems . The importance of nanomaterials in façades lies in their ability to modify or reinforce the mechanical, chemical, and thermal properties of conventional construction materials. Their high surface area-to-volume ratio, tunable reactivity, and compatibility with both organic and inorganic matrices make them ideal for developing multifunctional coatings, composites, and surface treatments. Nanomaterials such as nano-silica, nano-titanium dioxide (TiO_2), nano-alumina, graphene oxide, carbon nanotubes (CNTs), and nanoclays are widely used to improve the strength, abrasion resistance, water resistance, anti-corrosive properties, and self-healing capabilities of façade elements. Whether incorporated into concrete, metal panels, paint systems, stone sealants, or glass coatings, these materials enhance structural performance while also protecting surfaces from deterioration caused by environmental exposure. A critical area where nanomaterials demonstrate their value is in protecting façades from moisture and chemical intrusion. Water ingress is one of the most damaging threats to building exteriors, leading to freeze-thaw damage, efflorescence, corrosion of steel reinforcements, mold growth, and eventual structural failure. Nanomaterials such as nano-silica and nano-clays improve the density and impermeability of concrete and plaster by filling microscopic pores and reducing capillary absorption . This helps prevent water penetration while maintaining the material's breathability-an essential characteristic for thermal regulation and indoor air quality. Nano-enhanced hydrophobic coatings further repel rain, pollutants, and biological contaminants, ensuring cleaner façades with less surface staining and discoloration over time. In metal façade systems, nanomaterials significantly improve corrosion resistance and surface integrity. Nanoparticle-infused primers and paints-often containing zinc oxide, cerium oxide, or graphene-based additives-act as powerful corrosion inhibitors that protect steel, aluminum, and copper claddings from rust, oxidation, and chemical degradation. These coatings work by forming dense, protective films at the micro and nanoscale, impeding the electrochemical reactions that cause corrosion. In marine or industrial environments where façades are constantly exposed to salt, acid rain, or airborne chemicals, nanomaterial-enhanced coatings extend the lifespan of metal surfaces, reduce repainting cycles, and maintain aesthetic uniformity (Mahmoud, 2023; Konbr & Mamdouh, 2022; Kuda & Yadav, 2022).

Another key benefit of nanomaterials is their role in self-cleaning and anti-biofouling functions, which directly contribute to reduced maintenance. Nanoparticles such as titanium dioxide (TiO_2) possess photocatalytic properties, breaking down organic pollutants, soot, and microbial growth under UV light. This makes them ideal for self-cleaning coatings on glass, concrete, and metal surfaces, particularly in urban environments with high levels of dust and air pollution. In humid or shaded areas prone to moss, algae, or mold, nano-silver and copper nanoparticles inhibit microbial colonization, ensuring surfaces remain clean, hygienic, and structurally sound. This minimizes the need for frequent pressure washing, chemical treatments, or repainting, resulting in substantial cost savings and environmental benefits over the building's lifecycle . Nanomaterials also contribute to UV resistance and thermal stability. Prolonged exposure to ultraviolet radiation can lead to fading, chalking, and embrittlement of façade coatings and polymers. By incorporating nano-titanium dioxide or nano-zinc oxide into surface finishes, architects can significantly enhance UV protection, ensuring color retention and surface integrity over time. Similarly, nanomaterials help stabilize façade components against temperature extremes and thermal cycling, reducing expansion-contraction stresses that often cause cracking, joint failure, or delamination in façade systems. Looking ahead, the future of nanomaterials in durable and low-maintenance façades is bright, driven by interdisciplinary research, digital integration, and increasing emphasis on sustainable design. Researchers are developing multifunctional nanocoatings that combine self-cleaning, corrosion resistance, thermal insulation, and anti-microbial properties in a single layer. Innovations in nano-encapsulation, bio-based nanomaterials, and AI-optimized formulations are expanding the performance and applicability of these materials across all building types and climates . As buildings become more intelligent and interconnected, smart façades equipped with nanosensors may be able to monitor surface integrity, predict maintenance needs, and dynamically adapt to environmental stresses, redefining how maintenance and durability are managed in the built environment. In conclusion, nanomaterials are reshaping the future of façade engineering by offering superior protection, extended durability, and simplified maintenance through advanced material science. Their ability to enhance structural performance, resist environmental degradation, inhibit microbial growth, and reduce upkeep requirements makes them an essential asset in the development of high-performance, resilient, and sustainable buildings. While technical and regulatory hurdles remain, the continued advancement of nanotechnology promises to deliver even more efficient, cost-effective, and environmentally conscious solutions for the architectural industry. Embracing nanomaterials in façade design is not only a pathway to longer-lasting buildings but a critical strategy for meeting the demands of tomorrow's cities (Fu & Gray, 2021; Dahmani, 2024; Garg et al., 2024).

10. ENERGY EFFICIENCY AND SUSTAINABILITY: BENEFITS AND ARCHITECTURAL SIGNIFICANCE

Energy efficiency and sustainability have become central pillars of 21st-century architecture and urban planning, reflecting an urgent global response to the environmental, economic, and social challenges posed by climate change, resource scarcity, and urbanization. In the context of the built environment, these principles are not merely theoretical ideals-they are practical, measurable strategies that significantly improve building performance, reduce greenhouse gas emissions, optimize energy use, lower operational costs, and enhance occupant well-being. The benefits of energy efficiency and sustainability extend far beyond cost savings; they contribute to the long-term resilience of cities, the health of ecosystems, and the creation of equitable, livable, and future-ready communities . As governments, organizations, and citizens strive toward decarbonization and environmental stewardship, energy-efficient and sustainable architecture has emerged as a transformative force capable of addressing global sustainability targets while redefining how we design, construct, and inhabit the built world. The most immediate and tangible benefit of energy-efficient design is the reduction in building energy consumption. Buildings account for approximately 30-40% of global energy use and a significant share of carbon dioxide (CO_2) emissions. By incorporating high-performance insulation, energy-efficient glazing, advanced HVAC systems, smart lighting, and passive design strategies, buildings can dramatically reduce their dependence on fossil fuels and electricity . This leads to lower operational costs, often resulting in energy savings of 30-70% compared to conventional buildings. Over time, these savings accumulate to significantly reduce total cost of ownership, making energy-efficient buildings financially attractive to both developers and occupants. In addition, energy efficiency reduces peak demand on electrical grids, especially during hot summers or cold winters, improving energy security and reducing the risk of blackouts or supply shortages. Energy-efficient and sustainable buildings are also essential for meeting national and international climate goals . Countries around the world have committed to reducing emissions under the Paris Agreement, with many adopting net-zero or near-zero energy building codes for new construction. Green building standards such as LEED (Leadership in Energy and Environmental Design), BREEAM (Building Research Establishment Environmental Assessment Method), WELL, EDGE, and Passive House provide frameworks for evaluating and certifying buildings that meet or exceed energy performance criteria. These certifications not only support compliance with climate policies but also enhance market value, brand reputation, and investor confidence. Buildings that meet high sustainability standards often command higher rents, have lower vacancy rates, and show improved asset performance over time, making them appealing for long-term investment (Ahmed & Gawad, 2022; Shang & Lv, 2023; González-Lezcano, 2021).

The integration of smart technologies with energy-efficient systems has further revolutionized building performance. Smart sensors, Internet of Things (IoT) devices, and AI-driven building management systems (BMS) enable real-time monitoring and control of energy usage, automatically adjusting lighting, HVAC, shading, and ventilation based on occupancy, weather conditions, and user behavior. This dynamic responsiveness enhances efficiency while providing greater comfort and customization for users. Predictive maintenance, energy modeling, and digital twins also allow building managers to identify inefficiencies and optimize operations continuously. The synergy between energy efficiency and digital innovation is creating buildings that are not only green but also intelligent, adaptive, and future-proof . In urban planning, energy-efficient and sustainable design principles scale beyond individual buildings to shape low-carbon neighborhoods, eco-districts, and smart cities. Sustainable urban design incorporates mixed land use, walkability, green mobility infrastructure, public spaces, and renewable energy systems to reduce per capita energy use and promote social inclusion. The transition to distributed energy systems, such as rooftop solar panels, district heating, and energy storage, enables communities to generate and manage their own clean energy, increasing local resilience and reducing reliance on centralized power grids. Urban sustainability also entails the protection of green infrastructure, biodiversity corridors, and natural water systems, helping to create livable, climate-adaptive urban environments. Looking forward, the future of energy efficiency and sustainability in the built environment is deeply promising. The emergence of net-zero energy buildings (NZEBs), positive energy buildings (PEBs), and regenerative design is shifting the focus from reducing harm to creating buildings that give back-producing more energy than they consume, restoring ecosystems, and uplifting communities . Innovations in materials science, such as smart façades, phase-change materials, and nanotechnology, are enhancing performance while reducing environmental impact. Meanwhile, the rise of circular economy models is pushing the industry to rethink material use, waste, and end-of-life strategies, aiming for closed-loop systems where resources are continually reused. In conclusion, energy efficiency and sustainability are not just technical strategies-they are foundational principles of responsible, future-oriented architecture. Their benefits span multiple dimensions: environmental protection, economic savings, occupant health, urban resilience, and social equity. As buildings become increasingly complex, interconnected, and challenged by climate and resource pressures, the need for high-performance, low-impact design has never been more critical. By embracing energy efficiency and sustainability at every stage-from concept to construction to operation-architects, engineers, and planners can lead the transformation toward a built environment that supports both planetary health and human prosperity for generations to come (Arowoiya et al., 2024; Zhang et al., 2022; Lee et al., 2021).

11. IMPROVED BUILDING PERFORMANCE AND COMFORT: IMPORTANCE AND BENEFITS IN THE BUILT ENVIRONMENT

In the evolving landscape of architecture, engineering, and urban design, improved building performance and occupant comfort have emerged as vital goals that underpin the effectiveness, sustainability, and long-term value of built environments. Building performance refers broadly to how well a structure fulfills its intended function while maintaining operational efficiency, structural integrity, environmental responsiveness, and user satisfaction. Comfort, in turn, represents the experiential dimension of building performance-encompassing thermal, acoustic, visual, and indoor air quality (IAQ) conditions that contribute to the health, productivity, and well-being of occupants. In a world increasingly shaped by climate change, technological innovation, resource scarcity, and changing human needs, the pursuit of higher-performing and more comfortable buildings is not merely desirable-it is essential for environmental responsibility, economic resilience, and human-centric design . One of the primary reasons improved building performance is so critical lies in its direct impact on energy efficiency and operational sustainability. Buildings that are well-designed and optimized for performance typically consume significantly less energy, reduce operational costs, and generate fewer greenhouse gas emissions. High-performance envelopes-integrating advanced insulation, thermal breaks, airtightness, solar shading, and high-performance glazing-help regulate internal temperatures and reduce reliance on active heating and cooling systems. Efficient HVAC systems, intelligent lighting, renewable energy integration, and demand-controlled ventilation all contribute to a reduction in energy use intensity (EUI) and improve overall environmental performance . As buildings account for nearly 40% of global energy use and carbon emissions, enhancing their performance plays a key role in meeting global sustainability goals and transitioning to a low-carbon future. Indoor air quality (IAQ) also plays a significant role in building performance and occupant well-being. High-performance buildings incorporate mechanical ventilation systems with HEPA filters, demand-controlled ventilation (DCV), and air purification technologies to maintain low levels of CO_2, VOCs (volatile organic compounds), particulate matter (PM), and humidity. Enhanced IAQ is especially vital in densely occupied spaces, healthcare facilities, and post-pandemic workplaces, where air cleanliness has become a public health imperative. A well-ventilated and pollutant-free indoor environment not only supports respiratory health but also reduces absenteeism and improves cognitive performance, particularly in educational and professional settings. Smart technologies and automation systems have revolutionized how building performance and comfort are achieved and managed . Building management systems (BMS), digital twins, IoT sensors, AI algorithms, and cloud-based analytics now allow real-time monitoring, control, and predictive maintenance of building systems. These technologies provide granular insights into energy use, occupant behavior, equipment status, and environmental conditions, allowing facility managers to make informed decisions and fine-tune performance dynamically. Occupants can also interact with building systems through mobile interfaces or voice controls to personalize their environment-adjusting temperature, lighting, or blinds to meet their comfort needs. The result is a more responsive and intelligent environment that adapts proactively to both external and internal stimuli (Bueno et al., 2021; Lameira et al., 2023; Lazar & Chithra, 2022).

Post-occupancy evaluations (POE) and performance benchmarking tools further support the continuous improvement of building performance and comfort. Standards such as ASHRAE 55 (thermal comfort), WELL Building Standard, RESET Air, LEED, BREEAM, and NABERS provide guidelines and metrics to assess and certify buildings against established comfort and performance criteria. These frameworks promote data-driven design and performance verification, ensuring that buildings deliver on their promises of efficiency and comfort. They also serve as tools for transparency and accountability, giving occupants confidence in the quality and sustainability of their indoor environment. Looking forward, the future of improved building performance and comfort will be shaped by innovation, sustainability, and human-centric design. The rise of net-zero energy and carbon-positive buildings, biophilic design, AI-driven adaptive systems, and materials with embedded intelligence will push the boundaries of what buildings can achieve . Future buildings will not only provide shelter but also actively enhance health, respond to users, harvest energy, and adapt to environmental changes. The growing emphasis on wellness in architecture, especially in response to the global health crisis, will further solidify comfort as a critical design criterion alongside energy and performance metrics. In conclusion, improved building performance and comfort are indispensable for creating sustainable, resilient, and livable spaces in today's rapidly changing world. By optimizing energy use, enhancing indoor environmental quality, integrating smart technologies, and prioritizing occupant well-being, high-performance buildings serve as a foundation for healthier cities and empowered communities. The convergence of design excellence, environmental responsibility, and human-centered innovation positions building performance and comfort not just as technical goals but as ethical imperatives in the journey toward a regenerative built environment (Bueno et al., 2021; Arowoiya et al., 2024; Zhang et al., 2022).

12. CONCLUSION

The integration of smart materials and nanomaterials in façade design represents one of the most transformative developments in contemporary architecture, redefining the building envelope from a static boundary to a dynamic, responsive, and multifunctional system. As buildings evolve to meet the challenges of the 21st century-including rapid urbanization, climate change, energy efficiency mandates, and heightened user expectations-façades are becoming critical platforms for innovation, sustainability, and performance optimization. Smart and nano-enabled façades offer a powerful convergence of form and function, seamlessly blending aesthetics with adaptive behavior, and delivering enhanced energy performance, occupant comfort, durability, and environmental responsibility. The applications of smart materials-such as thermochromic and electrochromic glazing, shape memory alloys, and phase change materials-have demonstrated how architectural envelopes can adapt in real time to external stimuli such as heat, light, humidity, or electrical current. These materials enable façades to regulate solar heat gain, modulate light transmission, and provide dynamic shading, thereby reducing HVAC loads and improving indoor environmental quality. Similarly, nanomaterials, including nanocoatings, aerogels, carbon nanotubes, and nanoparticle-enhanced composites, are enabling unprecedented advances in thermal insulation, self-cleaning properties, UV protection, water repellency, and antimicrobial surfaces. These enhancements contribute not only to reduced energy consumption and maintenance costs but also to greater building resilience and longevity. Looking ahead, the future of smart and nano façade materials will be marked by greater functionality, intelligence, and integration. Innovations will move beyond passive and reactive properties to materials that are self-sensing, self-healing, and even self-powering. Advanced nanostructures will enable façades to harvest solar energy more efficiently, purify urban air, regulate humidity, and provide real-time feedback on environmental conditions and material health. Combined with digital technologies such as the Internet of Things (IoT), Artificial Intelligence (AI), and Building Information Modeling (BIM), façades will evolve into cyber-physical systems-able to analyze, respond, and adapt to changes autonomously while communicating with other building systems in a smart urban ecosystem. Moreover, as sustainability becomes an imperative rather than an option, smart and nano materials will play a pivotal role in achieving net-zero energy goals, circular economy principles, and low-carbon construction. Their application supports green certifications like LEED, WELL, BREEAM, and Passive House, and helps designers meet stringent environmental performance standards across global regions. Bio-based and recyclable nanomaterials-such as cellulose nanofibers, biochar composites, or mycelium panels-will provide low-impact alternatives to conventional façade materials while offering comparable or superior performance. These innovations will not only reduce embodied carbon and operational emissions but also promote climate-responsive and regenerative building practices. Crucially, the user experience and occupant well-being will remain at the heart of future façade development. Materials that enhance natural light, control glare, reduce noise, ensure thermal comfort, and support circadian rhythms will contribute to healthier and more productive indoor environments. Smart façades will increasingly respond to user preferences and behavioral patterns, providing personalized microclimates and environmental settings. Real-time data from embedded sensors will allow for predictive maintenance, extending material life and reducing environmental impact from premature replacements. In conclusion, the future of façade design lies in the synergistic integration of smart materials and nanotechnology, offering a powerful toolkit for architects and engineers to meet the complex demands of a changing world. These materials hold the key to unlocking smarter, greener, more adaptable buildings that prioritize performance, human health, and environmental stewardship. With continued research, supportive policies, collaborative design practices, and a commitment to innovation, smart and nano-enabled façades will become the new standard in high-performance architecture-paving the way for a more sustainable, responsive, and resilient built environment for generations to come.

ACKNOWLEDGMENT

The author acknowledges the growing interdisciplinary scholarship on smart materials, nanotechnology, façade engineering, and sustainable building performance that informs this chapter.

REFERENCES

Abdelrady, A., Abdelhafez, M. H. H., & Ragab, A. (2021). Use of insulation based on nanomaterials to improve energy efficiency of residential buildings in a hot desert climate. *Sustainability (Basel)*, *13*(19), 10856. DOI: 10.3390/su13095266

Ahmed, M. G., & Gawad, M. A. (2022). Architecture sustainability and energy efficiency. *International Journal of Energy Production and Management*, *7*(4), 320–330. https://elar.urfu.ru/handle/10995/126075

Ahmed, S. F., Mofijur, M., Rafa, N., Chowdhury, A. T., Nuzhat, S., Rahman, S. A., Ong, H. C., & Mahlia, T. M. I. (2022). Green approaches in synthesising nanomaterials for environmental nanobioremediation: Technological advancements, applications, benefits and challenges. *Environmental Research*, *204*, 112320. DOI: 10.1016/j.envres.2021.111967 PMID: 34450159

Al-Yasiri, Q., & Szabó, M. (2021). Incorporation of phase change materials into building envelope for thermal comfort and energy saving: A comprehensive analysis. *Journal of Building Engineering*, *44*, 102943. DOI: 10.1016/j.jobe.2020.102122

Arnesano, M., Pandarese, G., Martarelli, M., Naspi, F., Gurunatha, K. L., Sol, C., Portnoi, M., Ramirez, F. V., Parkin, I. P., Papakonstantinou, I., & Revel, G. M. (2021). Optimization of the thermochromic glazing design for curtain wall buildings based on experimental measurements and dynamic simulation. *Solar Energy*, *230*, 155–168. DOI: 10.1016/j.solener.2021.01.013

Arowoiya, V. A., Moehler, R. C., & Fang, Y. (2024). Digital twin technology for thermal comfort and energy efficiency in buildings: A state-of-the-art and future directions. *Energy and Built Environment*, *5*(1), 1–19. DOI: 10.1016/j.enbenv.2023.05.004

Aziz, H. N. A., & Abdelall, M. I. (2021). Responsive façades design using nanomaterials for optimizing buildings' energy performance. *WIT Transactions on Ecology and the Environment*, *253*, 321–332. https://books.google.com. DOI: 10.2495/SC210331

Bagheri, M. (2025). Advancing architecture through shape memory alloy actuators. *Virginia Tech ETDs*. https://vtechworks.lib.vt.edu/handle/10919/119126

Brzezicki, M. (2021). A typology of adaptive façades. *An empirical study based on the morphology of glazed facades. Cogent Arts & Humanities*, *8*(1), 1960699. DOI: 10.1080/23311983.2021.1960699

Bueno, A. M., de Paula Xavier, A. A., & Broday, E. E. (2021). Evaluating the connection between thermal comfort and productivity in buildings: A systematic literature review. *Buildings (Basel, Switzerland)*, *11*(10), 446. DOI: 10.3390/buildings11060244

Dahmani, S. (2024). Sustainable approach to nanocomposites and nanocoatings: Research and industrial applications of hydrophobic nanocoatings. *Sustainable Approach to Protective Nanocoatings, IGI Global.* https://www.igi-global.com/chapter/sustainable-approach-to-nanocomposites-and-nanocoatings/330064

Faragalla, A. M. A., & Asadi, S. (2022). Biomimetic design for adaptive building façades: A paradigm shift towards environmentally conscious architecture. *Energies*, *15*(15), 5390. DOI: 10.3390/en15155390

Farid, M., Khudhair, A. M., Razack, S. A. K., & Al-Hallaj, S. (2004). A review on phase change energy storage: Materials and applications. *Energy Conversion and Management*, *45*(9-10), 1597–1615. DOI: 10.1016/j.enconman.2003.09.015

Fu, H., & Gray, K. A. (2021). The key to maximizing the benefits of antimicrobial and self-cleaning coatings is to fully determine their risks. *Current Opinion in Chemical Engineering*, *33*, 100715. DOI: 10.1016/j.coche.2021.100761 PMID: 36569284

Garg, R., Garg, R., & Eddy, N. O. (2024). Applications of nanocoatings: Smart and sustainable protection. *Sustainable Approach to Protective Nanocoatings, IGI Global.* https://www.igi-global.com/chapter/applications-of-nanocoatings/330071

Ghosh, J., Rupanty, N. S., Noor, T., Asif, T. R., & Islam, T. (2025). Functional coatings for textiles: Advancements in flame resistance, antimicrobial defense, and self-cleaning performance. *RSC Advances*, *15*(12), 9876–9891. DOI: 10.1039/D4RA08567B

Gonçalves, M., Figueiredo, A., Almeida, R., & Vicente, R. (2024). Dynamic façades in buildings: A systematic review across thermal comfort, energy efficiency and daylight performance. *Renewable & Sustainable Energy Reviews*, *182*, 114934. DOI: 10.1016/j.rser.2024.114474

González-Lezcano, R. A. (2021). Advancements in sustainable architecture and energy efficiency. *Google Books.* https://books.google.com

Grillo, E., Milardi, M., & Olivieri, F. (2022). A review of innovative materials for the design of adaptive biomimetic façades. *Advances in Urban Planning and Architectural Design, 67*, 231–245. DOI: 10.1007/978-3-031-16261-8_19

Hussein, H. S. (2023). The state of the art of nanomaterials and its applications in energy saving. *Bulletin of the National Research Center, 47*(1), 76. DOI: 10.1186/s42269-023-00984-4

Jamilu, G., Abdou, A., & Asif, M. (2024). Dynamic facades for sustainable buildings: A review of classification, applications, prospects and challenges. *Energy Reports, 10*, 4123–4140. DOI: 10.1016/j.egyr.2024.05.047

Konbr, U., & Mamdouh, H. (2022). A proposed strategy to evaluate nanomaterials in construction to boost sustainable architecture. *Civil Engineering and Architecture, 10*(6), 2329-2339. https://www.academia.edu/81868293

Kuda, A., & Yadav, M. (2022). Opportunities and challenges of using nanomaterials and nanotechnology in architecture: An overview. *Materials Today: Proceedings, 62*(3), 10094–10099. DOI: 10.1016/j.matpr.2022.07.052

Lameira, G., Ramos, R. J. G., Valentim, N., & Cruz, C. O. (2023). Good architecture matters: The architect's perspective on design for ageing and energy efficiency. *Buildings (Basel, Switzerland), 13*(4), 979. DOI: 10.3390/buildings13041067

Lazar, N., & Chithra, K. (2022). Role of culture in sustainable development and sustainable built environment: A review. *Environment, Development and Sustainability, 24*(5), 11544–11565. DOI: 10.1007/s10668-021-01691-8

Lee, J. H., Ostwald, M. J., & Kim, M. J. (2021). Characterizing smart environments as interactive and collective platforms: A review of the key behaviors of responsive architecture. *Sensors (Basel), 21*(10), 3417. DOI: 10.3390/s21103417 PMID: 34069004

Mahmoud, A. H. (2023). Studying the impact of the role of nanomaterials technology in developing the performance of buildings and their effect on the architectural formation. *Journal of Al-Azhar University Engineering Sector, 18*(68), 1371–1380. https://journals.ekb.eg/article_304211.html. DOI: 10.21608/auej.2023.283051

Malik, N., Nirupama, E., Selvakumar, P., Sankpal, U., Nathsharma, S. K., Sharma, M., & Manjunath, T. C. (2025). 2D Materials for Energy Storage and Conversion. In Mhadhbi, M. (Ed.), *Innovations in Next-Generation Energy Storage Solutions* (pp. 1–24). IGI Global Scientific Publishing., DOI: 10.4018/979-8-3693-9316-1.ch001

Matin, N. H., & Eydgahi, A. (2019). Factors affecting the design and development of responsive facades: A historical evolution. *Intelligent Buildings International, 12*(4), 257–270. DOI: 10.1080/17508975.2018.1562414

Naeem, N., Abdin, A., & Saleh, A. (2024). An approach to using shape memory alloys in kinetic façades to improve the thermal performance of office building spaces. *Civil Engineering and Architecture, 12*(3), 1067-1079. https://www.researchgate.net/publication/375012719

Perez, G., Sirvent, P., Sanchez-Garcia, J. A., & Guerrero, A. (2021). Improved methodology for the characterization of thermochromic coatings for adaptive façades. *Solar Energy, 221*, 310–319. DOI: 10.1016/j.solener.2021.10.062

Rashidzadeh, Z., & Matin, N. H. (2023). A comparative study on smart windows focusing on climate-based energy performance and users' comfort attributes. *Sustainability (Basel), 15*(3), 2294. DOI: 10.3390/su15032294

Rathore, P. K. S., Gupta, N. K., Yadav, D., Shukla, S. K., & Kaul, S. (2022). Thermal performance of the building envelope integrated with phase change material for thermal energy storage: An updated review. *Sustainable Cities and Society, 76*, 103445. DOI: 10.1016/j.scs.2022.103690

Saleh, A. N., Attar, A. A., Ahmed, O. K., & Mustafa, S. S. (2021). Improving the thermal insulation and mechanical properties of concrete using Nano-SiO2. *Results in Engineering, 12*, 100284. DOI: 10.1016/j.rineng.2021.100303

Shang, W. L., & Lv, Z. (2023). Low carbon technology for carbon neutrality in sustainable cities: A survey. *Sustainable Cities and Society, 93*, 104512. DOI: 10.1016/j.scs.2023.104489

Suresh, C., Hotta, T. K., & Saha, S. K. (2022). Phase change material incorporation techniques in building envelopes for enhancing the building thermal comfort: A review. *Energy and Building, 264*, 112086. DOI: 10.1016/j.enbuild.2022.112225

Tabrizikahou, A., Hadzima-Nyarko, M., Kuczma, M., & Lozančić, S. (2021). Application of shape memory alloys in retrofitting of masonry and heritage structures based on their vulnerability revealed in the Bam 2003 earthquake. *Materials (Basel), 14*(24), 7700. DOI: 10.3390/ma14164480 PMID: 34443003

Toader, T. P., Dico, C., & Mircea, A. C. (2024). Thermal insulation of buildings using innovative materials based on nanoparticles. *European Journal of Materials Science and Engineering, 9*(1), 1–10. https://www.ejmse.ro. DOI: 10.36868/ejmse.2024.09.01.063

Wang, X., Li, W., Luo, Z., Wang, K., & Shah, S. P. (2022). A critical review on phase change materials (PCM) for sustainable and energy efficient building: Design, characteristic, performance and application. *Energy and Building*, *256*, 111783. DOI: 10.1016/j.enbuild.2022.111923

Zhang, W., Wu, Y., & Calautit, J. K. (2022). A review on occupancy prediction through machine learning for enhancing energy efficiency, air quality and thermal comfort in the built environment. *Renewable & Sustainable Energy Reviews*, *167*, 112716. DOI: 10.1016/j.rser.2022.112704

Chapter 2
Innovative Materials for High-Performance Cladding:
The Future of Green Urban Facades

Leevesh Kumar
http://orcid.org/0000-0001-9476-0841
World University of Bangladesh, Bangladesh

ABSTRACT

This chapter reviews high-performance cladding materials and facade systems for greener, climate-responsive urban buildings. It covers bio-based composites, smart and responsive materials, building-integrated photovoltaics, and vegetated facades, evaluating thermal performance, durability, lifecycle impact, carbon reduction, and urban environmental quality. It also examines the growing role of digital tools, including AI-assisted analysis, BIM, and predictive monitoring, in material selection, performance optimization, and long-term facade management. Case studies such as Bosco Verticale, The Edge, and Al Bahar Towers demonstrate how integrated facade strategies enhance energy efficiency, environmental performance, and design quality. Key barriers include high upfront costs, regulatory uncertainty, technical integration challenges, and maintenance complexity. Overall, the chapter argues that combining innovative materials with AI-supported design can advance net-zero and resilient urban development when guided by lifecycle thinking, supportive policies, and effective implementation.

DOI: 10.4018/979-8-3373-6023-2.ch002

1. INTRODUCTION

Cities are developing at a pace never witnessed before with the projection that in 2050, over 70 percent of the world population will be living in towns and cities (United Nations, 2018). These gigantic urban sprawls put additional strain on infrastructure, energy as well as a built environment. The building sector is behind nearly 40 percent of the total energy-related carbon dioxide emissions worldwide because building consumes a huge amount of energy, as well as produces greenhouse gases (IEA, 2021b). Therefore, architecture, engineering and construction (AEC) sector is undergoing a paradigm shift in terms of sustainable business operation and processes and technology and facades and building envelope have received most attention.

Among the components of architectural design, the cladding system - and the facade in particular - plays a major role in shaping aesthetic character, thermal behavior, environmental impact, and weather resilience. Rather than functioning only as a protective outer layer, facade systems now operate as active interfaces between interior and exterior conditions (Addington & Schodek, 2005; Bran, 2014).

The current fashion of innovation in the design of the facade improves the implementation of the use of high-performance materials that are multifunctional: thermal efficient, energy generative, air purifying, durable and have low environmental costs. The following materials such as bio-composites, phase-change materials (PCMs), building-integrated photovoltaic (BIPV), and nano-engineered coatings have become the primary ingredients of the next generation of green architecture. They may find an application as the complementary steps in the broader measures of carbonization, circular economy, and climate-responsive design.

Where material science, environmental engineering, and architectural innovation intersect, facade design becomes a tool for urban ecological repair and climate-responsive performance. Advances in biomimicry, digital fabrication, and material innovation are widening the role of building skins as active environmental moderators rather than passive barriers (Ashby, 2013; Bran, 2014).

However, although the future of such advanced materials looks very promising, there is still a series of limitations to their large-scale implementation: economic affordability, technical demandingness, the absence of unified rules, and ignorance of long-term operation in various climates (Asdrubali et al., 2015). The only solution to narrow this gap is to emerge with a full-fledged knowledge about these new materials, their life-cycle effects, how to implement them on a practical basis, and how they can be integrated into current construction methods.

Concrete, glass and aluminum are traditional cladding materials that traditionally have dominated the design of the building envelope but these materials are highly lacking when analyzed in the context of modern-day sustainability objectives. Examples of materials with very high embodied energy and embodied carbon include concrete and aluminum, since the extraction and manufacture processes are energy-intensive, and aluminum has a high embodied energy and embodied carbon of over 10 kg CO2-eq / kilogram of material. Glass facades are architecturally attractive but do not regulate thermal well due to their high solar heat gain combined with poor thermal insulation ability hence excess cooling is required in hot climates. Furthermore, these traditional materials usually offer minimal flexibility to the dynamic environmental conditions and do not offer the multifunctional features like in-built shade, energy production, or air cleaning. With the increasing pace of climate change and the built environment in search of post-carbon approaches to the production of net-zeros, the perceived wastefulness of conventional systems of cladding highlights the impetus to locate renewable, adaptive and low-carbon facade materials.

This chapter will bridge such knowledge deficit by providing an interdisciplinary critical review of the innovative materials of high-performance cladding systems, and the use of such systems in converting the facades into green, intelligent, responsive interface in an urban context. This chapter is faced with the synthesis of recent research, case studies, and environmental assessment strategies and reflects how cladding can lead to efficient energy saving, climate adaptation, beauty, and eco-optimization in cities.

1.1 Relevance of the Facade in Sustainable Urbanism

There is nothing more significant than the role of building facades in urban energy performance. A substantial share of building energy exchange is governed by the envelope, particularly through heating and cooling loads (Attia et al., 2017). Concrete panels, aluminum, steel, and glass remain common cladding materials, but they are not always sufficiently adaptable or energy efficient in the context of a warming climate. Increased embodied carbon, low thermal performance in certain situations, and limited recyclability are also associated with these materials (Pomponi & Moncaster, 2016).

The development of the smart and sustainable facade material is an opportunity to minimize energy consumption in operations, use the passive approach toward design, and counter the impact of the urban heat island. In addition, high-performance cladding systems play an important role in green engineering accreditation programs like LEED, BREEAM, and WELL, which focus on material transparency, the environmental impact, and the health of the occupants (U.S. Green Building Council, 2020).

1.2 Research Questions and Scope

The research questions that guide this chapter include the following:

1) What existing and emerging high-performance cladding materials offer significant environmental and functional benefits?
2) How do these materials contribute to the thermal, ecological, and aesthetic performance of buildings across different climates?
3) How should lifecycle cost, carbon footprint, and durability be assessed when comparing innovative cladding systems?
4) How can cladding systems incorporate nature-based solutions such as vertical gardens and living walls?
5) What barriers limit wider adoption and what opportunities could support scaling up these innovations?

To answer these questions, the chapter examines material innovation through the following analytical lenses:

1) Thermal, acoustic, and energy-efficiency performance
2) Environmental effects, including embodied energy, recyclability, and LCA
3) Technical integrity, including design flexibility, fabrication, and resilience
4) Design expression and architectural integration
5) Regulatory and socio-economic issues

1.3 Research Methodology

This chapter adopts a qualitative, review-based methodology that combines literature synthesis, comparative material assessment, lifecycle-oriented evaluation, and case-study interpretation (Snyder, 2019). The framework presented in Figure 1 structures the review around the following complementary steps:

1) Systematic review of peer-reviewed journal articles, white papers, standards, and conference proceedings.
2) Comparative evaluation of indicators such as thermal performance, embodied carbon, cost efficiency, and durability.
3) Life cycle assessment (LCA) perspective to examine environmental impacts from material production to end-of-life, consistent with ISO 14040 principles (ISO, 2006).
4) International case-study analysis of projects such as Bosco Verticale, The Edge, and Al Bahar Towers to assess how innovative facade strategies perform in practice.
5) SWOT-based interpretation of advanced cladding systems in relation to climatic flexibility, implementation risk, and scalability.

Figure 1. Research methodology framework

1.4 Goals of the Chapter

Within the frames of the research scope, the current chapter endeavors to:

1) Determine and assess innovative cladding technologies that align with sustainability and performance objectives.
2) Explain the technical and functional characteristics of bio-composites, phase change materials, nanomaterials, and BIPV systems.

3) Evaluate the ecological value of green facade systems, including vertical gardens and living walls.
4) Interpret environmental trade-offs through life cycle assessment principles.
5) Highlight implementation challenges, policy implications, and pathways for broader global adoption.

In doing so the chapter not only helps advance the existing body of scholarly research but also serves a broader and unifying body of knowledge to architects, material scientists, engineers as well as policymakers in the formation of their intent, the creation of a resilient energy-positive urban environment.

2. CONVENTIONAL AND INNOVATIVE CLADDING MATERIALS

The cladding materials have generally been used to fulfill two functions; appearance and protection of the buildings against external factors. In the context of the climate crisis, rise in energy demands and environmental deterioration though, these have come to be viewed increasingly as capable of more demanding roles, not only in terms of thermal insulation but active energy generation as well. In this section, the author gives a comparative review of the common cladding materials versus the new high-performing ones through a critical examination considering the aspects of thermal performance, ecological footprint, and flexibility and cycle behaviors.

The development of facade technologies can be divided into several unique stages, which can be traced back to the general changes in architecture, technology, and environment. Early facade systems were mainly solid massive load bearing walls and these were used to provide structural stability and not energy efficiency. The introduction of steel and reinforced concrete at the beginning of the 20th century made it possible to use lightweight curtain walls and facade glazing systems, which became the significant technological turnover. Since the 1980s, growing concerns on energy consumption saw the emergence of double-skin facades, high-performance insulation systems and ventilated rainscreens, bringing in enhanced thermal performance and moisture control. Ventilated-cavity strategies and solar-chimney concepts later expanded the discussion by linking facade design with passive airflow and natural ventilation (Zemitis et al., 2016; He et al., 2021). The rise of green facades and living walls emerged in the early 2000s, in which vegetation is incorporated as a passive cooling system and to improve the environment. In the 2010s, there has been a focus on adaptive and dynamic facade systems, such as kinetic shading, smart coatings and building-integrated photovoltaics. The current trend of research has recently moved to the optimization of materials aided by AI, digital twins, and bioengineered materials, showing a shift in direction between rigidly fixed envelopes to multifunctional and smart facade ecosystems.

2.1 Conventional Cladding Materials: Strengths and Limitations

Concrete panels, glass, natural stone, brick, ceramic tiles and aluminum panel are the most frequently applied materials in the traditional architecture. The selection of materials has been due to durability, strength, ability to withstand weather conditions and monetary flexibility in the aesthetic appeal. To give an example, in the majority of commercial high-rise buildings, people use glass and aluminum composite panels (ACPs) due to their design (as it gives the building a sleek look) and because of their versatility (it is easy to install them).

Many conventional facade materials do not perform well when assessed against current sustainability goals. Glass can support daylighting and visual openness, yet poorly configured glazed facades may increase solar heat gain and cooling demand in warm climates (Attia et al., 2017). At the same time, aluminum and concrete carry substantial embodied-energy and embodied-carbon burdens across production and end-of-life stages (Pomponi & Moncaster, 2016).

The traditional cladding materials are also rather passive systems; they do not connect to the environment being changed dynamically. They are not capable of responding to the changing climatic conditions or even assist in the energy efficiency of a building unless complicated insulation layers or mechanical support substructures are employed. Consequently, it is necessary to add additional HVAC systems to the buildings that are outfitted with the conventional cladding, which also adds to the amount of energy needed to operate such buildings (Kibert, 2016).

2.2 Performance Gaps and the Rise of Innovative Materials

The shortcoming of traditional construction materials has spurred the growth of interest in the radical cladding technology, as sustainability becomes a fundamental determinant of a building. The new materials present multi-functionality, responsiveness along with major environmental advantages. In contrast to traditional materials, which are combined mainly inert, the high-performance materials can respond to stimuli in the exterior world, control their temperature, energy supplies, clean the air, or even modify the properties in accordance with the environmental settings.

Innovative cladding materials can improve operational energy performance when they are carefully integrated into the building envelope. Climate-adaptive facade strategies, responsive shells, and higher-performing envelope assemblies can reduce operational loads while expanding the design role of the facade beyond static enclosure (Loonen et al., 2015).

In the following pages, major types of such sophisticated materials will be presented, such as lightweight composites, climate adaptive coatings, phytochemical panels, and the materials with functional grading. All of these inventions are redefining traditional uses of cladding as well as extending the limits of architectural sustainability.

2.3 Lightweight Composite Panels

Lightweight composites, generally resin matrices reinforced with fibers, which could be glass, carbon or natural fiber are stronger in proportion to its weight and have design freedom. They are also not as heavy to install, have lower structural weight and embodied energy than heavy masonry or stone panels (Ashby, 2013). This innovation in bio-composites with a new generation of bio-composites made up of hemp, flax, or recycled plastics has gone further to the domain of the sustainable designs.

In addition, the composites can be thermally insulated, acoustically designed; adding to overall comfort in the indoors. Their modular construction permits control of form, depth and layering of materials in such a way that they can be done accurately, something that cannot be easily achieved with traditional materials.

2.4 Bio-Based Panels and Renewable Materials

Bio-based panels use naturally occurring materials e.g. lignin, cellulose and mycelium to form panels that are frequently locally sourced, biodegradable and have low embodied carbon. The following panels have the capacity to facilitate the objectives of a circular economy, through facilitating cradle-to-cradle circuits of materials. Other technology advancements like hemp Crete panels and fungal mycelium boards promise to be thermally insulating, non-toxic and provide moisture control (Jones et al., 2020).

Living components of Phytochemical panels may be applied as well which will not only serve the purpose of air purification but also harmonize with the surrounding nature. They are especially helpful in green facades and may be used to give environmental value in towns which are extensively developed where very little vegetation exists.

2.5 Adaptive and Functionally Graded Coatings

Construction Functionally graded materials (FGMs) are fabricated in a gradual change in composition and structure, being able to provide differentiated responses to say, solar radiation, temperature, or moisture factors in the environment. Case in point, to optimize a panel, the side facing the sunlight could become reflective and retain thermal mass on the inside side (Zhang et al., 2021).

Also under this category are smart coatings that include photochromic, thermo chromic, and hydrophobic nano coating. These materials dynamically react to the alterations in light, temperature or even humidity. Applied to the structure's facade, they may change the reflectivity, self-clean, or repel the pollution, helping to save on maintenance and improve the performance of the building (Rabajczyk et al., 2021).

2.6 Climate-Sensitive Coatings and Active Systems

In contrast, however, to static traditional systems, climate-responsive coatings and dynamic cladding materials change their characteristics in real-time, in order to maximize interior comfort. One of them is phase change materials (PCMs) that can store and release heat during melting and solidification processes. In cases of cladding systems, PCMs improve this situation by minimizing or avoiding use of mechanical cooling or heating (Cabeza et al., 2011).

Additionally, kinetic facades such as panels that move or rotate in response to solar angles represent a new frontier in building adaptability. Though technologically complex and still relatively expensive, these systems offer high potential for climate-adaptive design, particularly in extreme climates (Sadegh et al., 2022).

2.7 Comparative Summary

The comparative performance characteristics of various cladding materials are summarized in Table 1, Performance comparison of different cladding materials and Figure 2, Material adaptability in construction systems, extends this comparison by visualizing how each material varies in terms of adaptability.

Table 1. Performance comparison of different cladding materials

Material Type	Thermal Performance	Embodied Carbon	Adaptability	Aesthetic Potential
Concrete Panels	Low	High	Low	Medium
Glass Façade	Very Low	Medium	Low	High
Aluminum Panels	Medium	High	Low	High
Lightweight Composites	Medium-High	Medium	Medium	High
Bio-Based Panels	Medium	Low	Medium	Medium-High
Smart/Responsive Materials	High	Medium	High	High

Source: Author's synthesis based on the cited literature and case studies

Figure 2. Material adaptability in construction systems

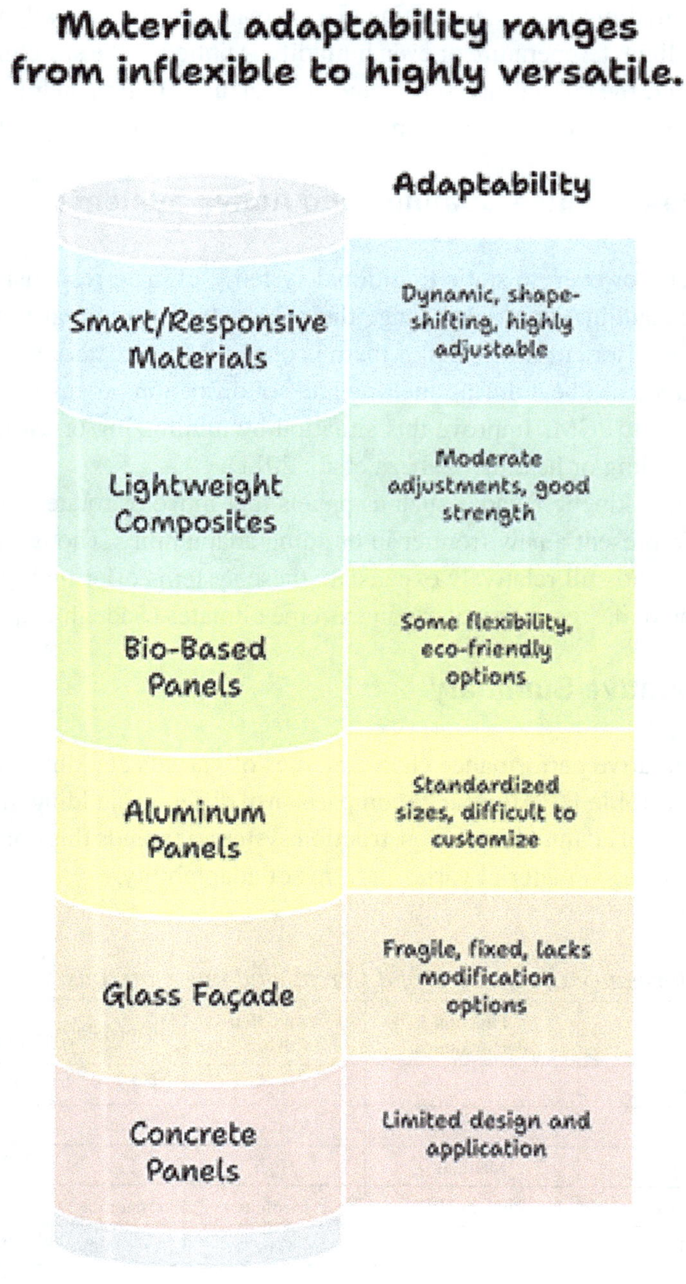

2.8 Concluding Remarks

Adaptability, low thermal performance, poor environmental cost; these are some of the factors that make the traditional cladding materials not adaptable overall, which lends credence to the underwhelming potentiality of alternatives of new generations. Revolutionary materials come in with a package of advantages which restructure the functionality of building facades. High-performance materials represent a way forward to resilient, flexible and visually engaging urban cities through the transition of static and extractive to dynamic and regenerative and isolated to integrated.

3. HIGH-PERFORMANCE CLADDING MATERIAL INNOVATIONS

3.1 Recycled and Bio-Based Materials

With growing urbanization and environmental awareness, the construction sector is increasingly exploring bio-based and recycled materials as substitutes for more carbon-intensive cladding products. These materials often derive from agricultural by-products, plant fibers, or recovered industrial waste streams, and they can offer lower embodied energy, carbon-storage potential, and improved circularity compared with conventional systems (Pittau et al., 2018; Jones et al., 2020; Zhang et al., 2021).

Hemp Crete, a mix of hemp fiber, lime, and water, is one of the thoroughly examined natural cavity fillers of a cladding system. Hemp Crete panels offer weak insulation, fire proof and moisture control and carbon dioxide sequester during its lifecycle (Arnaud & Gourlay, 2012). They are mostly used in regulating indoor humidity levels because of their porous nature which increases the thermal comfort without the use of mechanical systems. Mycelium based panels is another promising category and they are prepared by growing fungal mycelia on organic materials e.g. saw dust or straw. The weight of these panels is low, they are biodegradable, and have good thermal and acoustic insulation qualities (Jones et al., 2020). The aesthetic capabilities of mycelium composites have been tested by the research as well, and it has been discovered that they may be shaped to create architecturally interested constructions and may be touched up with natural dyes and resins to look better. The use of bamboo can be seen in the cladding systems especially in both the tropical and temperate zones. Bamboo is a very quick-growing renewable resource with an amazing tensile strength and dimensional stability when treated well. Composite prefabricated version is structural, and fireproof material applicable in facades modules (Sharma et al., 2015).

Another potential route toward sustainability is the use of recycled plastic composite panels, especially where post-consumer polymers are combined with natural fibers. These materials can offer durability and moisture resistance for exterior applications while lowering lifecycle burdens relative to more carbon-intensive alternatives. Nevertheless, uptake remains limited because of uneven performance data, fewer structural ratings, and concerns about long-term durability (Bribian et al., 2010). Moisture sensitivity, UV degradation, and biological attack also need to be addressed through coatings, detailing, or hybrid assemblies.

However, as material testing techniques, digital design and performance models, evolve, so do the credibility of these materials. Such integration in building codes and certification systems such as LEED, DGNB and WELL will be critical steps towards achieving the massive adoption in the urban construction.

3.2 Intelligent and Responsive Materials

Smart and responsive materials represent an important category of contemporary cladding technologies because they can adapt to variations in light, temperature, or pollution. Compared with static systems, these materials add adaptive behavior to the building envelope, helping to moderate thermal loads, improve occupant comfort, and reduce HVAC demand (Aelenei et al., 2016; Sadegh et al., 2022).

Phase change materials (PCMs) are among the most studied technologies in this area. Integrated into facade assemblies, PCMs absorb and release thermal energy during phase transitions, helping to buffer indoor temperature fluctuations, especially where diurnal thermal variation is high (Zhou et al., 2012; Giro-Paloma et al., 2015).

Even though intelligent and sensitive materials provide significant performance benefits, long-term monitoring shows that significant limitations should be considered. Experiments with phase change materials (PCMs) show that the thermal storage capacity can reduce by 8-15 percent of its original value after 2,000-3,000 melting-solidification cycles and leakage danger is more severe in such materials, which are not well encapsulated. Coatings that make use of thermochromics have been reported to become visible faded or low switching efficiency after 5-7 years of UV exposure, particularly in tropical conditions. Similarly, adaptive facades operated by sensors also rely on electronic elements and the industry statistics indicate sensor failure rates 2-5 annually, a failure that necessitates periodic adjustment. These findings underscore the importance of ensuring that smart materials, though offering new solutions, have to be maintained in the long term by employing effective strategies of maintenance, lifecycle testing and use of predictive monitoring systems.

Another important group of intelligent cladding technologies includes thermochromic and photochromic coatings. These materials alter optical properties such as transmittance and reflectance in response to environmental change, supporting passive control of solar heat gain in adaptive envelopes (Loonen et al., 2015). Self-cleaning nano-engineered surfaces, often based on titanium dioxide, also contribute to facade performance by limiting dirt accumulation and supporting photocatalytic removal of some air pollutants in urban environments (Rabajczyk et al., 2021).

Besides, kinetic facades/responsive shading devices are in development: shape-memory alloys (SMAs), hydrogel-based composites are being tested to be used in the future. These materials deform in a reversible manner with environmental changes and architecture elements can move or change shape without motor and electronic devices (Addington & Schodek, 2005). Despite these positive trends, smart materials have a number of problems such as high cost of installation, sorting out the intricate interconnection, and sustainability, in terms of long-term serviceability in real-life applications. They depend on the accurate model, lifecycle analysis and post-installation monitoring of their performance (Stazi et al., 2016).

These materials are, however, considered as a very important milestone in the development of climate-responsive buildings that would be able to lower down the energy demand but have a futuristic design.

3.3 Photovoltaic and Energy-Generating Facades

The next generation of building facade systems: Photovoltaic (PV) and energy-generating facade systems are transforming the synergy between buildings and energy infrastructures. The cladding technology, taking care of building integration of solar energy in the form of solar energy harvesting technologies, allows buildings to become partial or full energy generators. These Solar Built Surface (in short Building Integrated Photovoltaic-BIPV) systems not only combine utility and beauty but also provide clean energy production at no cost to architecture (Peng et al., 2011).

BIPV systems differ by replacing conventional facade material with PV modules, whether crystalline silicon or thin-film solar cell, embedded in panels, glazing or shading. Such systems are especially favorable to high-rise and business buildings, where vertical surface area is excessive, and there is minimum rooftop area (Jelle et al., 2012). Large amounts of energy can be generated by facade-mounted PV especially on south- or west-facing facades in sun-rich areas. Such technological innovations as the invention of semi-transparent PV glass, flexible solar films, and color-customized panels have permitted architectures to balance the design objectives and energy performance (Hong et al., 2018). Additionally, the hybrid facade systems (combination of PV with thermal insulation, ventilated cavity or responsive coatings) can be used to augment the energy efficiency of the building.

Despite these advantages, integrating PV into facades involves several technical constraints, including heat build-up, reduced efficiency under non-optimal orientation, increased structural load, and higher upfront cost. Designers must also account for wiring, shading analysis, maintenance access, and whole-envelope integration (Jelle et al., 2012; Mangherini et al., 2023). At the same time, BIPV supports low-carbon construction and net-zero building agendas, especially when combined with adaptive facades, smart controls, and digital performance management.

4. GREEN FACADE SYSTEMS

Integrating green facade systems brings architecture, ecology, and urban sustainability into direct interaction. These systems introduce vegetation - climbing plants, planted panels, and living-wall assemblies - onto the outer skin of buildings. Beyond aesthetic value, green facades can generate economic, environmental, and psychological benefits, while also improving urban microclimates and strengthening residents' access to restorative green space (Perini & Rosasco, 2013; Alexandri & Jones, 2008; Ulrich et al., 1991; He et al., 2017).

Figure 3. Advantages and disadvantages of green facade applications

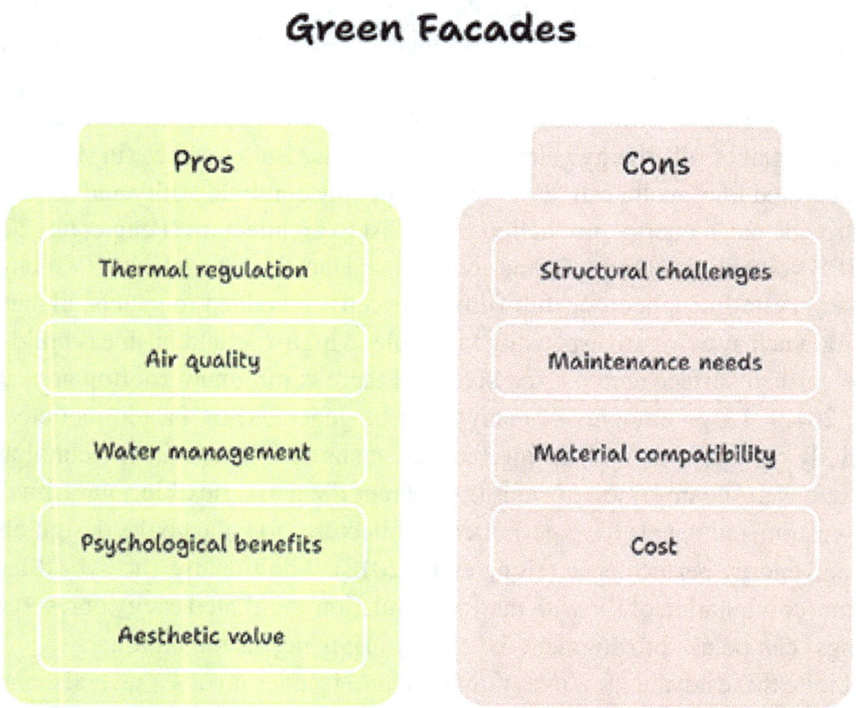

5. LIFE CYCLE ASSESSMENT AND ENVIRONMENTAL EFFECTS

The use of new cladding and green facade systems is beginning to gain popularity as the cities turn to greenery and become more sustainable and climate-resistant. Nonetheless, the choice of such materials should not just base its selection criteria on performance basis and aesthetics only, but on their entire environmental effect through the lifecycle. Here Life Cycle Assessment (LCA) is an essential analyzing tool. LCA assesses environmental impacts on all the phases of a product, the raw material extraction, the production process, transportation, installation, the working life cycle, the maintenance of the said property and disposal or rebirth (ISO, 2006).

The LCA, by measuring the effect of global warming potential (GWP), embodied energy, resource depletion, acidification, and toxicity, allows making informed material choices and design recommendations that would complement sustainable development and low-carbon construction (Cabeza et al., 2014).

Alongside regional frameworks such as Level(s), LEED, and related green-building standards, international climate commitments are increasingly shaping the adoption of low-carbon and renewable facade materials. The Paris Agreement established a global decarbonization framework that has encouraged signatory states to strengthen low-carbon construction policies and performance requirements for buildings (UNFCCC, 2015). These regulatory shifts reinforce the case for adaptive, renewable, and lifecycle-conscious cladding systems.

5.1 Embodied Energy and Carbon Footprint

Embodied energy is the overall energy required to manufacture and transport a substance to the construction site and embodied carbon is an equivalent greenhouse gas emission. Aluminum and glass which are traditional cladding materials are characterized by high embodied energy since they have energy intensive production processes. An example is that of aluminum that produces more than 10 kg-CO2 -eq kg-1 of material, one of the busiest facade materials (Pomponi & Moncaster, 2016).

Bio-based products, including hempcrete, timber composites, and mycelium panels, on the contrary, have much less embodied energy and, in their growth and manufacturing phases, can even serve as carbon sinks (Churkina et al., 2020). Such materials trap carbon dioxide in the air partially compensating other lifecycle stages. In a comparative study, Pittau et al. (2018) revealed that replacing traditional cladding with their bio-based counterparts would lead to a 30-70 reduction of the CO 2 emissions involved with the facade.

5.2 Operational Performance and Lifecycle Effects

While embodied impacts dominate the early lifecycle stages of a material, operational performance - including thermal insulation, solar control, and heat-delay behavior - strongly influences long-term environmental outcomes. High-performance envelope systems such as ventilated facades, insulated sandwich panels, and PCM-enhanced assemblies can reduce heating and cooling loads over extended service periods (Yucer & Hepbasli, 2010; Zhou et al., 2012).

Such example is the application of phase change materials (PCMs) within the facades, which has the potential to produce the reduction of peak indoor temperatures by 2-5C, therefore, reducing the need to use air conditioning systems and lowering carbon emissions in the use phase (Zhou et al., 2012). In the same way, green facades can be used to improve passive cooling by inducing shading and evapotranspiration, which over 30 years can have higher climate impacts than more embodying high-tech systems (Manso & Castro-Gomes, 2015).

5.3 Maintenance, Durability and Longevity

Longevity material Longevity material has direct effects on environmental performance within the lifecycle. Sturdy materials minimize the replacements, increasing the service life of cladding systems and decreasing the overall effect. As an example, self-cleaning materials and nano-coatings extend the functional life of the facades as they prevent dirt build-up and protect the surface against degradation processes, therefore, reducing the necessity of conducting repairs and replacing the components prematurely (Stazi et al., 2016).

However, biological degradation, root intrusion, and damage due to moisture may adversely affect the integrity of the structure in green facades in the long run. Performance and water-balance studies of vegetated envelope systems also show that substrate moisture, irrigation strategy, and detailing can significantly affect long-term behavior and maintenance demand (Jim & Peng, 2012; Fassman-Beck et al., 2013). Most LCA-based studies propose that because of support frameworks and irrigation schemes, the early life cycle effects of green walls might be worse than ordinary greenery, but the operational advantages and overall benefits of ecosystem services, pollution regulation, biodiversity and lowering the urban climatic temperature, compensate the embodied burdens in the longer run (Perini & Rosasco, 2013).

5.4 End-of-Life and Recycling

The least considered aspect of the end of life (EoL) of cladding material is the EoL itself and it has significant ramifications about sustainability. The end of a closed-loop circular economy is a material that can be recycled, upcycled, or even degrade itself, and thus less virgin material has to be extracted and directed to disposal. As an example, aluminum panels can be recycled few times with a relatively low amount of energy being used, although there prevail problems regarding their separation out of composite layers of a mixed-material facade (Ashby, 2013; Zhang et al., 2021).

Bio-based and biodegradable materials have the benefit of being compostable or allow low impact biodegradation in EoL management. Composite products such as Mycelium-based cladding can be put straight back into soil with little processing or a modular facade system can be extensively reassembled and reused in another place. Nevertheless, their implementation may demand the use of design for disassembly (DfD) strategies, since they are currently not the universal aspect of building codes.

LCA-based Environmental Product Declaration (EPDs) are increasingly being used in the comparison of EoL scenarios to stimulate circular practices of material sourcing.

5.5 Decision-Making Tool and Tools of Comparative Analysis

The modeling of the differences between the environmental impacts of cladding materials can be carried out with the help of various database and LCA software platforms, which include Sima Pro, GaBi, Open LCA, and convent. The tools also facilitate multi-criteria decisions since they are a combination of environment, performance, and economic-based measurements. The most innovative models actually combine the Building Information Modeling (BIM) and LCA data to facilitate real-time impacts at design time (Heigermoser et. al. 2019). Similar systems such as Cradle-to-Cradle (C2C) certification, BRE Environmental Profiles and LEED v4 Materials and Resources credit systems further standardize LCA usage in sustainable building assessment.

5.6 Market Trends and Regulatory Context

Within Europe market places, reporting of LCA is becoming compulsory, as it is happening globally. Other countries such as Netherlands, Sweden, and France have used some LCA metrics to be incorporated in the building codes, making the utilization of low-impact materials encourage able. The Level(s) framework of the EU is also encouraging lifecycle thinking and material transparency of the public procurement and design guidance.

In the meantime, material suppliers are on their way to releasing product-specific LCA data and EPDs (that are third party-verified) to meet the market requirements of sustainability certification credentials. This openness enables designers to make decisions based on data to address the future of carbon neutrality and resilience of urban plans.

6. CASE STUDIES

The performance of innovative cladding materials and green facade systems is best understood through built examples. The projects discussed here illustrate how bio-based materials, smart facade technologies, vegetated envelopes, and energy-generating skins can be combined in practice to support sustainability, energy performance, and architectural quality.

- Bosco Verticale, Milan, Italy (Boeri, 2015)
- The Edge, Amsterdam, Netherlands
- Al Bahar Towers, Abu Dhabi, UAE
- One Central Park, Sydney, Australia
- Parkroyal Collection Pickering, Singapore

6.1 Lessons Learned

1) Combination of Multiple Systems

The majority of the projects in question have been able to integrate several facade technologies, including BIPV, green walls, and dynamic shading among others in order to achieve optimal energy, environmental, and aesthetic performance. The Edge in Amsterdam is able to combine solar energy harvesting with smart thinking about the double-skin facade and Bosco Verticale can be considered as a combination of a biophilic building and interesting irrigation designs.

2) Climate-Specific Design

The projects that are successful are those that are designed with regard to the local climatic condition. Al Bahar Towers provide a movement shelter against severe sun of Abu Dhabi whereas Bosco Verticale is calibrated to temperate weather and use of dense vegetation in Milan.

3) Maintenance and Longevity

Maintenance strategies are very important in all case studies. Green facades, especially, avoid to be left without irrigation, pruning and pest management to ensure long-time effectiveness. Sophisticated surveillance systems like those used in One Central Park disrupt manual transmission of water and nutrients since the systems provide automatic delivery of water and nutrients.

4) Economic and Social Value

The preliminary cost of investing in innovational facades can be great, but with time-consequences such as decreased cost of operations, enhancement of property value and overall welfare of occupants, the cost is vital to the overall benefits. An example of such energy efficient building is The Edge, which not only saves huge cost in energy but is also used as a model of corporate sustainability.

5) Policy and Certification Support

The green building frameworks like LEED, BREEAM, or Green Star that prompted the utilization of new materials and facade technologies have played to the advantage of the majority of the case studies. Moreover, incentives and governmental regulations that are market-driven were instrumental in the effective implementation of solutions in projects such as One Central Park and the Park royal Collection Pickering. The principles behind these innovative designs of facade are illustrated in Figure 4, Innovative facade design approaches.

Figure 4. Innovative facade design approaches

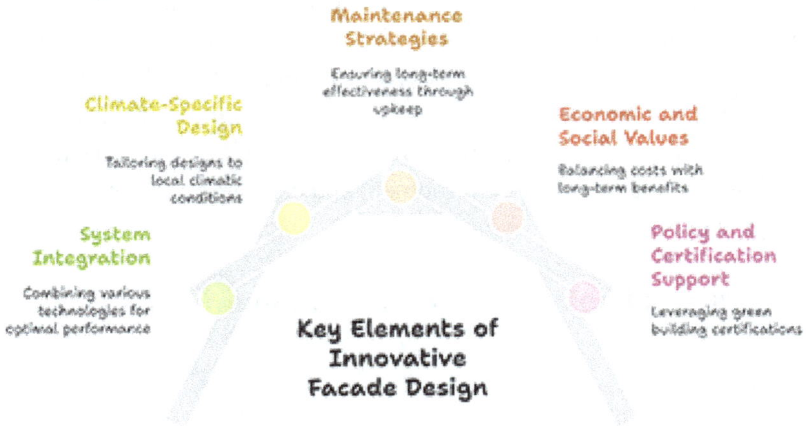

6.2 Comparative Analysis

The examples above take place in the world and underline the revolutionary possibilities of integrated facade solutions. Although the innovations of material used in each project are dissimilar, they all get dedicated in making sure that the operation energy has been reduced, better outcomes to the ecology, and visually powerful architecture. These different forms of green technologies feeding, smart materials, and renewable energy systems are essential sources of synergy in providing new standards of building performance in cities.

The durability measurement in innovative facade systems is supported by long-term performance data teased out of world case studies. In one case, Bosco Vertical reported an average of 3-5 plant replacement rate after the building was occupied (post-occupancy evaluation), mostly because of the exposure to wind at high levels. In the Al Bahar Towers, kinetic shading panels proved to have a 97% success rate of operation in the first 5 years although periodic recalibration of the actuators was necessary. According to the statistics of One Central Park, hydroponic living wall modules would need maintenance after 6-8 months to avoid nutrient discrepancies and pump malfunctions. These empirical results will be critical in determining the viability in the long run and the stakeholder issues on maintenance overheads and system reliability.

7. CHALLENGES AND OPPORTUNITIES

The inclusion of innovative materials in cladding, intelligent facade systems, and green envelopes marks a substantial shift toward climate-responsive and sustainability-oriented architecture. Despite clear environmental and design advantages, large-scale uptake remains constrained by cost, maintenance requirements, regulatory uncertainty, and durability concerns. Similar implementation barriers have also been noted in sustainable urban-development and climate-policy transitions (Bran, 2014; Reckien et al., 2019).

7.1 Barriers and Implementation Difficulties

1) High upfront cost
2) Maintenance and operational complexity
3) Technical and climatic constraints
4) Regulatory and certification barriers
5) Risk aversion and limited awareness

Figure 5. Challenges in adopting advanced cladding systems

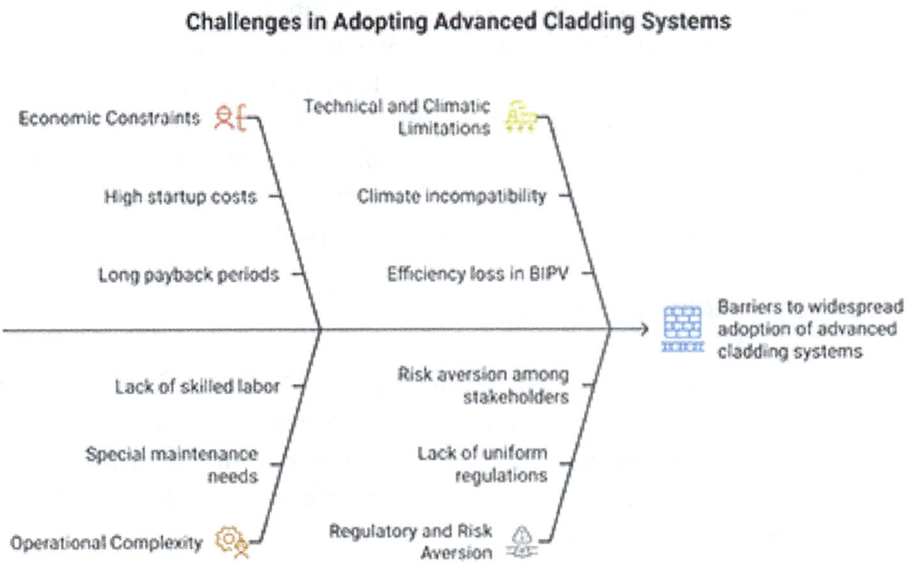

7.2 Opportunities and Emerging Trends

7.2.1 Increasing Demand for Low-Carbon Solutions

Growing commitment to net-zero buildings and low-carbon construction is increasing demand for innovative cladding materials. International roadmaps for building-sector decarbonization and performance improvement reinforce the importance of envelope innovation in reducing both operational and embodied impacts (IEA, 2021a; IEA, 2021b).

7.2.2 Digital Technology and Artificial Intelligence

Digital twin platforms, BIM-based workflows, and AI-assisted analytical tools are increasingly shaping cladding design, evaluation, and maintenance practice. Recent review studies show that AI in architecture and BIM is being applied to performance optimization, scenario testing, predictive maintenance, and design decision support (Bolek et al., 2023; Bassir et al., 2023). In facade contexts, these tools can help designers compare materials, anticipate operational behavior, and respond more effectively to climate and occupancy conditions.

7.2.3 Optimizing Performance and Modulation

With reduced wastage of material and new possibilities of customized cladding solutions, 3D printing and additive manufacturing are creating the next frontier of explorative manufacturing. Computer controllable advanced materials such as biodegradable panels or functional graded materials can be accurately manufactured to suit precise thermal, acoustic or structural requirements (Wu et al., 2016). Such technology also allows creating more complex geometries and biomimetic patterns that typically are hard to replicate with conventional methods.

7.2.4 Green Building Certifications

Green-building certification systems such as LEED, BREEAM, and WELL increasingly encourage developers to adopt facade materials with stronger energy, health, and recyclability performance. These frameworks also help structure market demand for BIPV systems, green facades, and lower-carbon envelope assemblies (U.S. Green Building Council, 2020).

7.2.5 Nature-Technology Synergy

Synergy between nature-based and technology-based facade strategies becomes most valuable when ecological and engineering functions are integrated rather than treated separately.

Combining green facades or living walls with dynamic shading, smart controls, or PV-integrated systems can produce more complete urban-design solutions by linking ecological benefits such as cooling and biodiversity support with energy and performance objectives (Manso & Castro-Gomes, 2015).

7.3 Future Outlook

Future cladding systems are likely to become increasingly multi-functional, adaptive, and data-informed. Research is already exploring self-healing bio-based composites, algae-integrated facade concepts, transparent photovoltaic glazing, and digitally assisted design workflows that support fabrication, monitoring, and circularity (Wu et al., 2016; Bolek et al., 2023).

The involvement of both material scientists, architects, urban planners, and policymakers is going to be decisive in avoiding the technical, economic, and regulatory wells. New generation cladding systems will become a fundamental part of sustainable cities as digital technologies evolve and carbon responsibility rises in the list of priorities.

8. CONCLUSIONS AND RECOMMENDATIONS

8.1 Conclusions

Rapid urbanization, climate pressures, and the need for more sustainable buildings have transformed the role of cladding and facade systems in contemporary architecture. Facades are no longer treated only as protective outer skins; they are increasingly expected to contribute directly to energy efficiency, environmental performance, and urban resilience.

This chapter has examined a broad range of innovative cladding strategies, including bio-based composites, phase change materials, building-integrated photovoltaics, and green facade systems. Taken together, these approaches show how facades can reduce operational energy use, improve thermal comfort, support urban biodiversity, and enhance architectural expression.

A central lesson is the importance of lifecycle thinking. Materials that perform well in operation must also be evaluated in terms of embodied carbon, maintenance demand, durability, and end-of-life recovery if they are to support genuinely sustainable building practice.

The analysis also shows that adoption still faces practical barriers, including high upfront cost, technical integration challenges, limited standardization, and maintenance complexity, particularly in living, responsive, or kinetic facade systems.

Even so, the long-term outlook remains strong. As digital tools, lifecycle assessment methods, advanced materials, and policy incentives continue to mature, high-performance cladding is well positioned to support net-zero and climate-responsive urban development.

8.2 Recommendations

Based on the analysis presented in this chapter, the following recommendations are proposed for researchers, practitioners, and policymakers:

8.2.1 Lifecycle Assessment in Early Design Stages

Lifecycle assessment should be introduced early in design so that architects and engineers can compare cladding options not only on operational performance, but also on embodied carbon and end-of-life implications. Environmental product declarations and LCA tools can strengthen these early-stage decisions (Heigermoser et al., 2019).

8.2.2 Hybrid Facade System Promotion

Single-material cladding solutions rarely satisfy all performance goals. Hybrid assemblies - for example, combining BIPV with smart coatings, insulation, or vegetated facade components - can deliver more balanced outcomes across energy generation, shading, durability, and aesthetics.

8.2.3 Standardization and Policy Incentives

Governments ought to avail privileges, subsidies, tax rebates on covering up with new cladding, particularly those that have been shown to have an improved effect on the environment. Meanwhile, normalization of performance tests (i.e., fire-resistance, thermal performance, and structural integrity) will also help eliminate ambiguity among developers. EU countries, by way of example, have gone ahead and incorporated LCA metrics into their building codes through the Level(s) framework.

8.2.4 Digital and AI-Based Designing Tools

AI-assisted optimization and BIM-based performance modeling can support facade design by improving decisions about energy generation, daylighting, thermal comfort, and material efficiency. Digital twin approaches are also useful for long-term monitoring and maintenance planning in complex facade systems (Bolek et al., 2023; Bassir et al., 2023).

8.2.5 Training and Capacity Building

Advanced cladding systems also demand skilled persons to implement. There needs to be developed training program running to architects, contractors and maintenance teams so that they can learn how to install, monitor and maintain the smart facade, BIPV and vertical gardens. The universities and professional organizations can be significant in terms of filling this knowledge gap.

8.2.6 Performance Tracking in the Long-Term

Performance data of innovative materials in the real world are few and limited. It is suggested that pilot projects and demonstration buildings should receive monitoring systems in order to find the energy savings, maintenance costs and durability with time. Such information will be essential to assert, amend design criteria, and enhance subsequent versions of cladding technologies.

8.2.7 Circular Design by Intent

Material selection should also reflect the potential for reuse, repair, and recycling. Design for disassembly, modular assemblies, and circular-design principles can make innovative cladding systems easier to recover and repurpose at the end of their service life, especially when facade decisions are considered across long-life building stock transitions (Ashby, 2013; Kohler & Hassler, 2002).

8.3 Future Directions

Future facade development is likely to move toward more adaptive systems that respond to real-time environmental data. Emerging technologies such as self-healing bio-based composites, algae-based panels, and transparent photovoltaic glass suggest that the next generation of facade systems will combine environmental responsiveness with stronger energy performance, while also needing to be evaluated in relation to retrofit pathways across the wider building stock (Kohler & Hassler, 2002).

Nature-based strategies integrated with digital intelligence are likely to become increasingly central to urban facade design. Real progress, however, will depend on collaboration among material scientists, architects, engineers, urban planners, manufacturers, and policymakers so that innovative cladding solutions remain scalable, affordable, and durable, and are supported by coherent urban-climate planning frameworks (Reckien et al., 2019).

ACKNOWLEDGMENT

The author acknowledges the scholarly literature, technical standards, and case-study sources that informed this chapter. Appreciation is also extended to the wider interdisciplinary research community working at the intersection of facade engineering, sustainable materials, and climate-responsive design.

REFERENCES

Addington, M., & Schodek, D. (2005). *Smart Materials and Technologies in Architecture*. Architectural Press.

Aelenei, D., Gonçalves, H., & Pereira, R. (2016). Adaptive facades for sustainable buildings. *Energy Procedia*, *91*, 269–275. DOI: 10.1016/j.egypro.2016.06.218

Alexandri, E., & Jones, P. (2008). Temperature decreases in urban canyons due to green walls and green roofs in diverse climates. *Building and Environment*, *43*(4), 480–493. DOI: 10.1016/j.buildenv.2006.10.055

Arnaud, L., & Gourlay, E. (2012). Experimental study of parameters influencing mechanical properties of hemp concrete. *Construction & Building Materials*, *28*(1), 50–56. DOI: 10.1016/j.conbuildmat.2011.07.052

Asdrubali, F., Ferracuti, B., Lombardi, L., Guattari, C., Evangelisti, L., & Grazieschi, G. (2015). A review of structural, thermo-physical, acoustical, and environmental properties of wooden materials for building applications. *Building and Environment*, *114*, 307–332. DOI: 10.1016/j.buildenv.2016.12.033

Ashby, M. F. (2013). *Materials and the Environment: Eco-informed Material Choice* (2nd ed.). Butterworth-Heinemann.

Attia, S., Eleftheriou, P., Xeni, F., Morlot, R., Ménézo, C., Kostopoulos, V., Betsi, M., Kalaitzoglou, I., Pagliano, L., Cellura, M., Almeida, M., Ferreira, M., Baracu, T., Badescu, V., Crutescu, R., & Hidalgo-Betanzos, J. M. (2017). Overview and future challenges of nearly zero energy buildings (nZEB) design in Southern Europe. *Energy and Building*, *155*, 439–458. DOI: 10.1016/j.enbuild.2017.09.043

Bassir, D., Lodge, H., Chang, H., & Celik, T. (2023). Application of artificial intelligence and machine learning for BIM [Review]. *International Journal for Simulation and Multidisciplinary Design Optimization*, *14*, 5. DOI: 10.1051/smdo/2023005

Boeri, S. (2015). *Bosco Verticale: A home for trees, humans, and birds*. Corraini Edizioni.

Bolek, B., Tutal, O., & Ozbasaran, H. (2023). A systematic review on artificial intelligence applications in architecture. *Journal of Design for Resilience in Architecture and Planning*, *4*(1), 91–104. DOI: 10.47818/DRArch.2023.v4i1085

Bran, F. (2014). *Green infrastructure for sustainable urban development.* Urbanism. Architecture. *Constructions*, *5*(2), 29–38.

Bribian, I. Z., Capilla, A. V., & Usón, A. A. (2010). Life cycle assessment of building materials: Comparative analysis of energy and environmental impacts and evaluation of the eco-efficiency improvement potential. *Building and Environment*, *46*(5), 1133–1140. DOI: 10.1016/j.buildenv.2010.12.002

Cabeza, L., Castell, A., Barreneche, C., De Gracia, A., & Fernández, A. (2011). Materials used as PCM in thermal energy storage in buildings: A review. *Renewable & Sustainable Energy Reviews*, *15*(3), 1675–1695. DOI: 10.1016/j.rser.2010.11.018

Cabeza, L. F., Rincón, L., Vilariño, V., Pérez, G., & Castell, A. (2014). Life cycle assessment (LCA) and life cycle energy analysis (LCEA) of buildings and the building sector: A review. *Renewable & Sustainable Energy Reviews*, *29*, 394–416. DOI: 10.1016/j.rser.2013.08.037

Churkina, G., Organschi, A., Reyer, C. P. O., Ruff, A., Vinke, K., Liu, Z., & Schellnhuber, H. J. (2020). Buildings as a global carbon sink. *Nature Sustainability*, *3*(4), 269–276. DOI: 10.1038/s41893-019-0462-4

Fassman-Beck, E. A., Simcock, R., Wang, S., & Gruner, D. (2013). 4.6 Vegetated stormwater biofilters: Green roofs and living walls. In *Handbook of Water Sensitive Planning and Design.*

Giro-Paloma, J., Martínez, M., Cabeza, L. F., & Fernández, A. I. (2015). Types, methods, techniques, and applications for microencapsulated phase change materials (MPCM): A review. *Renewable & Sustainable Energy Reviews*, *53*, 1059–1075. DOI: 10.1016/j.rser.2015.09.040

He, G., Wu, Q., Li, Z., Ge, W., Lv, D., & Cong, L. (2021). Ventilation performance of solar chimney in a test house: Field measurement and validation of plume model. *Building and Environment*, *193*, 107648. DOI: 10.1016/j.buildenv.2021.107648

He, Y., Yu, H., Ozaki, A., Dong, N., & Zheng, S. (2017). Long-term thermal performance evaluation of green roof system based on two new indexes: A case study in Shanghai area. *Building and Environment*, *120*, 13–28. DOI: 10.1016/j.buildenv.2017.04.001

Heigermoser, D., De Soto, B. G., Abbott, E. L. S., & Chua, D. K. H. (2019). BIM-based Last Planner System tool for improving construction project management. *Automation in Construction*, *104*, 246–254. DOI: 10.1016/j.autcon.2019.03.019

Hong, S., Choi, A., & Sung, M. (2018). Impact of bi-directional PV blind control method on lighting, heating and cooling energy consumption in mock-up rooms. *Energy and Building*, *176*, 1–16. DOI: 10.1016/j.enbuild.2018.07.022

International Energy Agency (IEA). (2021a). *Net Zero by 2050: A Roadmap for the Global Energy Sector.* https://www.iea.org/reports/net-zero-by-2050

International Energy Agency (IEA). (2021b). *Tracking Buildings 2021.* https://www.iea.org/reports/tracking-buildings-2021

ISO. (2006). *ISO 14040:2006 Environmental management Life cycle assessment Principles and framework.* International Organization for Standardization.

Jelle, B. P., Breivik, C., & Drolsum Rokenes, H. (2012). Building integrated photovoltaic products: A state-of-the-art review and future research opportunities. *Solar Energy Materials and Solar Cells*, *100*, 69–96. DOI: 10.1016/j.solmat.2011.12.016

Jim, C., & Peng, L. L. (2012). Substrate moisture effect on water balance and thermal regime of a tropical extensive green roof. *Ecological Engineering*, *47*, 9–23. DOI: 10.1016/j.ecoleng.2012.06.020

Jim, C., & Peng, L. L. (2012). Substrate moisture effect on water balance and thermal regime of a tropical extensive green roof. *Ecological Engineering*, *47*, 9–23. DOI: 10.1016/j.ecoleng.2012.06.020

Jones, M., Huynh, T., Dekiwadia, C., Daver, F., & John, S. (2020). Mycelium composites: A review of engineering characteristics and growth kinetics. *Journal of Bionic Engineering*, *17*(5), 843–872.

Kibert, C. J. (2016). *Sustainable Construction: Green Building Design and Delivery* (4th ed.). Wiley.

Kohler, N., & Hassler, U. (2002). The building stock as a research object. *Building Research and Information*, *30*(4), 226–236. DOI: 10.1080/09613210110102238

Loonen, R. C., Trčka, M., Cóstola, D., & Hensen, J. L. (2015). Climate adaptive building shells: State-of-the-art and future challenges. *Renewable & Sustainable Energy Reviews*, *25*, 483–493. DOI: 10.1016/j.rser.2013.04.016

Mangherini, G., Diolaiti, V., Bernardoni, P., Andreoli, A., & Vincenzi, D. (2023). Review of Façade Photovoltaic Solutions for Less Energy-Hungry Buildings. *Energies*, *16*(19), 6901. DOI: 10.3390/en16196901

Manso, M., & Castro-Gomes, J. (2015). Green wall systems: A review of their characteristics. *Renewable & Sustainable Energy Reviews*, *41*, 863–871. DOI: 10.1016/j.rser.2014.07.203

Peng, J., Lu, L., & Yang, H. (2011). Review on life cycle assessment of energy payback and greenhouse gas emission of solar photovoltaic systems. *Renewable & Sustainable Energy Reviews*, *19*, 255–274. DOI: 10.1016/j.rser.2012.11.035

Perini, K., & Rosasco, P. (2013). Cost–benefit analysis for green facades and living wall systems. *Building and Environment*, *70*, 110–121. DOI: 10.1016/j.buildenv.2013.08.012

Pittau, F., Krause, F., Lumia, G., & Habert, G. (2018). Fast-growing bio-based materials as an opportunity for storing carbon in exterior walls. *Building and Environment*, *129*, 117–129. DOI: 10.1016/j.buildenv.2017.12.006

Pomponi, F., & Moncaster, A. (2016). Embodied carbon mitigation and reduction in the built environment – What does the evidence say? *Journal of Environmental Management*, *181*, 687–700. DOI: 10.1016/j.jenvman.2016.08.036 PMID: 27558830

Rabajczyk, A., Zielecka, M., Klapsa, W., & Dziechciarz, A. (2021). Self-Cleaning coatings and surfaces of modern building materials for the removal of some air pollutants. *Materials (Basel)*, *14*(9), 2161. DOI: 10.3390/ma14092161 PMID: 33922766

Reckien, D., Salvia, M., Pietrapertosa, F., Simoes, S., Olazabal, M., De Gregorio Hurtado, S., Geneletti, D., Lorencova, E. K., D'Alonzo, V., Krook-Riekkola, A., Fokaides, P., Ioannou, B., Foley, A., Orru, H., Orru, K., Wejs, A., Flacke, J., Church, J., Feliu, E., & Heidrich, O. (2019). Dedicated versus mainstreaming approaches in local climate plans in Europe. *Renewable & Sustainable Energy Reviews*, *112*, 948–959. DOI: 10.1016/j.rser.2019.05.014

Sadegh, S. O., Gasparri, E., Brambilla, A., & Globa, A. (2022). Kinetic facades: An evolutionary-based performance evaluation framework. *Journal of Building Engineering*, *53*, 104408. DOI: 10.1016/j.jobe.2022.104408

Sharma, B., Gatóo, A., Bock, M., & Ramage, M. (2015). Engineered bamboo for structural applications. *Construction & Building Materials*, *81*, 66–73. DOI: 10.1016/j.conbuildmat.2015.01.077

Snyder, H. (2019). Literature review as a research methodology: An overview and guidelines. *Journal of Business Research*, *104*, 333–339. DOI: 10.1016/j.jbusres.2019.07.039

Stazi, F., Tomassoni, E., & Di Perna, C. (2016). Super-insulated wooden envelopes in Mediterranean climate: Summer overheating, thermal comfort optimization, environmental impact on an Italian case study. *Energy and Building*, *138*, 716–732. DOI: 10.1016/j.enbuild.2016.12.042

Ulrich, R. S., Simons, R. F., Losito, B. D., Fiorito, E., Miles, M. A., & Zelson, M. (1991). Stress recovery during exposure to natural and urban environments. *Journal of Environmental Psychology*, *11*(3), 201–230. DOI: 10.1016/S0272-4944(05)80184-7

United Nations. (2018). *World Urbanization Prospects: The 2018 Revision.* https://population.un.org/wup

United Nations Framework Convention on Climate Change. (2015). *Paris Agreement.* https://unfccc.int/process-and-meetings/the-paris-agreement/the-paris-agreement

U.S. Green Building Council. (2020). *LEED v4.1 Building Design and Construction.* https://www.usgbc.org/leed/v41

Wu, P., Wang, X., & Wang, J. (2016). A critical review of the use of 3D printing in construction. *Automation in Construction*, *68*, 21–31. DOI: 10.1016/j.autcon.2016.04.005

Yucer, C. T., & Hepbasli, A. (2010). Thermodynamic analysis of a building using energy analysis method. *Energy and Building*, *43*(2–3), 536–542. DOI: 10.1016/j.enbuild.2010.10.019

Zemitis, J., Borodinecs, A., Geikins, A., Kalamees, T., & Kuusk, K. (2016). Ventilation system design in three European geo cluster. *Energy Procedia*, *96*, 285–293. DOI: 10.1016/j.egypro.2016.09.151

Zhang, S., Zhao, Y., Ding, H., Qiu, J., & Guo, Z. (2021). Recycling flue gas desulfurization gypsum and phosphogypsum for cemented paste backfill and its acid resistance. *Construction & Building Materials*, *275*, 122170. DOI: 10.1016/j.conbuildmat.2020.122170

Zhou, D., Zhao, C. Y., & Tian, Y. (2012). Review on thermal energy storage with phase change materials (PCMs) in building applications. *Applied Energy*, *92*, 593–605. DOI: 10.1016/j.apenergy.2011.08.025

Chapter 3
Technical Textiles in Building Facades:
An Exploration of Structural and Aesthetic Potential

Sushila Sahani
http://orcid.org/0000-0002-5739-144X
Lovely Professional University, India

ABSTRACT

Technical textiles have the potential to transform building façade design by uniting structural performance with aesthetic creativity. This chapter investigates the diverse functions of advanced textile materials in contemporary architecture, emphasizing their mechanical properties, environmental adaptability, and visual impact. This analysis assesses the tensile strength, thermal performance, and weather resistance of textiles such as ETFE, PTFE, and composite textiles when used in façade systems. The efficiency and design flexibility of these materials are compared to those of conventional cladding methods. Additionally, it addresses the challenges related to durability, cost, and scalability in practical applications. By analyzing case studies and synthesizing research on potential materials, the chapter provides a comprehensive overview of innovative options, highlights their advantages and limitations, and offers recommendations for future applications of technical textiles in the field.

1. INTRODUCTION

Technical textiles have become a revolutionary type of material in modern architecture, especially when it comes to designing and engineering façades. Their incorporation into architectural practice signifies an increasing focus on sustain-

DOI: 10.4018/979-8-3373-6023-2.ch003

able, versatile, and aesthetically groundbreaking solutions for construction related sustainability challenges. As the construction industry throughout the world deals with issues such as energy efficiency, urban retrofitting, and environmental effect, textile-based façades provide a unique way to do things that combines lightness, flexibility, and usefulness (Trencher et al., 2016; Procaccini & Monticelli, 2024). This integration is not just a technological breakthrough; it is also a reaction to urgent structural and ecological needs. It marks a major change in how materials can affect the look and performance of modern structures. In the European Union, buildings account for nearly 40% of total energy consumption and a similar share of carbon emissions (Solis-Guzman et al., 2014). Much of this infrastructure is aging and energy-inefficient, requiring urgent renovation or retrofitting to align with contemporary sustainability standards. Conventional retrofitting approaches, however, are often costly, time-consuming, and disruptive to occupants (Almeida et al., 2025). In contrast, textile-based façades offer a lightweight, non-invasive, and cost-effective solution that can enhance both the energy performance and aesthetic character of buildings, making them particularly well-suited for sustainable retrofitting applications (Iturralde Carrera et al., 2025).

These materials can be engineered to perform multiple architectural functions, including solar shading, thermal regulation, acoustic optimization, and even structural reinforcement (Singh, 2021; Procaccini & Monticelli, 2024; Jalil, 2020). Structurally, technical textiles can be employed in diverse configurations such as tensioned membrane and pneumatic systems that enable complex geometric forms while maintaining lightweight support (Zagubień et al., 2025). They can also be integrated into composites and reinforced mortars to strengthen existing materials like concrete or timber, thereby extending the life of heritage or seismic-prone buildings (Kouris & Triantafillou, 2018; Corradi et al., 2023). Moreover, when incorporated into laminated glass units or double-skin façades, textiles can improve mechanical resilience and thermal regulation while preserving transparency and visual fluidity (Gibilisco et al., 2024; Denz et al., 2021). In safety-focused designs, high-strength textile claddings and blast curtains provide protective barriers against impact, fire, and explosion, adding a critical layer of resilience to modern façades (Bedon & Rajčić, 2019; Houtman, 2015).

Beyond their functional advantages, technical textiles open new aesthetic horizons for architectural expression. Their pliability enables architects to experiment with free-form geometries that are difficult to achieve with rigid materials. The ability to modulate light and transparency allows façades to function as dynamic filters, mediating between interior comfort and exterior appearance. The vast range of available colors, textures, and translucencies facilitates designs that are both contextually responsive and visually engaging. Recent innovations in adaptive and responsive textiles further expand this potential, enabling façades that adjust in real time to environmental stimuli such as sunlight, temperature, and wind, thereby optimizing energy performance while enhancing user comfort (Procaccini & Monticelli, 2024; Jalil, 2020).

In façade retrofitting, textile membranes have proven particularly advantageous due to their lightweight nature and low environmental impact. Their modularity and reusability align with principles of circular design, allowing for easy disassembly, replacement, and material recovery (Niazy et al., 2023; Mottahedeh, 2023). They contribute to energy savings by improving insulation and solar shading, while also supporting innovative retrofit strategies such as wrapping existing façades, adding ventilated layers, or creating double-skin systems that function as thermal buffers (Procaccini & Monticelli, 2024; Procaccini, 2025). These qualities make textile façades a practical and sustainable solution for extending the lifespan and performance of existing buildings without compromising their architectural integrity.

However, the widespread adoption of textile façades is not without challenges. The absence of standardized design codes and performance guidelines limits broader acceptance within the construction industry. Durability remains a key concern, particularly regarding resistance to UV radiation, pollution, and mechanical stress over extended lifespan. Additionally, achieving seamless integration between textile systems and conventional building technologies is a complex process. It requires interdisciplinary collaboration among architects, engineers, and material scientists. Economic factors, including lifecycle cost assessment and maintenance considerations, also influence adoption decisions. Moreover, the future of textile architecture lies in multifunctional façades capable of harvesting solar energy, collecting water from fog, and dynamically regulating internal environments. These innovations transform façades from passive enclosures into active, intelligent systems that embody the principles of sustainable and responsive architecture.

2. KEY TEXTILE MATERIALS USED IN BUILDING FACADES

Textile-based façade systems can be broadly categorized based on the type of fiber reinforcement and surface coating applied. Both factors play a crucial role in determining the mechanical, optical, and environmental performance of the fabric. The key textile materials used in building façades include coated fabrics, polymer foils, and advanced smart textiles. These materials are engineered to provide superior structural integrity, environmental adaptability, and aesthetic versatility. These high-performance materials are characterized by their lightweight nature, exceptional tensile strength, weather resistance, and energy efficiency. Such properties make them particularly suitable for both new constructions and façade retrofit projects. In architectural practice, these materials are utilized in different configurations depending on functional requirements. The most common textile materials in façade construction have been discussed below. This overview examines a range of textile materials used in contemporary building façade applications, analyzing their composition, structural and functional properties, and sustainability potential.

2.1 Polymer-Coated Fabrics

- PVC-coated Polyester (PES-PVC) consists of a polyester-yarn fabric base coated with polyvinyl chloride (PVC), often finished with PVDF topcoats for extra durability. It is lightweight (1000-1400 g/m^2), offers tensile strength of 82-100 kN/m depending upon the weave structure of resulting fabric (Uhlemann, 2016; Asadi et al., 2018) and is UV-resistant, waterproof, and weldable/thermoformable with fire ratings B1-B3. Architects deploy it in tensile façades, roofing canopies, shading screens, cladding wraps, secondary façade layers, and temporary pavilions. Mechanically recyclable via Texyloop, it delivers a moderate 20-25-year lifespan and cost-effective retrofit compatibility (Monticelli & Zanelli, 2022;Temme // Obermeier GmbH, n.d.).
- PTFE-coated Glass Fiber (GF-PTFE) uses woven glass-fiber fabric coated on one or both sides with polytetrafluoroethylene (PTFE). It achieves high tensile strength but low tear strength, excellent UV resistance, chemical inertness, and temperature stability from -200 °C to +260 °C (Lyons, 2013; Ghorbanalavi, 2018; Zhang et al., 2017)). The low-surface-energy coating yields self-cleaning and non-combustible behavior with moderate translucency. Typical uses include tensile membranes for roofs and façades, double-skin systems, lightweight envelopes, and acoustic insulation layers. With a lifespan exceeding 25 years and full recyclability, its low weight cuts transport and installation impacts; it also supports smart-textile enhancements (Cherif et al., 2016; Asadi et al., 2021; Winter, 2022).
- Silicone-coated Glass Fiber pairs glass-fiber fabric with silicone-based polymers, delivering high weather/UV/chemical resistance, fire retardancy, and waterproof performance (Schwark & Müller, 1996; Chruściel, 2022). However, glass fibers are inherently brittle and difficult to fold, which can result in a decline in their mechanical performance during handling, transportation, or construction (Junhao et al., 2017). It appears in building envelopes, roofing membranes, and façade cladding for extreme climates. Long-lasting and chemically inert, it retains integrity across temperature swings and pairs well with self-cleaning or UV-protective smart coatings.
- PVDF-coated Membrane builds on a polyester substrate finished with polyvinylidene fluoride (Yan et al., 2025). It exhibits high tensile strength (especial-

ly at high strain rates), strong energy absorption, low mechanical anisotropy, and strain-rate sensitivity (Zheng et al., 2020). The PVDF layer extends lifespan, cuts maintenance, and enables integration into smart/functional systems, indirectly lowering environmental impact through durability.

- PTFE-coated Aramid (Kevlar-PTFE) combines aramid-fiber mesh with PTFE polymer, yielding very high tensile/impact resistance, fire resistance, chemical/UV stability, and light weight. ANF (aramid nanofibers) reinforcement enhances Kevlar/PTFE-phenolic composites by increasing tensile strength by 32.1%, bonding strength by 30.32%, and reducing wear rate by 76.1%, demonstrating superior mechanical and tribological performance (Zhao & Ma, 2022; Yang et al., 2020). Reinforcement in composite façade panels and claddings for high-wind-load buildings is common. Ballistic-grade resistance and long urban-exposure life are achievable.

2.2 Fluoropolymer Films & Membranes

- ETFE Foil is a thin ethylene tetrafluoroethylene (ETFE) fluoropolymer film supplied in single- or multilayer pneumatic cushions. It can transmit 94-97% visible light in three-layer configurations and offers strong UV and chemical resistance (Robinson, 2005). Three-layer cushions can reach U-values of approximately 1.96 W/m^2K (Robinson, 2005). Iconic pneumatic-cushion façades, adaptive or kinetic systems, sun-shading membranes, and free-form roofs employ ETFE because of its low weight and long service life (Runge et al., 2025; Kuda et al., 2025; Şahin & Okur, 2025). Lifespans can exceed 30 years, and the material is recyclable and substantially lighter than glass (Solanki et al., 2024; Charbonneau et al., 2014).

2.3 Textile-Reinforced Concretes

- Textile-Reinforced Reactive Powder Concrete (TRRPC) sandwiches two 25 mm facings of reactive powder concrete (RPC) reinforced with epoxy-coated carbon-textile grids around a 150 mm foam-concrete (FC) core, linked by GFRP connectors. Compressive strength reaches 147 MPa, tensile ~5 MPa (RPC), with thin, lightweight, ductile post-cracking behavior and minimal cracking under wind/anchor loads. Non-load-bearing sandwich façade panels, prefabricated cladding, and vertical inter-story spans benefit from its flexural and suction/pressure resistance. Lower embodied energy (especially with FC core), corrosion-free textile/carbon reinforcement, and easy integration of thermal/smart functions improve life-cycle performance (Flansbjer et al., 2018).

- Glass Fiber Reinforced Concrete with Textile Mesh (GFRC) embeds alkali-resistant glass-fiber textile mesh in a cementitious reactive powder concrete matrix (Oestreich et al., 2024). It has been noted that high-performance concrete demonstrates brittle characteristics under ultimate load conditions; however, the inclusion of fibers improves tensile capacity, flexural response, and ductility while also functioning as a crack inhibitor (Yuan & Jia, 2021). It was reported that the addition of fibers from 0.0% to 1.2% enhanced the split tensile strength from 3.06 to 4.92 MPa, whereas flexural strength increased from 4.84 to 7.27 MPa (Atewi et al., 2019). Prefabricated cladding, ventilated façades, and slim architectural panels use this approach. It can also reduce embodied energy, CO_2 emissions, and transport or installation impacts when compared with steel-reinforced concrete (Nishanth et al., 2023).

2.4 Bio-Based & Circular Composites

- Bio-Based Composites (Nabasco 8010) are bulk-moulding-compound (BMC) panels made from waste fibers (recycled toilet paper, roadside grass, reed, textiles), calcium-carbonate filler (drinking-water waste), and 50% bio-based polyester resin (from biodiesel/glycol/animal/frying-fat waste). It has Fire class rating of B-s1,d0, density range runs from 1.7-1.9 g/cm^3; 3-point bending strength 35-40 MPa; and modulus is between 8.8-10.2 GPa (Bottger, 2019; NPSP BV, 2019).

2.5 Nanocoated & Functional Smart Textiles

- Nanocoated smart textiles can be produced through several methods depending on the intended end use, including the application of TiO_2, Ag, or ZnO nanoparticles via sol-gel treatment, dip-coating, or plasma functionalization onto textile fibers such as cotton, wool, polyester, and cellulose. They deliver photocatalytic self-cleaning behavior, antimicrobial action, high UV protection, hydrophobic or superhydrophobic surfaces, improved mechanical strength, and optional flame retardancy with wash-durable cross-linking. Self-cleaning façade membranes, UV-protective shields, antimicrobial linings, thermal or fire-retardant layers, and kinetic textile systems can all benefit from these coatings. Energy savings may also arise through reduced cleaning demand, lower chemical use, improved indoor air quality, lower microbial risk, and potential infrared reflection (Pakdel et al., 2018; Feizbahr & Pourzanjani, 2024).
- Piezoelectric & Shape-Memory Textiles integrate nickel - titanium shape memory alloy (Materials World, 2004), electroactive polymers, or piezo-

ceramic nanoparticles into flexible fabrics (Mishra et al., 2019). This enables functions such as stimuli-triggered shape recovery, piezoelectric pressure response, and electrical/thermal actuation enable kinetic shading, self-regulating solar control, moveable membranes, and energy-harvesting layers (Raudaschl et al., 2022).

- Fog-Harvesting Meshes are polypropylene Raschel (35% shade) multilayer textiles optimized for droplet capture via pore size and hydrophilic/hydrophobic balance (Otafy et al., 2025). UV-resistant and lightweight, they integrate into double-skin façades, rooftops, and urban green walls (Di Bitonto & Torres, 2025).
- Aerogel-composite membranes embed silica aerogel, with thermal conductivity of approximately 0.012 W/mK at 25 C, within textile composites, yielding ultra-light, translucent, high-performance insulation for thin façade skins and multilayer systems (Kim & Koo, 2024).
- Phase-change textile layers microencapsulate paraffin wax or other PCMs into textile membranes, providing reversible solid-liquid thermal mass for temperature stabilization. Adaptive thermal façades and dynamic insulation can therefore lower peak loads passively within sustainable envelope systems (Zwanzig et al., 2013; Jha et al., 2024).

3. TEXTILE MATERIALS IN ARCHITECTURE: SYSTEMS, PROPERTIES, AND FUNCTIONS

3.1 Façade Construction Systems

Three dominant textile-based systems enable the creation of lightweight, tensioned envelopes that combine structural efficiency with expressive architectural form:

- Pneumatic Façades: Air-pressurized cushions, most commonly fabricated from ETFE (ethylene tetrafluoroethylene) foils create transparent, adaptable surfaces that provide excellent daylight transmission, multi-layer insulation, and high strength-to-weight ratios while requiring minimal supporting framework (Silver et al., 2015). Such systems are frequently used in stadiums, atria, and climate-responsive façades due to their ability to modulate solar gain and internal comfort dynamically.
- Double-Curved Membranes: These rely on pre-tensioned fabrics such as PTFE- or silicone-coated glass fibers to form complex, self-supporting geometries. The double curvature enhances both stiffness and aesthetic appeal, enabling the structure to efficiently distribute loads along optimized stress

trajectories. These membranes often define iconic architectural forms while maintaining durability and resistance to environmental degradation (Chilton & Velasco, 2005; Procaccini et al., 2023).

- Flat Tensioned Façades: Constructed from PVC-PES composite fabrics stretched over lightweight frames, these systems serve as planar shading devices, secondary cladding layers, or retrofit skins that improve building performance without extensive structural alteration. Their modularity and adaptability make them particularly suitable for façade renovation and sustainable retrofitting applications (Van Dongen et al., 2022; Procaccini et al., 2023).

These systems collectively underpin Textile Façade Retrofit Strategies (TFRS) (Procaccini & Monticelli, 2024), a set of adaptable approaches aimed at enhancing building performance through lightweight, reversible interventions:

- Replace: This strategy involves a complete substitution of existing opaque or glazed façades with advanced membrane or pneumatic cushion systems. The new textile-based envelope significantly reduces structural load while improving daylight quality, thermal insulation, and visual permeability, making it ideal for large-span or aging building façades requiring comprehensive modernization.
- Add: In this approach, textile elements are integrated as additional layers over existing façades to enhance insulation, solar control, or ventilation performance. These supplemental systems can function as dynamic shading membranes, ventilated skins, or seasonal screens, offering flexibility in managing solar gain and energy efficiency without major structural alterations.
- Wrap: The wrapping strategy encloses the building in a continuous textile-based double-skin or buffer-zone envelope, creating an intermediate climatic layer. This configuration improves thermal comfort, acoustic performance, and façade longevity while allowing the original architecture to remain intact beneath a protective, energy-regulating shell.

3.2 Design-Driving Properties (System-Level Synthesis)

3.2.1 Mechanical & Structural

The structural and material performance of textile-based façades derives from several distinctive mechanical attributes:

- Tension-Only Behavior: Textile membranes primarily carry loads through tensile stresses, exhibiting exceptionally high tensile strength combined with near-zero bending stiffness. This allows forces to be evenly distributed across the surface, minimizing localized stress concentrations and reducing the need for heavy supporting structures.

- Lightweight Resilience: Due to their minimal self-weight, textile systems significantly lower the overall dead load on the building, allowing for smaller foundations and reduced seismic demand. The lightweight nature also facilitates ease of handling and transportation, contributing to both economic and environmental efficiency.
- Form Adaptability: Textile materials can be shaped into free-form, kinetic, or pneumatic geometries, supporting innovative architectural expressions. Their flexibility enables rapid prefabrication, modular assembly, and on-site adaptability, making them ideal for both new constructions and retrofit applications.
- Durability under Dynamics: The inherent elasticity and energy-absorbing capacity of textile composites provide resilience against dynamic actions such as wind, impact, and blast loads. These characteristics enhance occupant safety and structural longevity, particularly in environments exposed to fluctuating or extreme conditions.

3.2.2 Environmental & Performance

Beyond their structural advantages, textile façades exhibit distinct environmental performance attributes that contribute to energy efficiency and sustainability:

- Thermal Modulation: The integration of air cavities, multi-layer cushions, or ventilated gaps within textile façades achieves low U-values and effective passive solar regulation. These systems can buffer external temperature fluctuations, optimize daylight transmission, and reduce reliance on active heating or cooling, thereby improving overall building energy performance.
- Weather Durability: Advanced coatings and surface treatments, such as fluoropolymer finishes or titanium dioxide-based self-cleaning layers, enhance UV stability, chemical resistance, and hydrophobicity. This ensures long-term material integrity and aesthetic retention even under extreme climatic exposure, significantly extending the service life of the façade.
- Material Efficiency: Textile-based systems are inherently resource-efficient, characterized by low embodied energy, high recyclability, and modular construction principles. Their ease of disassembly and reconfiguration aligns with circular retrofit frameworks, enabling reuse, minimal waste generation, and adaptive life-cycle extension.

3.2.3 Aesthetic & Experiential

In addition to structural and environmental performance, textile façades offer aesthetic and responsive design potential, allowing architecture to become both expressive and adaptive:

- Light Modulation: Textile membranes can achieve tunable translucency levels ranging from approximately 10% to 96%, enabling precise control over daylight diffusion, glare reduction, and visual privacy. This capacity to balance transparency and opacity allows designers to tailor interior lighting conditions while maintaining visual continuity with the exterior environment.
- Surface Variability: Advanced fabrication and finishing technologies allow textile façades to serve as dynamic visual interfaces. Surfaces can be digitally printed, color-shifting, or integrated with LED illumination to support branding, wayfinding, or contextual expression. Such variability transforms façades into communicative skins that respond to cultural, environmental, or programmatic cues.
- Kinetic Potential: By embedding sensors, actuators, or stimuli-responsive polymers within the membrane system, façades can dynamically adjust to external factors such as solar intensity, temperature, or wind. These responsive systems enable adaptive shading, natural ventilation, and real-time performance optimization, bridging architecture and smart material technologies.

3.3 Functional Roles in Façades

Textile façades serve diverse functional roles that enhance building performance, safety, and aesthetics. They act as protective layers, shielding interiors from wind, rain, UV radiation, and thermal extremes using PTFE-glass membranes or ETFE cushions in exterior cladding, secondary skins, or insulation systems exemplified by the Allianz Arena (Iftikhar & Aslam, 2025). In structural support, double-curved PTFE-glass fabrics provide tension-based load-bearing for primary membranes and lightweight roofs (Göppert & Paech, 2015). Adaptive environmental modulation enables dynamic control of light, heat, and airflow through sensor-integrated membranes in kinetic shading or responsive ventilation systems (Barozzi et al., 2016). Textiles also offer reinforcement and safety enhancement by embedding mesh in concrete or glass to improve crack, impact, and seismic resistance in retrofit wraps or blast curtains (Maity et al., 2012). For thermal and acoustic regulation, multi-layer ETFE cushions and nested membranes create insulated, noise-dampening sub-spaces (Sluyts et al., 2024). Solar and visual management is achieved with translucent foils and mesh screens that reduce glare, optimize daylight, and ensure privacy (Strasser et al., 2018). Self-cleaning surfaces leverage photocatalytic coatings on PTFE-glass to resist pollution and microbial growth, minimizing maintenance (Maheshwari et al., 2025). Kinetic and responsive systems use shape-memory alloys or electroactive polymers to adjust geometry, transparency, or shading in real time (Chamilothori et al., 2013). Energy systems integrate PV or LED elements into fabrics for power generation, smart shading, and illumination, such as in photovoltaic curtains (Liu et al., 2023). Aesthetically, color-dynamic and LED-illuminated textiles enhance visual identity and branding (Chilton & Lau, 2015).

4. COMPARISON OF TECHNICAL TEXTILE AND TRADITIONAL CLADDING SYSTEMS

Table 1. Comparison of technical textile and traditional cladding systems

Category	Aspect	Technical Textiles	Traditional Cladding Systems
Functional Comparison	Construction Logic	Lightweight tensioned membranes or multi-layer composites; pure tension; high tensile strength, coatings, layered configurations; adaptable to freeform geometries	Heavy, inert materials (stone, metal, glass); compression or bending; rigid, load-bearing structures; restricted form flexibility
	Substructure Needs	Minimal; framed or cable-supported assemblies; reduced foundation and steel requirements	Rigid frames, substantial anchors, deeper foundations
	Installation	Off-site fabrication; dry methods (welding, mechanical fixation); rapid, modular replacement	Wet or semi-permanent; longer construction times, greater labor demands
	Maintenance	Self-cleaning TiO_2 or PVDF coatings; resist pollution and microbial growth	Periodic sealing or washing; prone to chemical degradation
	Aesthetics	Translucency, dynamic color modulation, three-dimensional flexibility	Largely static; costly for complex geometries
	Durability	20–35 years (PTFE- or silicone-glass composites)	30–50 years (stone or metal)

continued on following page

Table 1. Continued

Category	Aspect	Technical Textiles	Traditional Cladding Systems
	Sustainability	Lightweight, low-carbon, recyclable (especially ETFE and PVC-PES)	Higher embodied energy, partial recyclability, demolition-intensive replacement cycles
	Architectural Philosophy	Tension-based adaptability, material efficiency, flexibility, modularity, environmental responsiveness	Permanence, structural solidity, durability through weight and inertia
Mechanical Comparison	Tensile Strength-to-Weight Ratio	Vary depending upon the material	High strength-to-weight ratio
	Elasticity & Movement	Accommodates thermal movement; reduces deformation risks	Rigid behavior
	Fire Rating	Class A2 (PTFE, silicone-glass); B1–B3 (PVC-based, depending on formulation)	varies by material
	Acoustic/Thermal/Solar Performance	Achieved via pneumatic cushions, double-skin membranes, or adaptive layers	Thicker layers and added insulation
Environmental and Energy Performance	Embodied Carbon	Cut by up to 70%; lightweight, recyclable polymer matrices	Higher due to materials
	Thermal Insulation	Enhanced by air cavities or pneumatic structures (ETFE, PTFE); improved U-values	Thermal bridging; greater HVAC dependence
	Energy Control	Smart membranes with photothermal coatings, diffusive layers, semi-transparent PV films	N/A
	Air Purification	Photocatalytic finishes (TiO_2 or ZnO) promote pollutant breakdown	N/A
	Operational Energy	Lower due to reduced HVAC dependence	Higher

continued on following page

Table 1. Continued

Category	Aspect	Technical Textiles	Traditional Cladding Systems
Economic and Practical Comparison	Installation Costs	Lower by up to 50%; prefabrication, modularity, tension-based mounting	High material, manufacturing, and installation costs
	Maintenance Costs	Minimized by self-cleaning surfaces and lightweight assemblies	Higher
	Retrofitting	Replace, Add, or Wrap methods; enhances performance without full dismantling; supports removability, repairability, upgradability	N/A
	Service Life	N/A	40–60 years
	Environmental Impacts	Lower	Significant
Integration Outcome	Strengths	Adaptability, sustainability, weight efficiency, flexibility, energy performance, visual dynamism	Long-term durability, structural capacity
	Hybrid Potential	Combine PTFE-glass or ETFE membranes with steel frames or integrated PV	Unite with textile flexibility
	Overall Paradigm	Redefine envelope performance in circular construction; next-generation retrofits, kinetic façades, adaptive skins	Transcend static limitations via integration

5. CASE STUDIES OF INNOVATIVE APPLICATIONS

5.1 Hazza Bin Zayed Stadium

Hazza Bin Zayed Stadium, located in Al Ain, UAE and completed in 2014. The facade is designed using a steel diagrid structure combined with 612 individually oriented membrane panels. These panels are made of pre-stressed glass-PTFE membrane supported on steel frames, developed specifically to provide maximum shading and allow to maintain views from inside the building. The design achieves comfort cooling in specific stadium areas using additional thermal envelopes made of insulated glass or aluminum facade material. The textile membrane panels are tailored with varied inclinations and installed in the diagrid frame using simple bolted clamp details, which is possible due to their lightweight nature. At night, the membrane panels are illuminated by LED units, providing dynamic color effects and a striking visual identity. These materials are highly valued in architecture for their lightweight properties, flexibility in forming complex free-form geometries, and the economic advantages of reduced dead loads, facilitating easier installation and replacement (Almohaimeed & Pérez-Villalba, 2025; Paech et al., 2014).

5.2 BC Place Stadium

BC Place Stadium in Vancouver, Canada, completed in 2012, features a single-layer ETFE foil façade system supported by a steel substructure. It is a prominent example of the integration of technical textiles into architectural design, combining membrane and foil technologies to create a lightweight, sustainable, and visually dynamic building envelope. Its most innovative feature is the approximately 9,500 m^2 ETFE façade, supported by steel rings and vertical steel arches that create a three-dimensional saddle geometry. ETFE foils in this application provide high transparency, low weight, weather resistance, and customizable visual effects through printed fritting with variable density. The façade is also illuminated by LED lights mounted on the horizontal steel rings, enabling dynamic nighttime effects while maintaining daylight transparency during the day (Masubuchi, 2013; Paech et al., 2014; Runge et al., 2025).

5.3 Allianz Arena

Allianz Arena, completed in Munich, Germany in 2005, is globally recognized for its ETFE pillows, which are inflated cushions forming the façade. The façade consists of air-inflated ETFE cushion elements that create a lightweight double-skin system capable of dynamic visual expression. These cushions offer high durability, UV resistance, and weather protection while weighing substantially less than traditional glass systems. The double-skin effect improves thermal performance by reducing heat transfer, while the low surface tension of ETFE contributes to self-cleaning behavior and long-term durability (Alioglu & Sirel, 2018; Mansy, 2019; Iftikhar & Aslam, 2025).

5.4 Coca-Cola Beatbox Pavilion

The Coca-Cola Beatbox Pavilion is a leading example of the use of technical textiles in contemporary architectural applications, specifically during the London 2012 Olympics. It demonstrates how advanced materials such as ETFE are leveraged for both aesthetic and interactive functions in temporary structures. The pavilion incorporated 230 LED-illuminated ETFE cushions, 40 of which featured built-in audio speakers and motion sensors, allowing the structure to become an interactive sound installation responding to visitor movement and touch. As a temporary pavilion, it showcased the adaptability and removable nature of textile-based architecture for high-profile events. The pavilion served both as a visual icon, glowing at night due to integrated LEDs, and as an acoustic participatory artwork, highlighting new possibilities for interactive architectural envelopes using smart and technical textiles (Architektur & Der Technik, 2017; Özen, 2022).

5.5 Soft House

The Softhouse, developed by Sheila Kennedy and Kennedy Violich Architecture, transforms household curtains and building membranes into energy-harvesting and light-emitting textiles that power lighting and portable devices in the home. The exterior uses flexible photovoltaic textile "twisters" that bend in response to the sun's angle or open for views, representing a soft, adaptive alternative to traditional solar panels. These solar textiles and smart curtains are integrated to address both energy production and comfort, highlighting the capacity of textiles to perform architectural functions beyond mere decoration. Projects like Soft House demonstrate the potential for mobile, fabric-based dwellings to provide adaptable, sustainable solutions for energy consumption, post-disaster housing, and environmental awareness (Robertson, 2012; Heinzel, 2018).

5.6 Cheops Technology Building

Cheops Technology Building in Canejan, France was finalized in 2019. The building's facade is covered using a single layer of Solits FT381 textile, which is stretched and wrapped around three sides of the building. This textile is fixed to extruded profiles made of a special aluminum alloy and digitally printed with a protective polyurethane (PU) varnish layer. The main functions provided by this textile facade include insulation and sun protection, effectively controlling the temperature inside the building (Jalil, 2020).

6. CHALLENGES OF TECHNICAL TEXTILES IN BUILDING FACADES

Several challenges significantly limit the widespread adoption and optimal performance of technical textiles in building facades. Below are the key issues and barriers, supported by detailed insights from recent literature and empirical studies:

Scale, System, and Time Dilemmas

Architects encounter four principal dilemmas which include scale, system, time, and conception, when considering the integration of textiles in architectural practice. The scale dilemma is rooted in the disparity between the dimensions and durability expectations of textiles and traditional building materials, making trials and adoption in full-scale construction risky and less predictable. The system dilemma involves the challenge of integrating textiles into larger building systems, which often requires complicated supporting structures and sophisticated joining solutions for both new and existing materials. Time relates to the temporary nature and limited lifespan of textiles, raising concerns about longevity and maintenance frequency compared to more permanent facade materials. Lastly, both architects and clients may hold negative conceptions about the aesthetic and functional reliability of textiles, viewing them as "soft," "impermanent," or "inadvisable" for core building envelope functions.

Mechanical and Structural Vulnerabilities

Textile membranes present specific mechanical challenges, including vulnerability to mechanical stresses and limited durability of coatings. Their effectiveness depends on pre-existing tension and the ability to recover from temporary deflections due to external loads like wind or temperature fluctuations. There are also concerns about fire resistance, the risk of thermal bridges, and the need for specialized labor all of which complicate the design and maintenance of textile-based facades. Compared to traditional rigid materials, textiles may suffer from a higher rate of degradation and less predictable long-term performance under environmental exposure.

Regulatory, Cost, and Execution Constraints

One of the most common challenges cited by architects is the lack of clear regulations, guidelines and standards for the specification, testing, modeling, and installation of technical textiles in facades. This regulatory uncertainty often discourages innovation, increases costs, and raises the perceived risk. Practical challenges include the high investment required for durable facade membrane systems, difficulty securing high-skilled labor, and the potential for aesthetic changes that may not be acceptable for historical or preserved structures. The disruption to building occupants during installation, the physical execution constraints, and exposure to weather during facade retrofits are additional barriers.

Unique Material and Performance Limitations

While textiles excel in offering lightweight, flexible solutions for shading, insulation, and rapid retrofitting, they may lack adequate protection from external environmental forces if not properly engineered. Issues such as poor thermal inertia, the need for effective waterproofing, overheating, reduction in interior or exterior space, and vulnerability to moisture-related problems must be carefully addressed. Furthermore, the complexity of construction detailing for tensioned or pneumatic membranes and the challenge of achieving optimal acoustic and UV resistance add layers of technical difficulty. While modularity and ease of replacement offer potential benefits, they also compound the challenge of ensuring long-term sustainability and resilience in architectural applications.

7. FUTURE DIRECTIONS

- Development of Standards and Codes: Concerted international efforts are ongoing to create verified structural design codes specifically for tensile textile membranes, ensuring safety and harmonization across jurisdictions (Uhlemann, 2016; Asadi et al., 2018; Runge et al., 2025).
- Advanced Functional Coatings: Research in nanotechnology-enabled self-cleaning, antimicrobial, UV-resistant, thermochromic, and self-healing coatings will enhance durability and environmental interaction in textile façades (Pakdel et al., 2018; Chruściel, 2022; Maheshwari et al., 2025).
- Integration of Smart Textiles: The rise of e-textiles and active materials capable of sensing, responding, and adapting, such as shape-memory polymers, electrochromic inks, and embedded photovoltaics, will shift façades from passive skins to dynamic, energy-optimizing, and data-responsive systems (Denz et al., 2021; Heinzel, 2018; Raudaschl et al., 2022).
- Multi-Functional Composite Design: New composite membranes combining improved mechanical strength, translucency, thermal insulation, and recyclability are expected to emerge, pushing performance toward net-zero energy building envelopes (Yang et al., 2020; Zhao & Ma, 2022; Kim & Koo, 2024).
- Digital Fabrication and Modular Construction: Increased use of parametric design, 3D printing, and prefabrication will enhance design flexibility, reduce waste, and enable rapid and reversible installation of textile façades (Masubuchi, 2013; Cherif et al., 2016; Gibilisco et al., 2024).
- Sustainable and Circular Practices: Emphasis on eco-design principles promotes the use of recyclable polymers, reusable modular components, and

life-cycle thinking to reduce environmental footprint (Monticelli & Zanelli, 2022; Mottahedeh, 2023; Procaccini et al., 2023).

- Hybrid Systems Development: Combining textiles with conventional materials such as glass, metals, or concrete can balance durability and functional versatility, enabling broader architectural integration (Bedon & Rajčić, 2019; Atewi et al., 2019; Liu et al., 2023).
- Energy Harvesting and Environmental Monitoring: Textile façades are projected to incorporate photovoltaic modules and sensor networks to enhance building energy independence and improve indoor environmental quality in real time (Van Dongen et al., 2022; Raudaschl et al., 2022; Iturralde Carrera et al., 2025).
- Public Awareness and Policy Support: Advocacy, demonstration projects, and evolving regulatory frameworks will boost confidence in textile façades and encourage their incorporation in large-scale retrofits and new buildings (Procaccini, 2025; Trencher et al., 2016; Almeida et al., 2025).

8. CONCLUSION

Textile-based façade systems represent a paradigm shift in contemporary architecture where materials are no longer static enclosures but dynamic interfaces between environment, structure, and human. Technical textiles, with their lightweight nature, mechanical adaptability, and multifunctional performance, are redefining the aesthetics and functionality of building envelopes. These materials include polymer-coated fabrics such as PTFE-glass and ETFE foils, along with advanced composites, smart nanocoatings, and bio-based reinforcements. Together, they offer innovative solutions to global challenges in energy efficiency, urban retrofitting, and environmental resilience.

By integrating these systems into architectural façades, designers are achieving unprecedented levels of environmental responsiveness, optimizing daylight, moderating thermal conditions, and improving acoustic comfort while drastically reducing embodied energy and construction waste. Compared with conventional cladding, textile façades deliver efficiency through modularity, recyclability, and reversible assembly, aligning perfectly with global priorities of circular construction and carbon neutrality. The case studies discussed, including the Allianz Arena, BC Place Stadium, and the Coca-Cola Beatbox Pavilion, underscore the real-world viability and expressive potential of these materials in both permanent and temporary structures.

However, realizing their full potential demands overcoming challenges related to standardization, durability, and perception. The absence of regulatory frameworks and long-term performance benchmarks continues to limit mainstream acceptance.

Addressing these gaps through research, material innovation, and interdisciplinary collaboration is crucial to bridging the divide between experimental architecture and practical application. Looking ahead, the fusion of smart materials, digital fabrication, and sustainable design principles will propel textile façades into the next generation of responsive architecture. As they evolve to harvest energy, self-regulate internal environments, and contribute to urban resilience, textile façades will no longer be viewed as supplementary skins but as integral components of high-performance, adaptive, and expressive building systems. By bridging material science with design ingenuity, they pave the way for a sustainable built environment that is lighter, smarter, and infinitely more adaptable urging architects, engineers, and policymakers to embrace this textile revolution for a greener tomorrow.

ACKNOWLEDGMENT

The author acknowledges the body of research and professional practice that has advanced the use of technical textiles, membrane structures, and lightweight façade systems in contemporary architecture.

REFERENCES

Alioglu, T. U. G. B. A., & Sirel, A. Y. S. E. (2018). The Use of Textile-Based Materials in Shell System Design in Architecture and an Evaluation in Terms of. *J. Contemp. Urban Aff, 2*(3), 88–94. DOI: 10.25034/ijcua.2018.4722

Almeida, R., Rodrigues, H., Gomes, M. G., & Furtado, A. (2025). Financial incentives for energy and seismic retrofitting of reinforced concrete buildings: A review of an integrated approach. *Innovative Infrastructure Solutions, 10*(8), 329. DOI: 10.1007/s41062-025-02123-6

Almohaimeed, N., & Pérez-Villalba, M. (2025). The UAE's Journey to Sports Sustainability: Commitment in Action.

Architektur & Der Technik. (2017). Textile-based architecture.

Asadi, H., Uhlemann, J., Stranghoener, N., & Ulbricht, M. (2021). Water influence on the uniaxial tensile behavior of polytetrafluoroethylene-coated glass fiber fabric. *Materials (Basel), 14*(4), 846. DOI: 10.3390/ma14040846 PMID: 33578878

Asadi, H., Uhlemann, J., & Stranghöner, N. (2018). Water infiltration impact on tensile strength and breaking strain of architectural fabrics. *Advances in Structural Engineering, 21*(16), 2605–2616. DOI: 10.1177/1369433218756005

Atewi, Y. R., Hasan, M. F., & Güneyisi, E. (2019). Fracture and permeability properties of glass fiber reinforced self-compacting concrete with and without nanosilica. *Construction & Building Materials, 226*, 993–1005. DOI: 10.1016/j.conbuildmat.2019.08.029

Barozzi, M., Lienhard, J., Zanelli, A., & Monticelli, C. (2016). The sustainability of adaptive envelopes: Developments of kinetic architecture. *Procedia Engineering, 155*, 275–284. DOI: 10.1016/j.proeng.2016.08.029

Bedon, C., & Rajčić, V. (2019). Textiles and fabrics for enhanced structural glass facades: Potentials and challenges. *Buildings (Basel, Switzerland), 9*(7), 156. DOI: 10.3390/buildings9070156

Bottger, W. (2019, June). Development of (interactive) façade elements on base of waste materials of (waste) water companies.

Chamilothori, K., Kampitaki, A.-M., & Oungrinis, K.-A. (2013, November). Climate-responsive shading systems with integrated shape memory alloys (SMA). In Proceedings of the 8th Energy Forum on Advanced Building Skins (pp. [page range if available]). Bressanone, Italy.

Charbonneau, L., Polak, M. A., & Penlidis, A. (2014). Mechanical properties of ETFE foils: Testing and modelling. *Construction & Building Materials*, *60*, 63–72. DOI: 10.1016/j.conbuildmat.2014.02.048

Cherif, C., Diestel, O., Engler, T., Hufnagl, E., & Weiland, S. (2016). Processing aspects and application examples. *In Textile Materials for Lightweight Constructions: Technologies-Methods-Materials-Properties (pp.* 599-661). Berlin, Heidelberg: Springer Berlin Heidelberg.

Chilton, J., & Lau, B. (2015). Lighting and the visual environment in architectural fabric structures. *In J.* I. de Llorens (Ed.), Fabric Structures in Architecture (pp. 203-219). Woodhead Publishing. DOI: 10.1016/B978-1-78242-233-4.00007-3

Chilton, J., & Velasco, R. (2005). Applications of textile composites in the construction industry. In *Design and manufacture of textile composites* (pp. 424–435). Woodhead Publishing. DOI: 10.1533/9781845690823.424

Chruściel, J. J. (2022). Modifications of textile materials with functional silanes, liquid silicone softeners, and silicone rubbers-A review. *Polymers*, *14*(20), 4382. DOI: 10.3390/polym14204382 PMID: 36297958

Corradi, M., Mustafaraj, E., & Speranzini, E. (2023). Sustainability considerations in remediation, retrofit, and seismic upgrading of historic masonry structures. *Environmental Science and Pollution Research International*, *30*(10), 25274–25286. DOI: 10.1007/s11356-021-17490-7 PMID: 34797539

Denz, P. R., Sauer, C., Waldhör, E. F., Schneider, M., & Vongsingha, P. (2021). Smart textile sun-shading: Development of functional adaptex prototypes. *Journal of Facade Design and Engineering*, *9*(1), 101–116.

Di Bitonto, M. G., & Torres, N. M. R. (2025). Energy Efficiency through Fog. *Sustainable Built Environment for People and Society, 129.*

Feizbahr, M., & Pourzanjani, P. (2024). Nanotechnology in Construction: Innovations, Applications, and Impacts. *environment, 1, 7.*

Flansbjer, M., Williams Portal, N., Vennetti, D., & Mueller, U. (2018). Composite behaviour of textile reinforced reactive powder concrete sandwich façade elements. *International Journal of Concrete Structures and Materials*, *12*(1), 71. DOI: 10.1186/s40069-018-0301-4

Ghorbanalavi, F. (2018). Performance characteristics and practical applications of common building polymeric materials for building envelope. *Journal of Energy and Economic Development*, *3*(2), 13–33.

Gibilisco, G., Monteleone, A., Rodonò, G., & Sapienza, V. (2024, October). Self-Sufficient and Responsive Textile. In Proceedings of the 11th International Conference of Ar. Tec.(Scientific Society of Architectural Engineering): Colloqui. AT. e 2024-Volume 2 (Vol. 611, p. 345). Springer Nature.

Heinzel, T. (2018). Reactive architecture, augmented textiles, domotics and soft architecture fabrication: On electronic and reactive textiles in domestic contexts. *Textile (Oxford)*, *16*(1), 34–61. DOI: 10.1080/14759756.2017.1332907

Houtman, R. (2015). Materials used for architectural fabric structures. In *Fabric Structures in Architecture* (pp. 101–121). Woodhead Publishing. DOI: 10.1016/B978-1-78242-233-4.00003-6

Iftikhar, H. B., & Aslam, S. (2025). Textile Architecture in Transition: Global Trends and Emerging Applications in Pakistan. [DSSR]. *Dialogue Social Science Review*, *3*(6), 612–628.

Iturralde Carrera, L. A., Díaz-Tato, L., Constantino-Robles, C. D., Garcia-Barajas, M. G., Zapatero-Gutiérrez, A., Álvarez-Alvarado, J. M., & Rodríguez-Reséndiz, J. (2025). Advances in Mounting Structures for Photovoltaic Systems: Sustainable Materials and Efficient Design. *Technologies*, *13*(5), 204. DOI: 10.3390/technologies13050204

Jalil, W. A. (2020, February). Smart textiles for the architectural façade. [). IOP Publishing]. *IOP Conference Series. Materials Science and Engineering*, *737*(1), 012078. DOI: 10.1088/1757-899X/737/1/012078

Jha, S. K., Sankar, A., Zhou, Y., & Ghosh, A. (2024). Incorporation of phase change materials in buildings. *Construction Materials*, *4*(4), 676–703. DOI: 10.3390/constrmater4040037

Junhao, X., Yingying, Z., & Jigang, X. (2017). Off-axial failure analysis of polytetrafluoroethylene-coated woven glass fibers under different loading rates. *Journal of Industrial Textiles*, *47*(3), 310–330. DOI: 10.1177/1528083716647198

Kim, S. D., & Koo, J. H. (2024). Development and material characterization of quartz-reinforced UHTR/aerogel composites: morphology, thermal, and flammability properties. *In AIAA SCITECH 2024 Forum (p.* 0364). Göppert, K., & Paech, C. (2015). Retractable roof solutions for sports stadiums and arenas. In P. Culley & J. Pascoe (Eds.), Stadium and arena design.

Kouris, L. A. S., & Triantafillou, T. C. (2018). State-of-the-art on strengthening of masonry structures with textile reinforced mortar (TRM). *Construction & Building Materials*, *188*, 1221–1233. DOI: 10.1016/j.conbuildmat.2018.08.039

Kuda, A., Yadav, M., & Ali, S. M. (2025). Performance Evaluation of Etfe and Auto Heal for Responsive Kinetic Facades: Improving Building Efficiency and Urban Built Environment.

Liu, M., Liu, C., Xie, H., Zhao, Z., Zhu, C., Lu, Y., & Bu, C. (2023). Analysis of the Impact of Photovoltaic Curtain Walls Replacing Glass Curtain Walls on the Whole Life Cycle Carbon Emission of Public Buildings Based on BIM Modeling Study. *Energies*, *16*(20), 7030. DOI: 10.3390/en16207030

Lyons, A. (2013). Polytetrafluoroethylene (PTFE) and Ethylene Tetrafluoroethylene Copolymer (ETFE). *In Construction Materials Reference Book (pp.* 269-278). Routledge.

Maheshwari, A., Bell, F., Gallo, E., Ibanescu, M., Robertson, E., Fairless, C., Logan, F., Schmuhl, M. J., Cheung, H., Rein, M., Danielescu, A., & Alistar, M. (2025). Self-Cleaning Textiles for Publicly Shared Soft Surfaces: An Accessible and Scalable Photocatalytic Polyester System with Embedded Light-Emitting Diodes. *ACS Omega*, *10*(4), 3978–3987. DOI: 10.1021/acsomega.4c09837 PMID: 39926538

Maity, S., Singha, K., & Singha, M. (2012). Textiles in earthquake-resistant constructions. *Journal of Safety Engineering*, *1*(2), 17–25. DOI: 10.5923/j.safety.20120102.01

Mansy, A. A. Y. (2019). Can Ethylene Tetra Fluoro Ethylene cushions double skin façade applications improve buildings envelope performance, reduce energy consumption and enhance visual and thermal comfort in UAE buildings? (Master's thesis, The British University in Dubai).

Masubuchi, M. (2013). Conceptual and structural design of adaptive membrane structures with spoked wheel principle-folding to the perimeter.

Mishra, S., Unnikrishnan, L., Nayak, S. K., & Mohanty, S. (2019). Advances in piezoelectric polymer composites for energy harvesting applications: A systematic review. *Macromolecular Materials and Engineering*, *304*(1), 1800463. DOI: 10.1002/mame.201800463

Monticelli, C., & Zanelli, A. (2022). *9. Design for Re-manufacturing (DfRem) of short chains from design-to-construction: the case of textile-based tertiary architecture*. ORKS FOR TERTIARY ARCHITECTURES.

Mottahedeh, R. (2023). Circularity and design: material reuse and sustainable practices in Architecture.

Niazy, D., Metwally, E. A., Rifat, M., Awad, M. I., & Elsabbagh, A. (2023). A conceptual design of circular adaptive façade module for reuse. *Scientific Reports*, *13*(1), 20552. DOI: 10.1038/s41598-023-47593-9 PMID: 37996556

Nishanth, L., Patil, N. N., & Ganesh, N. A. (2023). Mechanical, Durability and Embodied Energy Analysis of Geopolymer Concrete with Fly Ash, GGBFS and Glass Fiber. In *E3S Web of Conferences* (Vol. 455, p. 03006). EDP Sciences.

NPSP BV. (2019, October 19). Nabasco biocomposite with high fire safety class [News item]. Retrieved November 4, 2025, from https://www.npsp.nl/en/news-item/nabasco-biocomposite-high-fire-safety-class

Oestreich, C. R., Lazzari, P. M., Dutra, V. F. P., & Silva Filho, L. C. P. D. (2024). Bending reinforcement of reinforced concrete beams with alkali-resistant fiberglass textile-reinforced concrete with different surface treatments. *Revista IBRACON de Estruturas e Materiais*, *18*(1), e18102. DOI: 10.1590/s1983-41952025000800002

Otafy, S. S., Nassar, K., Abou-Taleb, E. M., & Aly, N. M. (2025). Sustainable Textile-Based Water Harvesting Systems: An Overview. *Egyptian Journal of Chemistry*, *68*(6), 147–164. DOI: 10.21608/ejchem.2024.304309.10014

Özen, G. (2022). Design and analysis of deployable reciprocal frames (Doctoral dissertation, Izmir Institute of Technology).

Paech, C., Göppert, K., & Wilbrenninck, S. (2014). Fast-track design and build of a FIFA compliant stadium: The Hazza Bin Zayed Stadium in Al Ain/UAE. *Stahlbau*, *83*(6), 394–399. DOI: 10.1002/stab.201410166

Pakdel, E., Fang, J., Fang, J., Sun, L., Wang, X., & Wang, X. (2018). Nanocoatings for smart textiles. *Smart textiles: wearable nanotechnology, 247-300.*

Procaccini, G. (2025). Textile materials and retrofit strategies. *New frontiers for energy efficiency. Techne : Journal of Technology for Architecture and Environment*, *29*(1), 299–308. DOI: 10.36253/techne-16600

Procaccini, G., & Monticelli, C. (2024). The application and potentialities of textile facade retrofit strategies for energy-efficient and resilient buildings. *In E3S Web of Conferences (Vol.* 3, pp. 1-5).

Procaccini, G., Prieto, A., Knaack, U., Monticelli, C., & Konstantinou, T. (2023). Textile Membrane for Façade Retrofitting: Exploring Fabric Potentialities for the Development of Innovative Strategies. *Buildings (Basel, Switzerland)*, *14*(1), 86. DOI: 10.3390/buildings14010086

Raudaschl, M., Levak, T., Riewe, R., Triantafyllidis, G., Drnda, E., Popek, S., Schlegl, D., Funke-Kaiser, D., & Lund, A. (2022, September). Piezoelectric Textile Facade for the energy supply of active sensor technology with regard to data management for circular economy in building construction. [). IOP Publishinging.]. *IOP Conference Series. Earth and Environmental Science*, *1078*(1), 012037. DOI: 10.1088/1755-1315/1078/1/012037

Robertson, K. (2012). *Felt Space: Responsive Textiles.* Fabric Dwellings and Precarious Housing.

Robinson, L. A. (2005). Structural Opportunities of ETFE (ethylene tetra fluoro ethylene) (Doctoral dissertation, Massachusetts Institute of Technology).

Runge, D., Surholt, F., Uhlemann, J., & Stranghöner, N. (2025). ETFE-foils in pneumatically retractable civil engineering structures. *Architectural Engineering and Design Management*, ●●●, 1–26. DOI: 10.1080/17452007.2025.2548549

Şahin, Y. D., & Okur, M. (2025). The types of textiles used in the façade and roofing systems of stadium facilities in Turkey. *Fibres and Textiles*, *32*(1), 96–101. DOI: 10.15240/tul/008/2025-1-019

Schwark, J., & Müller, J. (1996). High performance silicone-coated textiles: Developments and applications. *Journal of coated fabrics, 26(1), 65-77.*

Silver, P., Silver, P., & McLean, W. (2015). *Air Structures.*

Singh, S. (2021). ArchiTextile: A Review on Application of Textiles in Architecture. [JTATM]. *Journal of Textile & Apparel Technology & Management*, *12*(1).

Sluyts, Y., Rychtarikova, M., & Glorieux, C. (2024). Acoustic characterisation of ETFE cushions and their impact on the acoustic comfort of building users.

Solanki, K., Balster, T., Maywald, C., & Monticelli, C., & PADILLA MARCOS, M. Á. (2024). Evaluation of the wind load history of environmentally exposed ETFE foils. *In Proceedings of the IASS 2024 Symposium (pp.* 1-6).

Solis-Guzman, J. A. I. M. E., Martínez-Rocamora, A., & Marrero, M. (2014). *(pp* (Vol. 1). Methodology for determining the carbon footprint of the construction of residential buildings. *In Assessment of Carbon Footprint in Different Industrial Sectors*. Springer Singapore.

Strasser, D., Reber, A., Biesinger, A., & Eicker, U. (2018). Daylight Performance of a Translucent Textile Membrane Roof with Thermal Insulation. *Buildings (Basel, Switzerland)*, *8*(9), 118. DOI: 10.3390/buildings8090118

Temme // Obermeier GmbH. (n.d.). PVC/PES - PVC coated polyester fabrics - the all-rounder. Retrieved November 1, 2025, from https://www.to-experts.com/en/membranes/pvcpes/

Trencher, G., Broto, V. C., Takagi, T., Sprigings, Z., Nishida, Y., & Yarime, M. (2016). Innovative policy practices to advance building energy efficiency and retrofitting: Approaches, impacts and challenges in ten C40 cities. *Environmental Science & Policy*, *66*, 353–365. DOI: 10.1016/j.envsci.2016.06.021

Uhlemann, J. (2016). Elastic constants of architectural fabrics for design purposes (Doctoral dissertation, Duisburg, Essen, Universität Duisburg-Essen, Diss., 2016).

Van Dongen, P., Britton, E., Wetzel, A., Houtman, R., Ahmed, A. M., & Ramos, S. (2022). Suntex: Weaving solar energy into building skin. *Journal of Facade Design and Engineering*, *10*(2), 141–160. DOI: 10.47982/jfde.2022.powerskin.9

Winter, F. (2022). 3.3 Glass fibers. *From Construction Materials to Technical Gases, 55, 316.*

Yan, Y., Chen, W., Hu, J., & Tariq, I. (2025). Long-term testing study of tensile ETFE, PTFE and PVDF membrane structure integrated thin Si-PV. *Solar Energy*, *289*, 113307. DOI: 10.1016/j.solener.2025.113307

Yang, M., Zhang, Z., Yuan, J., Wu, L., Li, P., Zhao, X., & Men, X. (2020). Enhanced mechanical and tribological properties of Kevlar/PTFE-phenolic composites by improving interfacial properties by aramid nanofibers. *Polymer Composites*, *41*(10), 4192–4201. DOI: 10.1002/pc.25704

Yuan, Z., & Jia, Y. (2021). Mechanical properties and microstructure of glass fiber and polypropylene fiber reinforced concrete: An experimental study. *Construction & Building Materials*, *266*, 121048. DOI: 10.1016/j.conbuildmat.2020.121048

Zagubień, A., Wolniewicz, K., & Wesolowski, M. (2025). Overview of textile tensile surface structures with the largest uniform area. *Textile Research Journal*, *95*(7-8), 879–892. DOI: 10.1177/00405175241268836

Zhang, Y., Xu, S., Xue, J., & Zhang, Q. (2017). Anisotropic mechanical properties and constitutive relations of PTFE coated glass fibers. *Composite Structures*, *179*, 601–616. DOI: 10.1016/j.compstruct.2017.07.076

Zhao, J., & Ma, P. (2022). Comparative Study on the tribological properties of PTFE coated Kevlar fabric and hybrid PTFE/Kevlar fabric. *Fibers and Polymers*, *23*(4), 1111–1118. DOI: 10.1007/s12221-022-4421-8

Zheng, L., Ye, Y., Zhuang, J., & Zheng, Y. (2020). Impact tensile behaviors of PVDF building coated fabrics. *Advances in Civil Engineering*, *2020*(1), 1620760. DOI: 10.1155/2020/1620760

Zwanzig, S. D., Lian, Y., & Brehob, E. G. (2013). Numerical simulation of phase change material composite wallboard in a multi-layered building envelope. *Energy Conversion and Management*, *69*, 27–40. DOI: 10.1016/j.enconman.2013.02.003

Chapter 4
Built-Up Beam in Façade Engineering:
Analysis of Shear Force and Bending Moment of Z-Section and I-Section

Uma Shankar
http://orcid.org/0009-0001-4160-3380
Chandigarh University, India

Kedarisetti Pramoda Lochan
http://orcid.org/0009-0000-4456-2310
Chandigarh University, India

Tejaswini Yadav
Chandigarh University, India

ABSTRACT

Building structures must function according to façade engineering, integrating structural design with load analysis to ensure safe transmission of loads from facade and substructure to the primary structure while accommodating differential moments. To achieve efficient, strong systems, section modulus becomes crucial, particularly in primary and sub structural members of varied materials, where built-up beams play a vital role. This chapter analyzes structural behavior of different built-up beam designs for high-performance facades. Investigating properties such as center of gravity, moment of inertia, shear force, bending moment, bending and shear stresses enhances load-carrying capacity, material optimization, and adaptability to complex geometries. Further, connection strategies are evaluated for resisting wind loads

DOI: 10.4018/979-8-3373-6023-2.ch004

and supporting cladding systems. Beam modeling and FEA help assess stresses and optimize cross-sections for stiffness and strength, strengthening the role of built-up beams in façade engineering for durable and impactful structural integration.

INTRODUCTION

The built-up beams are made from two or more pieces of material joined together to form a single beam. Such beams can be constructed in a great variety of shapes to meet special architectural or structural needs and to provide larger cross-sections than are ordinarily available.

As illustrated in Figure 1, typical cross-sections of built-up beams include wooden box beams, glulam beams and plate girders.

Figure 1. Cross-sections of typical built-up beams: (a) Wooden box beam, (b) Glulam beam, (c) Plate girder

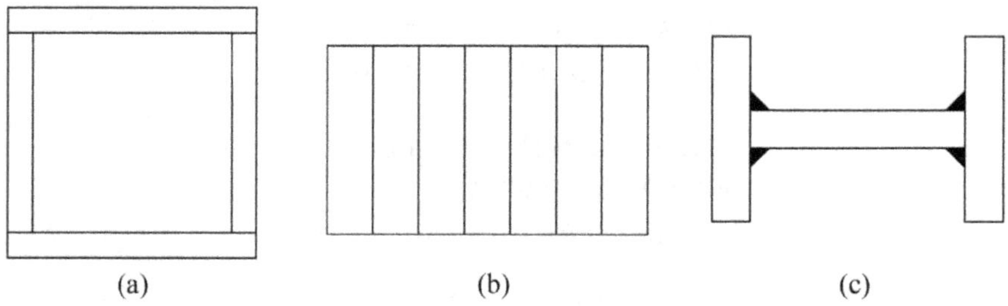

Figure 1(a) shows a wood box beam constructed of two planks, which serve as flanges, and two plywood webs. The pieces are joined together with nails, screws or glue in such a manner that the entire beam acts as a single unit. Box beams are also constructed of other materials, including steel, plastics and composites.

Figure 1(b) is a glued laminated beam (called a glulam beam) made of boards glued together to form a much larger longer beam than could be cut from a tree as a single member. Glulam beams are widely used in the construction of small buildings.

Figure 1(c) is a steel plate girder of the type commonly used in bridges and large buildings. These girders, consisting of three steel plates joined by welding, can be fabricated in much larger sizes than are available with ordinary wide flange or I-beams.

Built-up beams must be designed so that the beam behaves as a single member. The design calculations involve two phases. In the first phase, the beam is designed as if it is made up of one piece, taking into account both bending and shear stresses. In the second phase, the connection between the parts (such as nails, bolts, welds and glue) are designed to ensure that the beam does not behave as a single unit. The connections must be strong enough to transmit the horizontal shear forces acting between the parts of the beam. In order to obtain these forces, we have to use the concept of shear flow (American Institute of Steel Construction, 2016; Meza et al., 2020; Sujitha et al., 2022).

Shear Flow

Considering the Figure 2, we cut an element *mm, nn* from a beam and the horizontal equilibrium has been investigated, i.e., the element *mm, pp* as shown in Figure 2(b). From the horizontal equilibrium of the section, the force F_3 acting on its lower surface is:

Figure 2. Horizontal shear stresses and shear forces in a beam

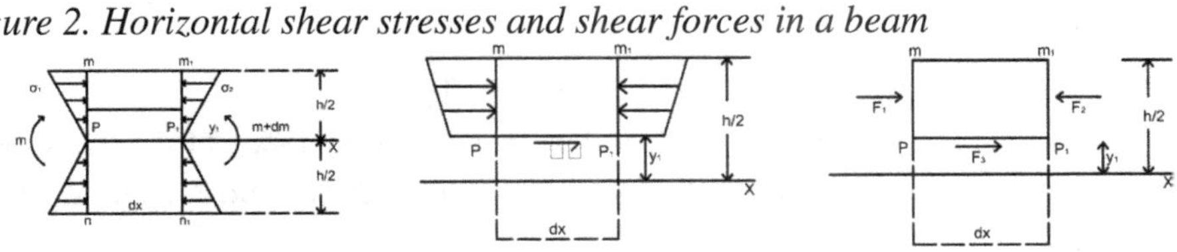

(a) Side view of element (b) Side view of element (c) Side view of element

$$\mathbf{F}_3 = \frac{\mathbf{dm}}{\mathbf{I}}\int \mathbf{y}\, dA\#(1)$$

Let: f = shear flow, which is the horizontal shear force per unit distance along the longitudinal axis of the beam.

Since, the force F_3 acts along the distance dx,

The shear force per unit distance is

$$f = \frac{F_3}{dx} = \frac{dm}{dx}\left(\frac{1}{I}\int y\, dA\right)$$

Let: shear force,

$$V = \frac{dm}{dx} = Q = \int y\, dA = \text{the first moment}$$

$$\text{Shear flow, } f = \frac{VQ}{I}\#(2)$$

This equation gives the shear flow acting on the horizontal plane PP_1, as shown in figure 2.

The shear stresses on plane *PQ* are uniformly distributed as we assumed for rectangular beams. Then the shear flow in this case becomes

$$f = \tau b$$

where b is the width.

In this case the shear flow formula reduces to a shear formula. As shown in equation (1) for the force F_3 does not involve any assumption about the distribution of the shear stresses in the beam. - Since, the force F_3 is the total horizontal shear force acting between the sub element and the rest of the beam, and it may be distributed anywhere over the sides of the sub element, not just on its lower surface. The same concept can be applied to the shear flow f.

Note:

1. V and I are not affected by the choice of sub element.
2. The first moment, Q, is a property of the cross-sectional face of the sub element (Elgaaly & Seshadri, 1998; Pasternak & Kubieniec, 2010; Luo & Edlund, 1996; Wu et al., 2020).
3. As shown in Figure 3(a) and (b) In order to have good stress distribution in a beam, the beam sections may be built-up from I-sections, angle sections, channel sections, plates etc.
4. While having a built-up section, the cost should be kept in mind.
5. Built-up girders form an important part of the beams in which the moment of inertia is normally varied by having different lengths of cover plates for the girders.

This ensures different values of Z at various sections.

Figure 3. Girder Sections (a) Compound girder section, (b) Plate girder section

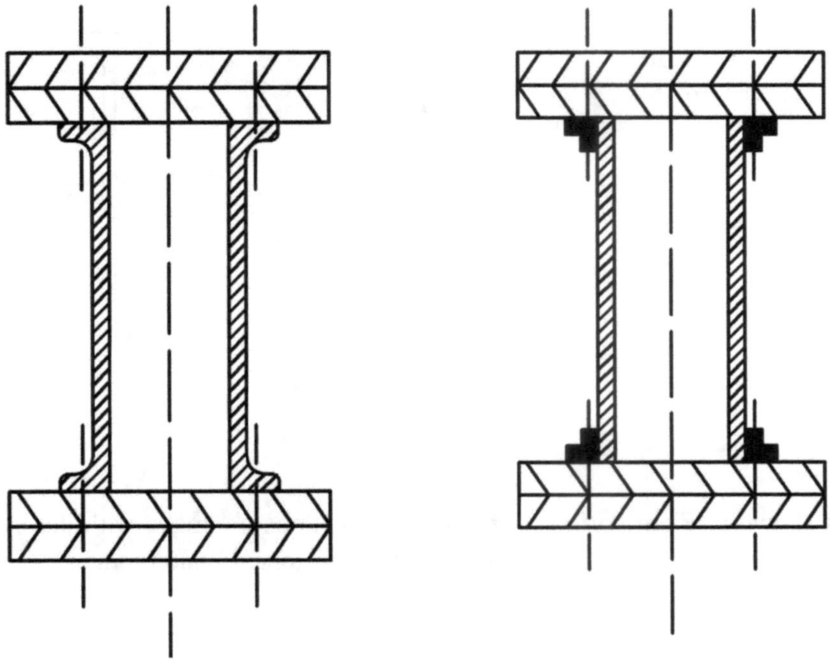

(a) (b)

6. To increase section modulus so as to improve the capacity of a beam to withstand high bending moment, a compound section is built up by using standard rolled sections as I, T, channel and angle (L) sections.
7. To reduce the effect of stress concentration at inner corners.and to eliminate sharp outer edges, the steel section of different shapes is rolled out by having fillet radius at inner corners and rounded outer edges as shown in Figure 4(a), 4(b) &4(c).

Figure 4. Rolled steel sections (a) L-section, (b) Channel Section, (c) I- section

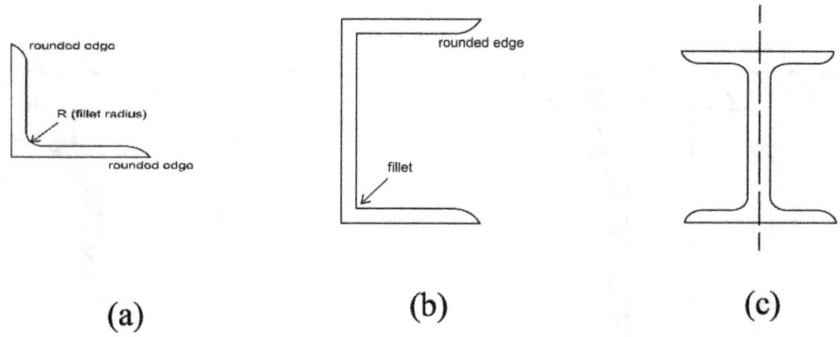

8. As shown in Figure 5(a) & 5(b) Built-up sections are made with the help of riveting or welding of plates with standard sections. C_x and moment of inertia of a built-up section has to be calculated.

Figure 5. Built-up sections (a) Box section, (b) Two channel section

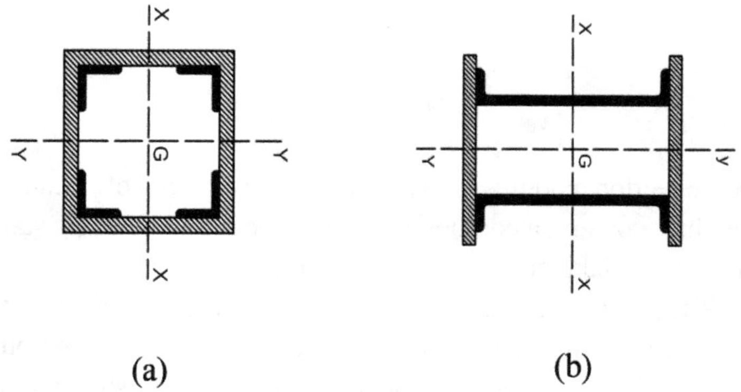

9. A built-up section is constructed from two or more parts joined together to form a single unit.

10. The capacity of this section to resist a moment, which will vary directly with its section modulus, since

$$S_{reqd} = \frac{M}{\sigma_{allow}}\left(S_{reqd} = Z\right)$$

where Z = section modulus.
If S_{reqd} is increased, then I is increased because

$$S_{reqd} = \frac{I}{C}\left(\text{where } c = y_{max}\right)$$

Due to this very reason, most of the material for a built-up section should be placed as far away from the neutral axis as practical. This is what makes a deep wide flange beam so efficient in resisting a moment (EN 1993-1-5, CEN, 2006; Dabaon et al., 2019; Narola et al., 2019; Both et al., 2023).

11. For a very large load, an available rolled steel section may not have a section modulus great enough to support the load. Hence, the engineers will usually build up a beam made from plates and angles.
12. A deep I-shaped section having this form is called a plate girder.
13. A steel plate girder can either be welded or, using angles, bolted to the web plate.
14. Wooden beams are also "built-up" usually in the form of a box beam.
15. For very large spans, "glulam beams" are used. These members are made from several boards glue-laminated together to form a single unit.
16. The design of a built-up section requires that the bending and shear stresses have to be checked.
17. In addition to this, shear stresses in the fasteners, such as weld, glue, nails etc. must be checked to be certain the beam performs as a single unit.

SOME IMPORTANT POINTS

1. Beam supports loadings that are applied perpendicular to their axes. So they are designed on the basis of strength: they must resist their allowable shear and bending stresses.
2. The maximum bending stress in a beam is assumed to be much greater than the localized stresses caused by the application of loadings on the surface of the beam.

3. Based on the discussion, the following procedure provides a rational method for the design of a beam on the basis of strength.

Shear and Moment Diagrams

Determine the maximum shear and moment in the beam, and this is done by constructing the beam's shear and moment diagrams.

Bending Stress

1. If the beam is relatively long, it is designed by finding its section modulus using the flexural formula:

$$S_{reqd} = \frac{M_{max}}{\sigma_{allow}}$$

2. Once S_{reqd} is determined, the cross-sectional dimensions for simple shapes can then be calculated, since:

$$S_{reqd} = \frac{I}{c}$$

3. If rolled steel sections are to be used, several possible beams can be selected that meet the requirement that:

$$S \geq S_{reqd}$$

Hence, choose the beam which is having smallest cross-sectional area because this beam has the least weight and thus is most economical.

4. Make sure that the selected section modulus, S, is slightly greater than S_{reqd} so that the additional moment created by the weight of the beam is considered.

Shear Stress

1. Normally, beams that are short and carry large loads, especially those made of wood, are first designed to resist shear and then later checked against the allowable bending stress requirement.
2. Using the shear formula, check that the allowable shear stress is not exceeded, i.e.,

$$T_{allow} \geq \frac{V_{max}Q}{It}$$

3. If the beam has a solid rectangular cross-section, the shear formula becomes

$$T_{allow} \geq 1.5\left(\frac{V_{max}}{A}\right)$$

and if the cross-section is a wide flange, it is generally appropriate to assume that the shear stress is constant over the cross-sectional area of the beam's web so that

$$T_{allow} \geq \frac{V_{max}}{A_{web}}$$

where A_{web} is determined from the product of the web's depth and its thickness.

ADEQUACY OF FASTENERS

The adequacy of fasteners used on built-up beams depends upon the shear stress the fasteners can resist. The required spacing of nails or bolts of a particular size is determined from the allowable shear flow, calculated at points on the cross-section where the fasteners are located (American Institute of Steel Construction, 2016; EN 1993-1-3, CEN, 2006; Meza et al., 2020).

$$q_{allow} = \frac{VQ}{It}$$

Worked Examples

Worked Example 1. The laminated wooden beam shown in the figure supports a uniform distributed loading of 12 KN/m. If the beam is to have a height to width ratio of 1.5, then determine its smallest width.

Take, $\sigma_{allow} = 9\text{MPa}$ and $\tau_{allow} = 0.6\text{MPa}$, Neglect the weight of the beam.

Solution: From the diagram:

$$V_{max} = 20KN$$

$$M_{max} = 10.67KNm$$

Bending Stress:

$$S_{regid} = \frac{M_{max}}{\sigma_{allow}}$$

$$= \frac{10.67 * 10^3}{9 * 10^6} = 0.00119 m^3$$

Assuming that the width is a and height is $1.5a$.

$$S_{regid} = \frac{I}{C}$$

$$= \frac{(a)(1.5a)^3/12}{(1.5a/2)} = 0.00119 m^3$$

$$\therefore a^3 = 0.003160 m^3$$

$$\boxed{\therefore a = 0.147m}$$

Figure 6. Laminated wooden beam with shear moment and bending moment diagram

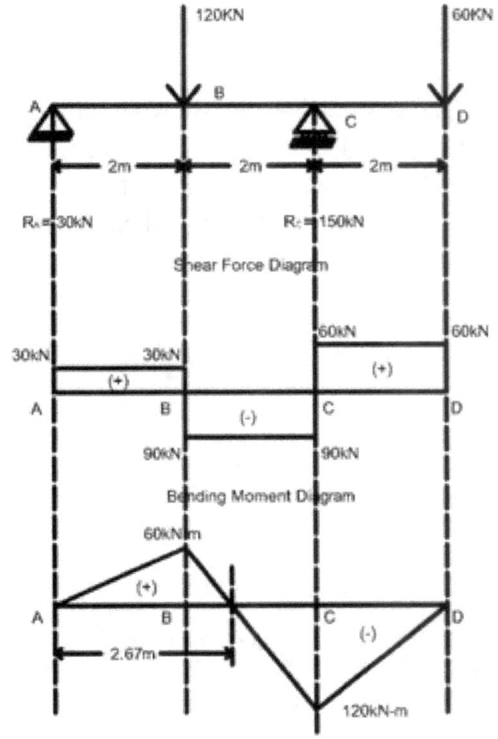

Figure 6 Shows the bending moment and shear force of the laminated wooden beam when it is subjected to the applied load.

Shear stress:

Applying the shear formula for a rectangular section

$$i.e.\tau_{max} = \frac{VQ}{It}$$

$$\therefore \tau_{max} = 1.5\frac{V_{max}}{A} = 1.5\frac{20*10^3}{0.147*1.5*0.147}$$

$$= 0.929MPa > 0.6Mpa$$

Since, the design based on bending fails the shear criterion, thus, the beam must be redesigned on the basis of shear

$$\therefore \tau_{allow} = 1.5\frac{V_{max}}{A}$$

$$600KN/m^2 = 1.5\frac{20*10^3}{a*1.5a}$$

$$\therefore a = 0.183m = 183mm$$

Worked Example 2. The wooden T-beam shown in the figure is made from two $20mm * 30mm$ boards. If $\sigma_{allow} = 12MPa$ and $\tau_{allow} = 0.8MPa$ determine if the beam can safely support the loading shown. Also specify, the maximum spacing of nails needed to hold the two boards together if each nail can safely resist $1.5KN$ shear.

Solution: From the diagram:

$$V_{max} = 1.5KN$$

$$M_{max} = 2KN - m$$

Bending stress:

$$\bar{y} = \frac{\sum A\bar{y}}{\sum A}$$

$$= \frac{(0.1m*0.03m*0.2m) + (0.215m*0.03m*0.2m)}{(0.03m*0.2m) + (0.03m*0.2m)} = 0.1575$$

Moment of Inertia, I_{xx}

$$= [\frac{1}{12}(0.03m)(0.2m)^3 + (0.03m)(0.2m)(0.1575m - 0.1m)^2]$$

$$+[\frac{1}{12}(0.2m)(0.03m)^3 + (0.03m)(0.2m)(0.215m - 0.1575m)^2$$

$$= 60.12s*10^{-6}m^4$$

$$\therefore c = 0.1575m(not\ 0.230m - 0.1575m = 0.0725m)$$

we require,

$$\sigma_{allow} \geq \frac{M_{max}C}{I}$$

$$\therefore 12*10^6 Pa \geq \frac{(2*10^3 Nm)\ (0.1575m)}{60.125*10^{-6}m^4}$$

$$= 5.24*10^6 Pa$$

Figure 7. T-beam geometry and loading diagram (a) Cross-section of the T-beam, (b) Loading and support conditions

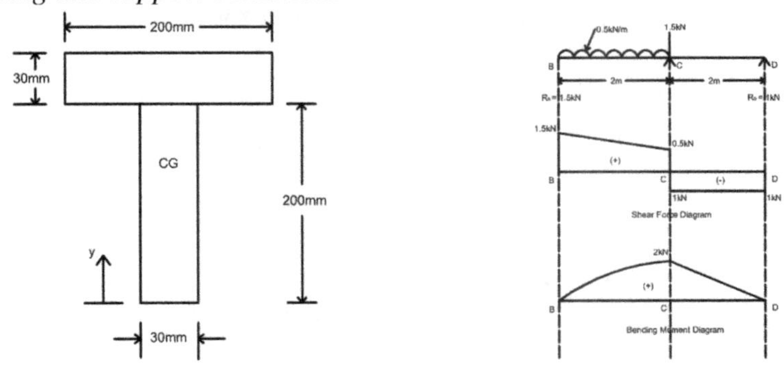

(a) (b)

Shear Stress:

Maximum shear stress in the beam depends upon the magnitude at Q and t as shown in figure 19. It occurs at the natural axis, since Q is maximum there and at the neutral axis the thickness $t = 0.03m$ is the smallest for the cross section.

$$Q = \bar{y}A' = \left(\frac{0.1575}{2}\right)[(0.1575m)(0.03m)$$

$$= 0.372 * 10^{-3} m^3$$

So that, $\tau_{allow} \geq \frac{V_{max} Q}{It}$

$$or\ 800 * 10^3 Pa \geq \frac{(1.5 * 10^3 N)(0.372 * 10^{-3} m^3)}{(60.125 * 10^{-6} m^4)(0.03m)}$$

$$\boxed{Q = 309 * 10^3 Pa}$$

Worked Example 3. The section of a welded plate girder as shown in figure 8 consists of two $450mm*25mm$ flange plates and a $1600mm*10mm$ web plate. Connection of the plates is made by four longitudinal welds running continuously throughout the length of the girder. If the girder is subjected to a shear force of $1400KN$, then determine the load per mm length each weld has to resist.

Solutions: M.I of the section

$$I = \frac{450 * 1650^3}{12} - \frac{440 * 1600^3}{12}$$

$= 1.8268 * 10^{10}\ mm^4$

Moment at the flange area about the neutral axis= ay

$= 450*25*812.5 = 9140625\ mm^3$

Figure 8. Welded plate girder

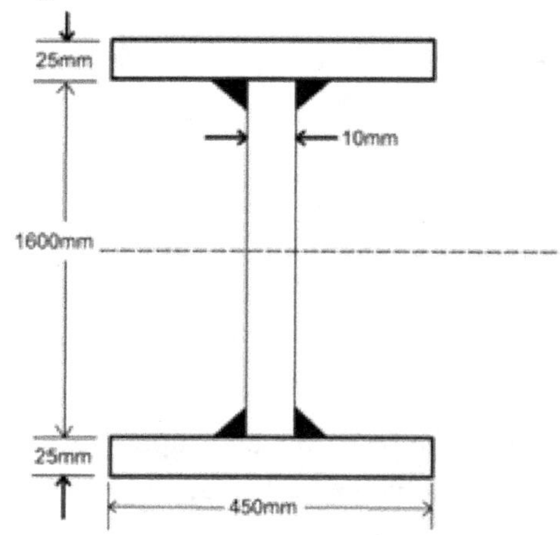

Longitudinal shear load per mm length at the junction of the flange and web$=\frac{Qay}{I}$

$$= \frac{1400 * 10^3 * 9140625}{1.8268 * 10^{10}} = \frac{700.5}{2}$$

$= 350.25 N/mm$

Worked Example 4. Figure shows a built-up beam section consisting of two ISMB 125 joints connected by 12 mmdiameter riverts as shown in figure 9, The beam is simply supported on a span at 2 $m$and carries a uniformly distributed load at such intensity that the maximum bending stress reached is 165 N/mm^2. Determine the intensity of loading on the beam and the spacing of rivets near the support. The safe average shear stress at the rivets is 100 N/mm^2 based on the gross diameter which shall be taken as 1.5 mm more than the nominal diameter.

Take for each joint ISMB 125 section. $A = 1660\ mm^2$, $I_{xx} = 449 * 10^4\ mm^4$.

Figure 9. Beam joined with the reverts

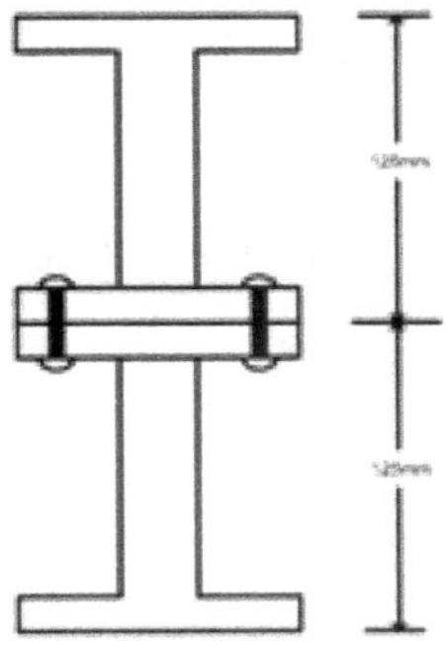

Answer: Moment of inertia at the built-up section about the neutral axis

$I = 2(449*10^4 + 1660*62.5^2)$

$= 21948750\ mm^4$

Section Modulus,

$z = \frac{21948750\ mm^3}{25}$

$= 175590\ mm^4$

moment resistance of the section,

$m = f.z = 165*175590 = 289723350\ Nmm$

$= 28.97235 KNm$

Let the loading on the beam be w KN/m
Maximum Bending moment,

$m = \frac{wl^2}{8} = \frac{w*2^2}{8} = 28.97235\ KNm$

$w = 57.9447\ KN/m$

Maximum shear force, $Q = \frac{57.9447*2}{2} = 57.9447\ KN$
Horizontal shear load per mm run near the ends at the level of the Neutral axis,

$$q_1 = \frac{Qay}{I} = \frac{57.9447*10^3*1660*62.5}{21948750}$$

$= 273.9\ N/mm$

Shearing strength at two rivets at 12*mm* diameter $= 2fQ\frac{\pi d^2}{4}$

$= 2*100*\pi*\frac{13.5^2}{4} = 28628\ N$

Let,
P be pitch of the rivets(mm)

$\therefore\ q_1 P = 28628$

$$\therefore\ P = \frac{28628}{273.9} = 104\ mm$$

Worked Example 5. A beam 10*m* long is simply supported at ends and carries a uniformly distributed load of 160 *KN/m* over the whole of its length as shown in figure 10. Determine a suitable built-up section made of standard I-beam with two plates at its upper flange and two more plates at its lower flange as shown in figure 23. All plates are to be of the same thickness. The overall depth of the section is not to exceed 800*mm* for the complete built-up section.

The plates are not to extend over the whole of the length of the beam but they are not to be cut along the length as shown in figure (10), so that requirements of resistance to bending are fulfilled with minimum length of plates. The maximum fiber stress is not to exceed $120MN/m^2$ and the weight of beam is to be neglected. Available standard plates have thicknesses of $10, 12, 16, 20, 25, 32$ *and* 40 *mm* and widths of $160, 200, 250, 320$ *and* 400 *mm*.

Figure 10. C/S of I-beam

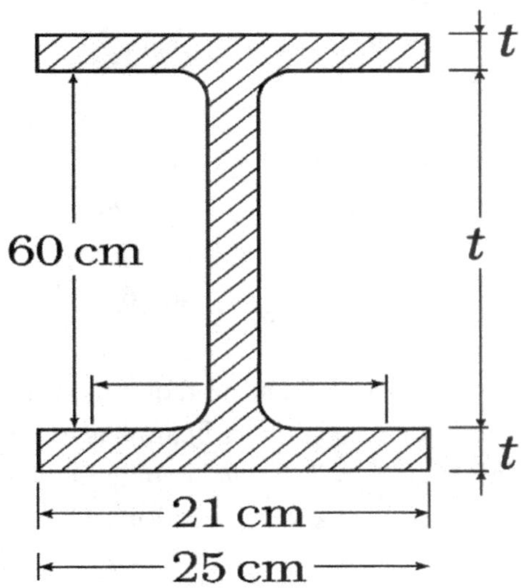

Answer:

Maximum Bending Moment = $\frac{Wl^2}{8}$ (at the center)

$$= \frac{160 * 10^2}{8} = 2000\ KNm$$

Required value at Z at the centre

$$= \frac{200 * 10^3}{120 * 10^6}\ m^3$$

$$= \frac{50}{3 * 10^3}\ m^3 = 16666.7\ cm^3$$

A beam of 600 *mm* depth (ISMB600) will be elected from the tables of properties of standard sections.

It has $I = 91813\ cm^4$ and its width is 210 *mm*. Use a plate of 250 *mm* width and t be its thickness.

$$I_{xx} = 91813 + 2t * 25 * 2(30tt)^2$$

(Neglecting the M.I of plates about their centroidal axis)

$$Z_{xx} = \frac{91813 + 100t(30tt)^2}{30 + 2t}$$

$$16666.7 = \frac{91813 + 100t(30)^2 t}{30}$$

$$\therefore\ t = 4.53\ cm = 45.3\ mm$$

The thickness 45.3 *mm* is only approximate. Let us calculate the exact value of Z_{xx} by taking the nearest available thickness of 40 *mm*.

$$I_{xx} = 91813\ + 2*4*25*2(34)^2\ + \frac{50*8^3}{12*38} = 14616\ cm^3$$

The calculated value of z_{xx} is less than the required section modulus so wider or thicker plates are required or some other I-beam has to be used. Let us provide plates of 320 *mm* in width instead of 250 *mm* keeping the I-beam of the same section. Thickness of plates is also kept same (i.e. 40 *mm*) The momentum of Inertia for the new section:

$$I_{xx} = 91813\ + 2*4*32*2(34)^2\ + \frac{2*32(8)^3}{12}$$

$$Z_{xx} = \frac{91813}{38}\ + \frac{512*34^2}{38}\ + \frac{64*8^3}{12*38} = 18082\ cm^3$$

This section is all right.

Note: Reader may try ISWB 600 with 250 *mm* wide plates 40 *mm* thick and find that,$Z_{xx} = 14996\ cm^3$, but if plates are made 320 *mm* wide and 40 *mm* thick and ISWB 600 is used. It can be seen by calculations that Z_{xx} is even greater than 18082 cm^3, Let us assume that only ISMB 600 is used without plates. It can resist the bending moment,

$$M = \sigma Z$$

$$= 120 \times 10^5 \times \frac{3060.4}{100^3}$$

$$\therefore\ M = 367.25 \times 10^3\ \text{Nm}$$

$$= 367.25\ \text{kNm}$$

The above moments are resisted by the ISMB 600 alone. This means ISMB 600 without plates can be used upto a point on the beam where B.M is 367.25 *KNm*. Let this point be at a distance of x from one of the supports as shown in Figure 11.

Figure 11. Simply supported beam with uniformly distributed load

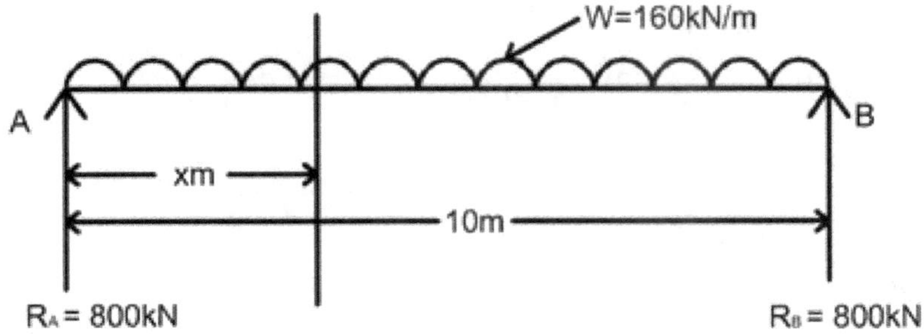

$$\therefore M_x = R_x - \frac{vx^2}{2}$$

$$= 800x - \frac{160}{2}x^2$$

$$= 800 = 80x^2$$

$$\therefore 367.25 = 800x - 80x^2 \ or \ 4.6 = 10x - x^2$$

$$\therefore x^2 - 10x + 4.6 = 0$$

$$\therefore x = \frac{10 \pm \sqrt{(100 - 18.4)}}{2} = 0.47 \ m \ or \ 9.53 \ m$$

So, the first plate should be present only over the beam for the central portion at 10 – 0.45*2 i.e.,9.1 m

Figure 12. Built-up beam made up of upper and lower flanges

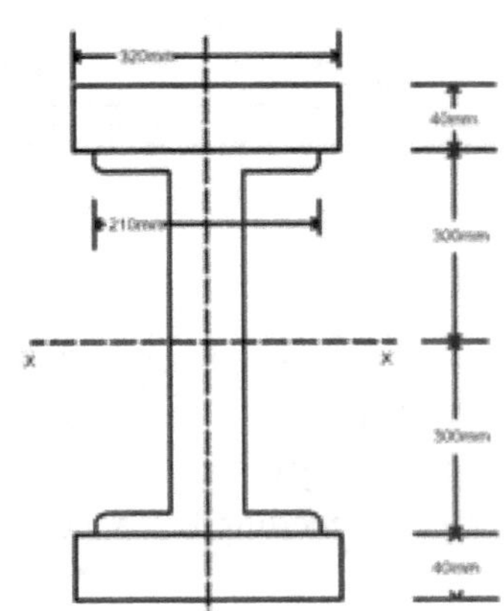

Now consider the I-beam with one plate on each side as shown in Figure 12:

$$I_{xx} = 91813 + 32*4*2(32)^2 + \frac{4^3*32*2}{12}$$

$$Z_{xx} = \frac{91813}{34} + \frac{256(32)^2}{34} + \frac{64*64}{12*34}$$

$$= 10490\ Cm^3$$

$$\therefore \text{ Resting moment } = \sigma Z_{xx}$$

$$= 120*10^6*\frac{10490}{106}$$

$$= 1258.80 \text{ KNm}$$

This means that I-beam with one plate on each flange can be used upto a point on the beam where the bending moment becomes 1258.80 KNm. Let the point be at a distance x from one end at the beam

$$\therefore\ 800x - 80x^2 = 1258.8$$

$$x^2 - 10x + 1574 = 0$$

$$x = \frac{10 \pm \sqrt{100 - 62.96}}{2} = 1.9608\ m$$

$\therefore$ It is more convenient to take, $x = 2\ m$ and $8\ m$

$\therefore$ The second plate should extend over a central beam length of ($10 - 2*2)\ m = 6m$ as shown in Figure 13.

Figure 13. C/S of the beam with shaded region

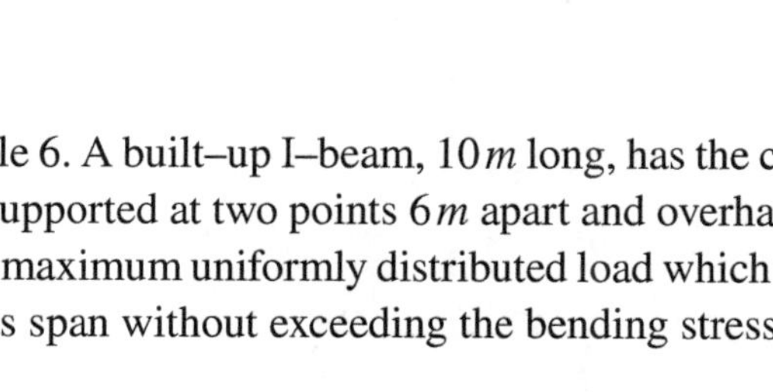

Worked Example 6. A built–up I–beam, $10m$ long, has the cross–section shown in Figure 14. It is supported at two points $6m$ apart and overhangs equally on both sides. Estimate the maximum uniformly distributed load which the beam can safely carry throughout its span without exceeding the bending stress of $40MPa$.

Solution:

Figure 14. Built–up I–section consisting of four angle sections

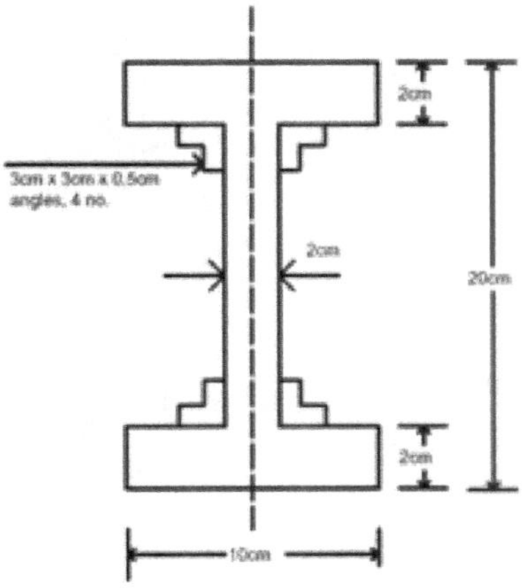

Section Properties:

$$A_1 = A_3 = 20 \text{ cm}^2, \; A_2 = 32 \text{ cm}^2, \; A_4 = 15 \text{ cm}^2$$

$$A_{\text{total}} = 83 \text{ cm}^2$$

Because of symmetry:

$$y = 10 \text{cm}$$

Moment of inertia:

$$I_{NA} = 4494.4 \text{ cm}^4$$

Reactions

For UDL w over the full $10m$ beam, supported at points $6m$ apart with $2m$ overhangs:

$$R_A \times 6 = w(6 \times 3 + 2 \times 1)$$

$$R_A = 5w, \; R_B = 5w$$

Bending Moment at a section

Let x be measured from left support:

$$M(x) = 5wx - \frac{wx^2}{2} - 2w(x+1)$$

Differentiate to find maximum:

$$\frac{dM}{dx} = w(5 - x - 2) = 0$$

$$x = 3m$$

Maximum Bending Moment

$$M_{max} = w\left[5(3) - \frac{3^2}{2} - 2\right] \tag{4}$$

$$M_{max} = w(15 - 4.5 - 8)$$

$$M_{max} = 2.5w \text{ (kNm)}$$

Stress Condition

$$\sigma = \frac{M_{max}y}{I_{NA}}$$

$$40 \times 10^6 = \frac{(2.5w)(0.10)}{4494.4 \times 10^{-8}}$$

$$w = 719.1 \text{ N/m}$$

$$\boxed{w_{max} \approx 720 \text{ N/m}}$$

This is the allowable uniformly distributed load over the entire length of the beam.

Worked Example 7. *A built-up steel column consists of two ISLC 400 channels and two plates, each* 400 mm × 18 mm *as shown in Figure* 15. *The column carries a vertical load whose line of action passes through the x–x axis at a distance of* 60 mm *from the centre. Determine the least distance between the backs of the channels if there is no tension anywhere in the section. For the section determined above, obtain on a sketch the area within which the load can act without causing tension in any part of the section.*

Solution:

Figure 15. Built-up column using two ISLC 400 channels and plates

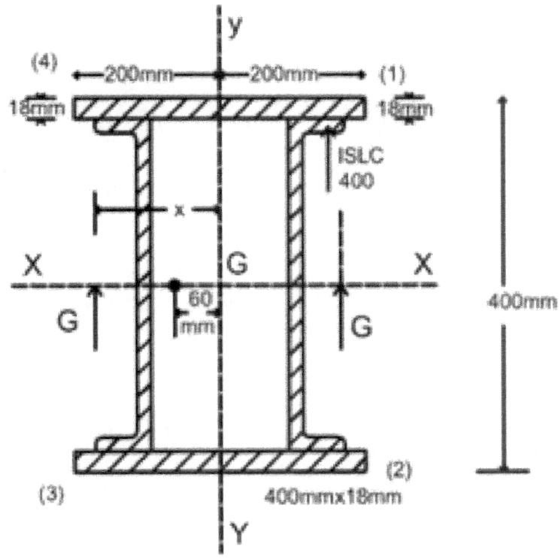

Given (for one ISLC 400 channel):

$$A_c = 58.25\ \text{cm}^2,\ I_{yy} = 460.4\ \text{cm}^4,$$

Centroid from back of channel:

$$c_1 = 23.6\,\text{mm}$$

Plates:

$$A_p = (400 \times 18)\ \text{mm}^2 = 7200\ \text{mm}^2 = 72\ \text{cm}^2$$

Total area:

$$A = 2(58.25) + 2(72) = 260.5\ \text{cm}^2$$

Let x_1 be the distance of the centroid of each channel from the y–y axis.

$$I_{yy} = 2\left[460.4 \times 10^4 + 58.25\, x_1^2 + 18 \times 400 \left(\tfrac{400}{2}\right)^2\right]$$

$$= 20120.8 \times 10^4 + 11650\, x_1^2\ (\text{mm}^4)$$

Location of centroid:

$$Z_{ey} = \frac{20120.8 \times 10^4 + 11650\, x_1^2}{260.5}$$

Condition to avoid tensile stress:
To just avoid tension anywhere:

$$\sigma_{(1)} = 0 = \frac{W}{A} - \frac{WMx_1}{I_{yy}}$$

Load acts at 60 mm from centroidal axis:

$$\frac{W}{260.5} = \frac{W \times 60\, x_1}{100.66 \times 10^4 + 58.25\, x_1^2}$$

After simplification:

$$100.66 \times 10^4 + 58.25\, x_1^2 = 156.3 \times 10^4$$

$$58.25\, x_1^2 = 55.64 \times 10^4$$

$$x_1 = 97.78\,\text{mm}$$

Distance between backs of channels:

$$D = 2(x_1 - 23.6)$$

$$D = 2(97.78 - 23.60) = 148.36\,\text{mm}$$

$$\boxed{D_{min} = 148.36\,\text{mm}}$$

This is the least spacing between the backs of the channels to avoid any tension in the built-up section.

Kern (Area Within Which Load Can Act)

Since:

$$y_{max} = 160\,\text{mm}$$

The load must lie within the kern boundary around the centroid such that:

$$e_{max} \leq \frac{I}{A\,y_{max}}$$

This region is shown in the following figure.

Worked Example 8. *The box beam shown in Figure* 16 *is made up of four* 15 cm × 2.5 cm *wooden planks connected by screws. Each screw can safely transmit a horizontal shear force of* 1250 N. *Determine the minimum necessary spacing (pitch) of screws along the length of the beam if the maximum shear force transmitted by the cross–section is* 15 kN. *Also sketch the corresponding shear stress distribution across the cross–section.*

Answer:

Figure 16. Box beam made of wooden planks connected by screws

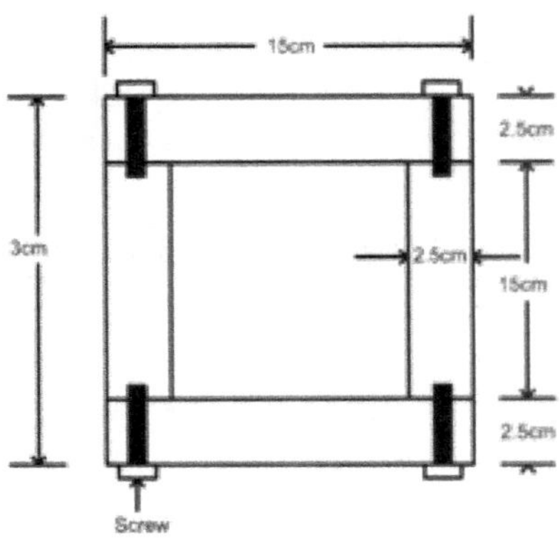

Moment of inertia

Each plank:

$$I = \frac{15 \times 2.5^3}{12}$$

Two vertical planks:

$$I_v = 2 \times \frac{15 \times 2.5^3}{12}$$

Two horizontal planks:

$$I_h = 2 \times \frac{10 \times 1.5^3}{12}$$

$$I = 7187.5 \text{ cm}^4$$

Shear stress in flanges

$$\tau_f = \frac{FA\bar{y}}{Ib}$$

$$\tau_f = \frac{5 \times 10^3(15 \times 2.5 \times 8.75 + 2 \times 2.5 \times 7.5 \times 3.75) \times 10^{-6}}{7187.5 \times 10^{-8} \times 15 \times 10^{-2}}$$

$$\tau_f = 0.152\,\text{MPa}$$

Shear stress in web

$$\tau_w = \frac{0.152 \times 15}{2.5 \times 2}$$

$$\tau_w = 0.456\,\text{MPa}$$

Average shear stress

$$\tau = \frac{5 \times 10^3(15 \times 2.5 \times 8.75 + 2 \times 2.5 \times 7.5 \times 3.75) \times 10^{-6}}{7187.5 \times 10^{-8} \times 5 \times 10^{-2}}$$

$$\tau = 0.652\,\text{MPa}$$

Pitch of screws
Let
P = pitch of screws (cm)
Horizontal shear force per pitch length:

$$V_p = 0.456 \times 2 \times 7.5 \times P \times 10^6 \times 10^{-4}$$

$$V_p = 228P\ \text{N}$$

Each screw safely carries:

1250 N

Thus:

$$228P = 2 \times 1250$$

$$P = 10.965\,\text{cm}$$

$$\boxed{P_{\min} \approx 11\,\text{cm}}$$

Shear Stress Distribution

The shear flow is highest in the vertical webs and lowest in the flanges. A parabolic distribution in the webs and nearly uniform in thin flanges is expected.

FACADE CLADDING AND GIRT BEAM DESIGN

Cladding Support and Girt design

The following girt-design procedure should be read as a preliminary engineering method for quick sizing and force estimation. It is useful for conceptual design and editorial illustration, but it does not replace a project-specific structural analysis performed under the governing design code.

Introduction

Cladding systems such as prefabricated panels, metal sidings, insulated sandwich panels, curtain wall units, and composite façade assemblies are among the most widely used components of contemporary building envelopes. These systems perform a critical role in separating the internal building environment from external climatic conditions while also contributing significantly to architectural expression, thermal efficiency, moisture protection, fire resistance, and acoustic insulation. Although these cladding elements are generally classified as non-load-bearing components because they are not intended to support the primary gravity loads of the building, they must nevertheless be capable of safely resisting and transferring their own self-weight, imposed maintenance loads, thermal movements, and environmental actions such as wind pressure and suction to the supporting structural framework (Klein, 2013; Paoletti & Nastri, 2024; Lee et al., 2018).

In modern façade engineering, the distinction between primary structural systems and secondary cladding support systems is important but should not diminish the structural significance of cladding. While the main frame—typically composed of reinforced concrete, structural steel, or composite systems—provides the principal resistance to gravity and lateral loads, the cladding and its support system must function as an integrated subsystem that ensures stability, durability, and serviceability under a wide range of operating conditions. Failures in cladding systems, even when they do not compromise the global stability of the building, can result in serious safety hazards, economic losses, water ingress, thermal inefficiency, and occupant discomfort. Therefore, façade support systems require careful structural analysis and detailing despite their non-primary classification.

The dead load of cladding refers primarily to the permanent self-weight of panels, framing members, insulation layers, sealants, glazing units, brackets, fasteners, and associated accessories. Depending on the material used, dead loads can vary significantly. For example, aluminum composite panels are relatively lightweight, whereas precast concrete façade panels impose substantially higher loads. These dead loads must be transferred efficiently through brackets, anchors, girts, rails, and mullions into the building's structural frame without causing excessive localized stresses or long-term deformation. The supporting elements must also accommodate installation tolerances and differential movements arising from creep, shrinkage, thermal expansion, and structural deflection.

In addition to dead load, wind loading is often the governing design consideration for cladding systems, especially in high-rise buildings, industrial structures, coastal regions, and open terrain where wind speeds are elevated. Wind acts on façades in both positive pressure and negative suction forms depending on building geometry, location, and flow separation characteristics. Positive pressure pushes the cladding inward toward the structure, while suction creates outward pull forces that can be particularly critical for connections and fasteners. The magnitude and distribution of these loads are influenced by factors such as building height, terrain category, topography, façade orientation, edge zones, corner effects, and openings in the building envelope. Localized pressures near roof edges and corners can be significantly higher than average façade pressures and often control connection design.

Because façade elements are directly exposed to these pressure fluctuations, the support system must provide a clear, continuous, and reliable load path from the cladding panels to the main structural frame. This load path typically begins at the panel level, where loads are transferred to clips, screws, anchors, or concealed fixing systems. From there, loads pass into secondary framing members such as girts, rails, mullions, and transoms, then into brackets or stand-off supports, and finally into columns, beams, slabs, or shear walls of the primary structure. Each component in this chain must possess adequate strength, stiffness, and ductility to prevent progressive failure. This requirement is central to both serviceability and safety assessment because failure at any intermediate stage can compromise the integrity of the entire façade system (Spence, 2006; Mazzucchelli et al., 2017; Nowak & Byrdy, 2018).

Serviceability considerations are particularly important in cladding design because excessive deflection can cause visible damage long before ultimate structural failure occurs. Excessive movement may lead to cracked sealants, broken glazing, joint opening, water penetration, vibration-induced discomfort, and aesthetic deterioration. For curtain wall systems, even small deformations can affect gasket performance and air-tightness. As a result, design standards often impose strict deflection limits such as span/180, span/240, or more restrictive criteria depending on material type and performance requirements. These limits are not merely aesthetic but are directly related to durability and occupant safety. For example, façade panel displacement under wind load must remain within tolerable limits to avoid glass breakage or fatigue in repetitive loading conditions.

Where cladding spans primarily in the vertical direction, backup framing may not always provide sufficient stiffness to control out-of-plane deformation under wind loading. This situation commonly arises in long-span metal cladding systems, insulated wall panels, warehouse envelopes, and industrial façades where panels are fixed between floor levels or between roof and base supports. If the primary framing spacing is too large, the cladding experiences increased bending moments, higher deflections, and elevated connection forces. In such cases, supplementary girts, intermediate rails, or support beams are introduced to reduce the effective span, decrease bending demand, and improve serviceability performance.

Girts are horizontal secondary structural members commonly used in steel-framed buildings to support wall cladding. They span between columns and provide intermediate restraint for vertically spanning panels. By reducing unsupported panel length, girts significantly reduce flexural stresses and deflection under wind pressure. Similarly, vertical mullions in curtain wall systems act as primary support members carrying wind loads to floor slabs, while transoms provide lateral support and help distribute localized panel loads. The design of these members depends on load combinations, connection stiffness, thermal movement requirements, and architectural constraints. Cold-formed steel sections are frequently used because of their high strength-to-weight ratio and ease of fabrication, although aluminum and hot-rolled steel sections are also common depending on exposure conditions and durability requirements.

Bracket design is another critical aspect of cladding support systems. Brackets serve as the interface between façade framing and the primary structure, often accommodating both vertical gravity loads and horizontal wind loads. They must resist bending, shear, pull-out, and local bearing stresses while also allowing for adjustment during installation. In ventilated façade systems and rainscreen assemblies, brackets often project outward from the structural wall to create a cavity for drainage and thermal insulation. This eccentricity increases moment demand and requires careful consideration of anchor capacity and thermal bridge reduction. Stainless steel, galvanized steel, and aluminum brackets are commonly selected based on corrosion resistance, structural demand, and compatibility with adjacent materials.

Connection detailing is frequently the most vulnerable part of façade systems because localized failures often originate at fasteners, anchors, or welds rather than within the panels themselves. Wind suction can generate uplift forces that exceed pull-out resistance if connections are inadequately designed. Repeated cyclic loading may also lead to fatigue, loosening, or progressive deterioration over time. Therefore, connection design must consider not only peak load resistance but also long-term performance, corrosion protection, and constructability. Bolted, screwed, riveted, and welded connections must be selected based on accessibility, maintenance requirements, and the potential need for replacement during the building's service life.

Thermal movement introduces another layer of complexity in cladding support design. Exterior façade elements are exposed to large temperature variations due to solar radiation, ambient climate changes, and differential heating between interior and exterior surfaces. Aluminum, in particular, exhibits significant thermal expansion and requires sliding or slotted connections to prevent restraint-induced stresses. If movement is not properly accommodated, thermal expansion may cause panel buckling, seal failure, cracking, or anchor distress. Thus, façade support systems must balance structural restraint with movement flexibility, ensuring that load transfer occurs without creating unintended stress concentrations.

Seismic considerations may also become relevant depending on geographic location and building importance. Although cladding is not part of the primary lateral-force-resisting system, it must remain stable during seismic events and should not detach in a way that creates falling hazards. Seismic drift between floors can impose significant demands on curtain walls and façade panels, particularly when rigid connections are used. Flexible anchorage systems and drift-compatible joints are therefore necessary in seismic design regions to allow inter-story movement without damaging the façade assembly.

Moisture management and durability further influence support system design. Water penetration through failed joints or improperly detailed connections can lead to corrosion, freeze-thaw damage, mold growth, and deterioration of insulation performance. Drainage cavities, pressure-equalized rainscreen principles, and proper flashing details are integrated into façade systems to mitigate these risks. Support members exposed to moisture must be protected through galvanization, coatings, or material selection to ensure long-term structural reliability. Galvanic corrosion between dissimilar metals must also be prevented through isolation layers and compatible detailing.

The increasing use of prefabricated façade systems in modern construction has further emphasized the importance of accurate support design. Prefabricated panels improve construction speed, quality control, and site safety, but they also require precise dimensional coordination and reliable load transfer mechanisms. Tolerance management becomes essential because even minor misalignment between the façade and structural frame can create installation difficulties and unintended stress concentrations. Digital modeling tools such as Building Information Modeling (BIM) and finite element analysis are increasingly used to coordinate façade geometry, simulate load paths, and optimize support member performance before fabrication and installation.

Sustainability considerations are also shaping cladding support strategies. Lightweight façade systems reduce embodied carbon by minimizing structural demand on the primary frame, while thermally broken brackets improve energy efficiency by reducing heat transfer across the envelope. Recyclable materials such as aluminum and steel support circular construction principles, but their structural behavior must still satisfy increasingly stringent performance expectations. The challenge for engineers is to balance sustainability goals with safety, durability, and cost-effectiveness.

Ultimately, cladding support systems should be viewed not as secondary details but as essential structural subsystems that directly influence building safety and long-term performance. Their design requires a multidisciplinary understanding of structural mechanics, material behavior, façade technology, environmental loading, and construction practice. The interaction between panels, framing members, brackets, and primary structural elements must be carefully evaluated to ensure that loads are transferred efficiently and that serviceability requirements are satisfied throughout the building's life cycle.

Where vertical spanning cladding lacks sufficient out-of-plane stiffness, the introduction of supplementary girts or support beams becomes not merely an enhancement but often a structural necessity. These elements reduce bending stresses, control deflection, protect connections, and improve façade resilience under wind loading. As façade systems become more complex and performance-driven, the importance of robust support design continues to increase. A reliable cladding support system is therefore fundamental to achieving safe, durable, and efficient contemporary building envelopes.

Figure 17 shows the detail of how such a cladding can be attached to the back-up system.

Figure 17. Typical arrangement showing cladding support and girt connection

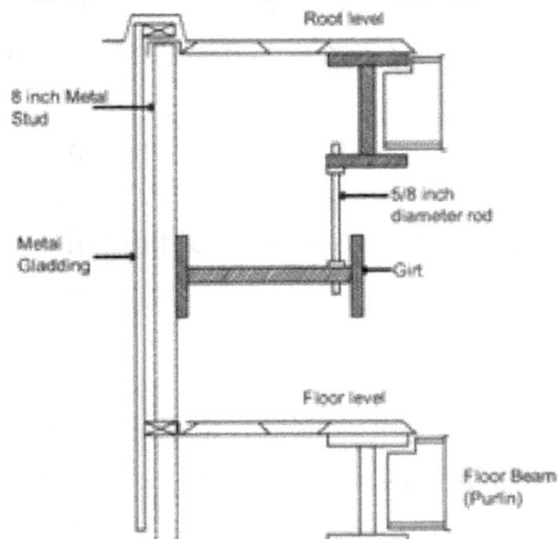

In Figure 17, the support for the cladding, in this case, is only at the two floor levels (or at the roof and the next floor below it).

When wind pressure becomes significant, the metal studs and cladding can bend in the vertical direction. A supplementary girt, commonly using a W-section, channel, square hollow section, or Z-profile, is then introduced to stiffen the system and improve load transfer.

DESIGN CONSIDERATIONS

Some claddings are able to support themselves and as such do not need girts. Examples of these are reinforced concrete precast facade panels.

Sag rods are commonly required to stabilize girts against lateral movement and weak-axis instability. Because girts frequently act as laterally restrained members under gravity but may become more vulnerable under wind suction or erection-stage conditions, their restraint details deserve explicit design attention (Spence, 2006; Lee et al., 2018).

For self-weight, a girt may often be idealized as part of a continuous system; however, for wind loading it is common and conservative to model the member as simply supported over the clear column spacing. The corresponding load path should be consistent with the tributary area of the cladding and the actual connection arrangement.

A sag rod is more critical for buckling and the compression force causing it. This compression force (as suggested by Prof. Joseph Yura of the University of *Texas*) is taken as 1% of the compression force that is developed in the flange of the girt as a result of the wind pressure.

This compression force is compensated by the tension in the sag rod because of the weight of the girt. If the compression force is not fully compensated with the tension force, then the difference must be smaller than the buckling strength of the sag rod.

In situations where the adjacent columns are far from each other, a large size girt will be needed. This makes the design very uneconomical and unreasonable. In these situations, it is more economical to add additional columns in the perimeter of the building to shorten the span length of girts. These additional columns are sometimes called wind (or soldier) columns.

In the foregoing details, we used a W-section positioned on its side; however, a channel (laid flat on its side), a hollow square section, or Z-profiles are also used as girts. In seismic areas, consult with a relevant local code to find out whether special provisions for design are required. However, most designers believe the design for wind is adequate enough.

The span length of a girt is 24 feet (spacing between the two adjacent columns). The tributary width for wind (spacing between girts in the vertical direction) is 18 feet, and the total wind pressure (including the sum of pressure and suction) is 20 psf. A W8×18 is suggested for the girt ($F_y = 50$ ksi); sag rods are $\frac{5}{8}$ inches in diameter, each with a length of 12.6 feet.

The following worked example is presented using an allowable-stress design format for continuity with the original submission. In current practice, section properties and flexural resistance should be verified against the applicable edition of the AISC Specification and the Steel Construction Manual (American Institute of Steel Construction, 2016; American Institute of Steel Construction, 2017).

The uniformly distributed wind load on the girt is given by:

$$w_u = 20 \times 18\,\text{ft} = 360\,\text{lb/ft}.$$

The applied bending moment is:

$$M = \frac{w_u L^2}{8}$$

Therefore,

$$M = \frac{360 \times (24.0)^2}{8} = 25,920\,\text{ft-lb}.$$

Because the girt is attached on its compression flange to the cladding in multiple places, we can assume that the compression flange is fully supported laterally. Hence, the bending resistance at this section per AISC (for ASD) is:

$$M_R = 26.5\,\text{ft-kips}$$

$$M_R = 26,500\,\text{ft-lb} > 25,920\,\text{ft-lb}.$$

Sag rods have a critical force equal to 1% of the compression force (F) in the flange of the girt due to the wind load.

$$F = \frac{M}{d} = \frac{25,920 \times 12}{8.0} = 38,880\,\text{lb}.$$

For two rods:

$$F = \frac{38,880}{2} = 19,440\,\text{lb per rod}.$$

Hence, the compression force in the sag rod is:

$$F_{\text{rod}} = 0.01 \times 19{,}440 = 194.4\,\text{lb.}$$

(Fig. 14) The tension in the rod is equal to the reaction of the three-span beam.

$$w_g = \text{Girt weight (lb/ft)} = 18\,\text{lb/ft}$$

$$\text{Tension in rod} = 1.19wL/3 = \frac{1.19 \times 18 \times 24}{3} = 158.4\,\text{lb}$$

The net compression in the rod is:

$$194.4 - 158.4 = 36.0\,\text{lb.}$$

This value must be smaller than the buckling strength of the rod.
Allowable Buckling Strength

$$P_{\text{allowable}} = F_a A_b$$

where

F_a = allowable stress, A_b = cross-sectional area.

For a $\frac{5}{8}$ in diameter rod:

$$A_b = 0.31\ \text{in}^2$$

The slenderness ratio kL/r for the rod is:

$$\frac{kL}{r} = \frac{1.0 \times (7.6 \times 12)}{0.16} = 456$$

Considering the rod as a connecting element per AISC Section 14.1, for this large slenderness ratio, the corresponding design provisions apply. In Section E (AISC) applies — first, we need to find F_e:

$$F_e = \frac{\pi^2 E}{(kL/r)^2}$$

Substitute:

$$E = 29{,}000{,}000\,\text{psi},\ (kL/r) = 445$$

$$F_e = \frac{\pi^2(29{,}000{,}000)}{(445)^2} = 580.5\,\text{psi}$$

Then, the allowable stress is:

$$F_a = \frac{0.877F_e}{\Omega_c} = \frac{0.877 \times 580.5}{3.02} = 168.3\,\text{psi}$$

Allowable Buckling Strength of Rod

$$P_{\text{allowable}} = F_a A_b = 168.3 \times 0.31 = 52.7\,\text{lb}$$

$$\therefore\ P_{\text{allowable}} = 52.7\,\text{lb} > 36.0\,\text{lb}$$

Hence, the rod is safe in buckling.

Causes and Signs of Damage

The following list summarizes typical causes of distress in girts and related facade-support members.

1. The main cause of damage in girts is an inappropriate design. This may stem from:
   - Lack of sag rods to prevent girt movement.
   - Using an inadequate structural member.
   - Inadequate connection of the girt to the supporting columns.
2. Excessive corrosion of the girt
3. Inadequate cladding connection to the girt.

Signs of damage are:

1. Deformed metal cladding, metal cladding span, or seams connecting elements missing.
2. Corrosion stains appearing on the cladding.
3. Crushed or displaced claddings.
4. Cracked and broken glass panes.

Repair Options

In many cases, portions of the facade, cladding, or curtain-wall system must be removed before the condition of the girt and its supporting members can be evaluated properly.

Because girts are typically proportioned for wind actions transferred from the cladding, any repair or strengthening scheme should reassess both the original demand and the current code-required load effects.

When facade panels or cladding units are removed for inspection, built-up beams composed of plates, angles, channels, or I-sections often become accessible for direct condition assessment. At this stage, deterioration, connection weakness, misalignment, and corrosion can be identified and addressed systematically (Mazzucchelli et al., 2017; Nečasová et al., 2020; Paoletti & Nastri, 2024).

1. Section Restoration Through Plate Replacement or Addition

Built-up beams typically rely on flange plates, web plates, and cover plates. When these plates undergo corrosion or deformation, the following repairs are applied:

1.1 Replacement of Corroded Plates

- Severely corroded flange or web plates are cut out using controlled gas cutting or plasma cutting.
- New plates of matching grade and thickness are welded in place.
- Continuous fillet or full-penetration butt welds are used based on design demand.

1.2 Addition of Cover Plates

- Used when partial section loss occurs, or strengthening is required.
- Plates are welded on the tension or compression flange.
- Welding is done in a staggered manner to avoid residual stresses.
- Length of cover plates is computed based on bending moment distribution.

1.3 Web Doubling or Side Plating

- Side plates are added to both sides of a weakened web.
- This increases shear capacity and restores local buckling resistance.
- Side plates are stitch-welded or fully welded depending on design loads.

2. Weld Restoration and Crack Repair

Built-up beams rely heavily on welded connections. Ageing, overload, or vibrations can initiate cracks.

2.1 Removal of Defective Welds

- Grind or gouge out cracked or porous welds.
- Re-weld with appropriate electrodes (E7018 or equivalent) following WPS/PQR.

2.2 Crack Arresting

- Stop-holes drilled at crack tips.
- Cracks filled by controlled weld buildup.

2.3 Full-Penetration Rewelding

- Required where cracks propagate through plates.
- Preheating may be needed for thicker sections to avoid HAZ cracking.

3. Bolted Splice or Bolted Reinforcement Plates

When welding onsite is difficult or thermal distortion must be avoided, bolting provides a practical solution.

3.1 Bolted Cover Plates

- High-strength friction-grip bolts (HSFG) are used.
- Reduces site welding hazards near façade elements.

3.2 Bolted Splice Joints

- Damaged beam segments are cut out.
- A new segment is inserted and connected with splice plates.

4. Replacement or Strengthening of Girt Connections

During façade removal, girt connections are frequently exposed.

4.1 Reinforced Clip Angles

- Corroded or under-designed clip angles are replaced with thicker angles.
- HSFG bolts are preferred for long-term performance.

4.2 Stiffener Installation

- Stiffeners added on the beam web at girt locations to avoid web crippling.

4.3 Bearing Pad Installation

- Introduced between girts and beams to reduce point load concentrations.

5. Strengthening Against Increased Wind Loads

If wind load requirements have changed under current codes, built-up beams may require strengthening:

5.1 Flange Enlargement

- Addition of new flange plates to increase moment capacity.

5.2 Web Stiffeners

- Spacing of stiffeners optimized to enhance shear capacity.

5.3 Lateral Bracing for Compression Flange

- Newly added bracing reduces lateral-torsional buckling risk.

6. Correction of Beam Alignment or Deflection

Prolonged façade loading can cause deflection or twist.

6.1 Heat-Straightening Technique

- Uses controlled heating to realign beams.
- Only permitted under strict engineering supervision.

6.2 Jacking and Realignment

- Hydraulic jacks lift or realign the beam.
- Reinforcing plates added afterward to maintain geometry.

7. Corrosion Control and Protective Treatment

Once the structural repairs are complete:

7.1 Surface Preparation

- Grit blasting to SA 2.5 / SSPC-SP10.
- Removal of mill scale, rust, and weld slag.

7.2 Application of Protective Coatings

- Zinc-rich primers.
- Two-layer epoxy system.
- Polyurethane topcoat for UV resistance.

7.3 Cathodic Protection (for severe environments)

- Used in coastal or industrial zones with aggressive corrosion.

8. Complete Replacement as a Last Resort

If section loss exceeds 40% or cracks are widespread:

8.1 Beam Removal

- Temporary shoring provided.
- Old member removed in segments to avoid destabilization.

8.2 New Built-Up Beam Installation

- Fabricated in workshop under controlled conditions.
- Installed with proper alignment and connection quality checks.

9. Quality Assurance Measures

Repair of built-up beams should follow stringent QA/QC:

- Ultrasonic testing (UT) of full-penetration welds
- Magnetic particle testing (MT) for surface cracks
- Radiographic testing (RT) where required
- Bolt torque verification for HSFG bolts
- Coating thickness measurement (DFT gauge)

CONCLUSION

The study of built-up sections is integral to modern structural engineering, providing an efficient solution for achieving higher load-carrying capacity and stiffness without excessively increasing material usage. Through the analysis of various built-up configurations—such as welded and bolted I-sections, plate girders, and cold-formed composite sections—it becomes evident that the performance of these assemblies is strongly influenced by connection detailing, fastener adequacy, and shear flow continuity between individual elements.

The exploration of shear flow demonstrated that maintaining uniform load transfer across flanges and webs is critical to ensuring composite action within the built-up member. Any discontinuity in shear transfer can lead to localized slip, stress concentration, and eventual structural inefficiency. The inclusion of sufficient welds, bolts, or mechanical fasteners, po- sitioned in accordance with design codes, helps maintain the structural integrity and stiffness required under bending and shear loading.

Finite element and experimental analyses available in literature consistently reveal that the deformation characteristics and ultimate strength of built-up beams are governed not only by geometry and material properties but also by the degree of interaction among the component plates. Moreover, optimization of connection layout, plate thickness, and web-flange propor- tions enhances performance and reduces material wastage, supporting sustainable construction practices.

In conclusion, built-up sections continue to serve as a versatile and economical alternative to rolled sections, offering superior adaptability in both conventional and composite structures. Understanding the principles of shear flow, connection behavior, and stiffness coupling is essen- tial for their effective design. Future developments in computational modeling and experimental validation are expected to further refine design approaches, leading to safer, lighter, and more sustainable built-up structural systems.

ACKNOWLEDGMENT

No specific funding statement or separate acknowledgment text was provided in the original submission. The revised chapter therefore retains a neutral acknowledgment section while preserving the technical content contributed by the authors.

REFERENCES

American Institute of Steel Construction. (2016). *ANSI/AISC 360-16: Specification for Structural Steel Buildings*. American Institute of Steel Construction.

American Institute of Steel Construction. (2017). *Steel Construction Manual* (15th ed.). American Institute of Steel Construction.

Both, I., Burcă, M., Benzar, ., & Ungureanu, V. (2023). Numerical study on the behaviour of built-up cold-formed steel corrugated web beams end connections. *Civil Engineering Journal, 9*(4), 770–786. DOI: 10.28991/CEJ-2023-09-04-01

CEN. (2006). *EN 1993-1-5 Eurocode 3: Design of Steel Structures, Part 1-5: Plated Structural Elements*. European Committee for Standardization.

Dabaon, M., El Hadidy, A. M., & Manaa, R. (2019). Shear behaviour of continuous tapered steel plate girders with corrugated webs. *Journal of Engineering Research, 3*(6), 60–65. DOI: 10.21608/erjeng.2019.125506

Elgaaly, M., & Seshadri, A. (1998). Steel built-up girders with trapezoidally corrugated webs. *Engineering Journal (New York), 35*(1), 1–11. DOI: 10.62913/engj.v35i1.694

Klein, T. (2013). *Integral Facade Construction*. TU Delft OPEN Publishing., DOI: 10.59490/ABE.2013.3.611

Lee, A. D., Shepherd, P., Evernden, M., & Metcalfe, D. (2018). Optimizing the architectural layouts and technical specifications of curtain walls to minimize use of aluminium. *Structures (Oxford), 13*, 8–25. DOI: 10.1016/j.istruc.2017.10.004

Luo, R., & Edlund, B. (1996). Shear capacity of plate girders with trapezoidally corrugated webs. *Thin-walled Structures, 26*(1), 19–44. DOI: 10.1016/0263-8231(96)00006-7

Mazzucchelli, E. S., Angelo, L., & Tattoni, S. (2017). Analysis and control of façade claddings structural issues. *Rivista Tema, 3*(1). Advance online publication. DOI: 10.30682/tema0301i

Meza, F. J., Becque, J., & Hajirasouliha, I. (2020). Experimental study of cold-formed steel built-up beams. *Journal of Structural Engineering*, *146*(7), 04020126. Advance online publication. DOI: 10.1061/(ASCE)ST.1943-541X.0002677

Narola, D., Meena, A., & Meena, K. (2019). Effect of corrugation angle in strength of trapezoidal corrugated web steel built-up beam. *Proceedings of the International Conference on Structural Engineering and Construction*. DOI: 10.11159/icsect19.143

Nečasová, B., Liška, P., Šlanhof, J., Sedlák, P., & Motyčka, V. (2020). Long adhesive joints in façade applications exposed to wind suction. *Proceedings of the Institution of Mechanical Engineers. Part E, Journal of Process Mechanical Engineering*, *234*(5), 488–498. DOI: 10.1177/0954408920945065

Nowak, K., & Byrdy, A. (2018). Effect of mounting brackets on thermal performance of buildings with ventilated facades. *Journal of Building Physics*, *43*(1), 46–56. DOI: 10.1177/1744259118790759

Paoletti, I., & Nastri, M. (2024). *Construction of the Façade Systems*. Springer Nature Switzerland., DOI: 10.1007/978-3-031-49608-0

Pasternak, H., & Kubieniec, G. (2010). Plate girders with corrugated webs. *Journal of Civil Engineering and Management*, *16*(2), 166–171. DOI: 10.3846/jcem.2010.17

Spence, S. M. J. (2006). *Database-Assisted Design for Wind*. National Institute of Standards and Technology., DOI: 10.6028/NIST.BSS.180

Sujitha, R., Sunmathi, N., Manikandan, R. K., Arunprasad, J., Rajkumar, S., Sharma, S., Sharma, K., Li, C., & Tag Eldin, E. M. (2022). Analytical and experimental study on cold-formed steel built-up sections for bending. *Materials (Basel)*, *15*(20), 7140. DOI: 10.3390/ma15207140 PMID: 36295206

Wu, C., Hassanein, M. F., Deng, H., Zhang, Y.-M., & Shao, Y.-B. (2020). Shear buckling response of S690 steel plate girders with corrugated webs. *Thin-walled Structures*, *157*, 107015. DOI: 10.1016/j.tws.2020.107015

Chapter 5

Lifecycle Assessment of Aluminium Composite Panel Façades in Tropical and Temperate Climates: A Durability-Informed Approach to Sustainable Design

Arkar Htet
http://orcid.org/0000-0003-1301-3604
Lincoln University College, Malaysia

Theingi Aung
http://orcid.org/0009-0003-5063-575X
Lincoln University College, Malayisa

Sui Reng Liana
Lincoln University College, Malayisa

Om Prakash Giri
Pokhara University, Nepal

Zakir Hossen Shaikh
http://orcid.org/0000-0003-4733-4166
Kingdom University, Bahrain

Yin Hlaing Min
http://orcid.org/0009-0004-2266-6935
Azteca University, Mexico

ABSTRACT

This chapter develops a durability-informed lifecycle assessment perspective for aluminium composite panel (ACP) façades in tropical and temperate climates. It argues that ACP sustainability cannot be judged by initial embodied carbon alone, because climatic exposure influences coating degradation, corrosion risk, sealant ageing, moisture ingress, maintenance frequency, service life, and end-of-life recovery. Tropical climates tend to intensify humidity-, rainfall-, UV-, and corrosion-related deterioration, while temperate climates present challenges associated with thermal movement, freeze-thaw exposure, sealant fatigue, and seasonal moisture manage-

DOI: 10.4018/979-8-3373-6023-2.ch005

ment. The chapter integrates lifecycle assessment, façade durability science, and green building certification implications to show that ACP façade sustainability is a dynamic whole-life performance outcome. It concludes that more reliable assessment requires climate-sensitive durability assumptions, preventive maintenance planning, realistic recycling pathways, and transparent environmental documentation.

1. INTRODUCTION

1.1 Role of Façade Systems in Sustainable Building Performance

Façade systems are central to the sustainability performance of contemporary buildings because they mediate heat flow, daylight access, weather resistance, indoor comfort, energy demand, and architectural identity. As the boundary between internal space and external climatic exposure, the façade is both a technical assembly and a long-term environmental asset. Recent concern over climate change, resource depletion, and the environmental burden of construction has therefore shifted façade evaluation beyond appearance and short-term functionality toward whole-life performance (Cabeza et al., 2014; Pomponi & Moncaster, 2017).

This shift is particularly important because reductions in operational energy demand through efficient mechanical systems, renewable energy integration, and improved controls increase the relative importance of embodied environmental impacts. Embodied carbon associated with extraction, manufacturing, transport, installation, maintenance, replacement, and disposal can represent a substantial share of total building-related emissions, especially in high-performance buildings (Pomponi & Moncaster, 2017). Façade materials such as glass, aluminium, cladding panels, insulation, sealants, and supporting framing systems therefore require careful assessment because they influence both embodied and operational performance.

Lifecycle thinking has consequently become a necessary part of façade engineering and material selection. Rather than evaluating façade systems only by initial cost, appearance, or short-term performance, designers are increasingly expected to consider environmental consequences across the service life of the envelope. Lifecycle assessment (LCA) provides a structured method for examining impacts associated with raw material extraction, production, transport, construction, maintenance, repair, replacement, and end-of-life treatment. Standards such as ISO 14040, ISO 14044, and EN 15978 provide the methodological basis for applying this logic in the building sector (International Organization for Standardization, 2006a, 2006b; European Committee for Standardization, 2011). When applied to façade systems, these frameworks allow engineers and architects to compare alternatives not only on technical and economic grounds, but also through measurable environmental performance (Asadi et al., 2012).

1.2 Aluminium Composite Panels in Contemporary Façade Design

Aluminium composite panels (ACPs) are widely used in contemporary façade design because they combine low weight, geometric flexibility, and efficient installation. Typically composed of two aluminium sheets bonded to a polymer or mineral-filled core, ACP systems are valued for their dimensional stability, surface finish options, and adaptability in commercial and high-rise applications. Their light weight can reduce structural demand and simplify transportation and handling compared with heavier façade materials (Efthymiou et al., 2010; Han et al., 2015; Khan et al., 2021).

From a sustainability perspective, however, ACP performance is more complex than these practical advantages suggest. Primary aluminium production is energy-intensive and contributes substantially to embodied impact, while the composite structure of ACPs can complicate end-of-life separation and material recovery. The environmental value of ACP systems therefore depends not only on production impacts but also on durability, maintenance demand, recoverability, and the feasibility of local recycling pathways across the building life cycle (Pomponi & Moncaster, 2017; European Commission, 2020; Trubina et al., 2024).

1.3 Climate Influence on Façade Performance

Climate strongly influences façade durability, maintenance demand, and long-term environmental performance. In tropical climates, high humidity, intense solar radiation, elevated temperature, and frequent rainfall can accelerate coating degradation, sealant ageing, corrosion-related processes, and biological growth. In temperate climates, the dominant concerns are more often thermal movement, freeze–thaw exposure, moisture ingress, and pollution-related weathering (Kumar & Imam, 2013; Straube & Burnett, 2005).

These exposure differences affect sustainability because reduced service life increases repair, replacement, and associated embodied impacts over time. For façade systems, this means that lifecycle assessment should account not only for material production but also for climate-sensitive durability performance throughout the service period (Cabeza et al., 2014; Pomponi & Moncaster, 2017).

1.4 Research Gap

Despite the growth of sustainability research in construction, three gaps remain particularly important for façade engineering. First, many LCA studies evaluate buildings or materials at a general level without isolating façade assemblies, even though façades experience distinct exposure conditions and performance demands. Second, durability and environmental impact are still often assessed separately, which can lead to misleading conclusions when service life varies by climate. Third, comparative studies of façade material sustainability under tropical and temperate conditions remain limited, especially for ACP systems (Cabeza et al., 2014; Riahinezhad et al., 2021; Park et al., 2018).

1.5 Aim, Objectives, and Chapter Structure

This chapter develops a durability-informed lifecycle sustainability assessment of ACP façades through a structured synthesis of secondary evidence from tropical and temperate climatic contexts. The purpose is to connect lifecycle environmental assessment with façade durability and climate-responsive engineering so that ACP sustainability is interpreted as a whole-life performance issue rather than as a production-stage material attribute (Pomponi & Moncaster, 2016, 2017).

The objectives are to:

(1) synthesize lifecycle assessment literature relevant to ACP façade systems;
(2) compare environmental performance trends across tropical and temperate climates;
(3) evaluate the influence of durability and service life on lifecycle outcomes;
(4) propose climate-responsive design and maintenance strategies; and
(5) consider implications for frameworks such as LEED and BREEAM.

The chapter proceeds from conceptual foundation and methodology to comparative analysis, engineering discussion, certification implications, practical recommendations, and final conclusions.

2. LITERATURE REVIEW AND CONCEPTUAL FOUNDATION

2.1 Lifecycle Sustainability Assessment in Façade Engineering

Lifecycle sustainability assessment has become increasingly important in construction as the sector shifts toward low-carbon and resource-efficient design. In façade engineering, this perspective is especially relevant because façades contribute to both embodied environmental impact and long-term building performance. As operational emissions decline in better-performing buildings, embodied carbon associated with envelope materials becomes a more important component of whole-life assessment (Cabeza et al., 2014; Pomponi & Moncaster, 2016).

Lifecycle assessment provides a structured method for examining impacts from material extraction to end-of-life management. However, façade-specific studies remain less developed than broader building-level research, and many assessments still treat the envelope as a generalized system rather than distinguishing among façade materials, service lives, exposure conditions, and replacement scenarios (Han et al., 2015; Huedo et al., 2016; Keles & Yazıcıoğlu, 2023; Park et al., 2018).

2.2 Environmental Impact Indicators Relevant to Façade Materials

The environmental assessment of façade materials commonly focuses on global warming potential, embodied energy, resource depletion, and end-of-life recovery. For materials such as aluminium, these indicators are particularly significant because manufacturing is energy-intensive, while recyclability can substantially affect whole-life performance. A meaningful sustainability assessment must therefore consider both initial production impacts and later-stage recovery potential (Dixit et al., 2010; Pomponi & Moncaster, 2017).

These indicators should not be interpreted in isolation. High embodied impact may be justified where longer service life, lower maintenance demand, or improved functional performance reduce cumulative impacts over time. This is why durability remains central to lifecycle interpretation in façade engineering (Jelle, 2011; Cabeza et al., 2014; Struhala et al., 2018).

2.3 ACP Material Characteristics Affecting Sustainability

ACP systems consist of aluminium face sheets bonded to a polymer or mineral-filled core and protected by surface coatings such as PVDF or polyester. Their low weight, finish flexibility, and ease of fabrication explain their widespread use in modern façade applications. However, recent façade material selection research indicates that such systems should be evaluated through technical, environmental, and performance criteria rather than cost and appearance alone (Naseer et al., 2024; Efthymiou et al., 2010). Sustainability performance therefore depends on more than these practical advantages: primary aluminium production is energy-intensive, and the composite structure complicates material separation, recycling, and fire-related material selection decisions (Efthymiou et al., 2010; European Commission, 2020).

Durability is especially important because longer service life reduces maintenance and replacement-related impacts. In practice, ACP performance is strongly influenced by coating quality, joint detailing, moisture management, thermal movement, and workmanship. For this reason, lifecycle sustainability assessment should treat ACP systems as climate-exposed assemblies rather than simply as materials with fixed environmental values (Riahinezhad et al., 2021; Straube & Burnett, 2005). Figure 2.1 presents the conceptual framework adopted in this chapter to link ACP material composition, climatic exposure, durability behaviour, lifecycle impact indicators, and overall sustainability outcomes.

Figure 1. Conceptual framework for durability-informed lifecycle sustainability assessment of ACP façade systems

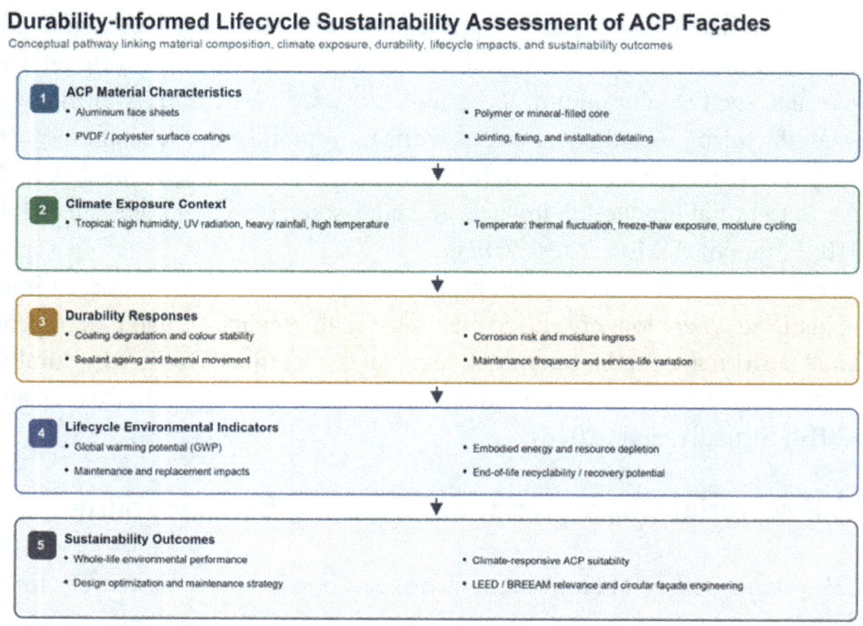

Source: Developed by the author based on Cabeza et al. (2014), Pomponi and Moncaster (2016, 2017), Han et al. (2015), and Struhala et al. (2018)

2.4 Durability Science of Façade Materials

Durability is a fundamental consideration in façade engineering because building envelopes are continuously exposed to complex and interacting environmental stressors. Service life planning standards, particularly ISO 15686, provide structured methodologies for predicting material longevity based on exposure conditions, material properties, and maintenance strategies. These frameworks emphasize that durability is not an intrinsic material property alone but a function of environmental context and design detailing (International Organization for Standardization, 2011; Straube & Burnett, 2005).

Environmental degradation mechanisms affecting façade materials include ultraviolet (UV) radiation, moisture ingress, temperature fluctuation, atmospheric pollutants, and biological growth. These stressors may act independently or synergistically to accelerate deterioration. External façade surfaces are particularly vulnerable because condensation, pollutant deposition, and cyclic environmental loading occur directly at the weathering interface (Kumar & Imam, 2013; Rabajczyk et al., 2021). For ACP façades, coating deterioration is a major durability concern because it affects corrosion resistance, aesthetic stability, and weatherproofing performance. Improper joint detailing, sealant failure, and inadequate drainage can further accelerate system-level deterioration, reinforcing the need to evaluate surface protection and maintenance as part of lifecycle interpretation (Búryová & Sedlák, 2021).

Integrating durability assessment with lifecycle environmental analysis is increasingly recognized as essential for accurate sustainability evaluation. A material with higher initial embodied impact may ultimately be more sustainable if it demonstrates longer service life and reduced maintenance requirements. Conversely, materials with lower embodied carbon but shorter durability may result in higher cumulative environmental burdens due to frequent replacement and maintenance cycles. This trade-off highlights the importance of combining durability science with lifecycle assessment in façade engineering practice (Pomponi & Moncaster, 2016; Cabeza et al., 2014).

2.5 Climate-Specific Façade Performance Considerations

Tropical Climate Context

Tropical climates present significant challenges for façade systems because high humidity, intense solar radiation, heavy rainfall, and elevated temperatures often act simultaneously. High moisture levels can accelerate corrosion processes, particularly in coastal or industrial environments, while persistent UV exposure contributes to coating degradation, colour fading, and polymer embrittlement. These conditions increase the likelihood of moisture ingress, biological growth, and sealant deterioration, all of which can adversely affect façade durability and performance (Kubba, 2012; Kumar & Imam, 2013).

Empirical observations and field studies in tropical regions indicate that façade maintenance cycles are often shorter than those in temperate climates. This can require more frequent inspection, recoating, and component replacement. Consequently, lifecycle environmental impacts may increase through additional material consumption, transport, and maintenance-related emissions. These findings highlight the importance of climate-responsive façade design strategies, including enhanced protective coatings, improved ventilation detailing, and effective drainage systems (Straube & Burnett, 2005; Rabajczyk et al., 2021).

Temperate Climate Context

In contrast, temperate climates subject façade systems to different environmental stress patterns. Seasonal temperature variations result in repeated thermal expansion and contraction, which can induce mechanical stresses in panels, joints, and fixing systems. Over time, these cyclic movements may lead to fatigue, sealant failure, and joint integrity issues. Additionally, freeze–thaw cycles in colder climates can exacerbate deterioration when moisture penetrates façade assemblies (Kumar & Imam, 2013; Straube & Burnett, 2005).

Although UV exposure may be less intense than in tropical climates, urban pollution and atmospheric contaminants can accelerate coating degradation and surface soiling. These factors contribute to reduced aesthetic performance and may require more frequent cleaning or maintenance interventions. The variation in environmental stress mechanisms between tropical and temperate climates underscores the necessity of region-specific façade design and material selection strategies. Ignoring climatic context in sustainability evaluation can result in misleading conclusions regarding long-term performance and environmental impact (Pomponi & Moncaster, 2017; Cabeza et al., 2014).

2.6 Conceptual Integration Framework for Durability-Informed Lifecycle Assessment

Based on the reviewed literature, a conceptual framework can be established that integrates three primary dimensions: lifecycle environmental assessment, façade durability science, and climatic exposure conditions. This integrated perspective recognizes that environmental performance is not determined solely by initial material production impacts but is significantly influenced by service life, maintenance frequency, and end-of-life management strategies (Pomponi & Moncaster, 2016; Antypa et al., 2022).

Figure 2. Integrated framework illustrating how climatic conditions influence façade durability mechanisms, service life, maintenance cycles, and lifecycle environmental impacts

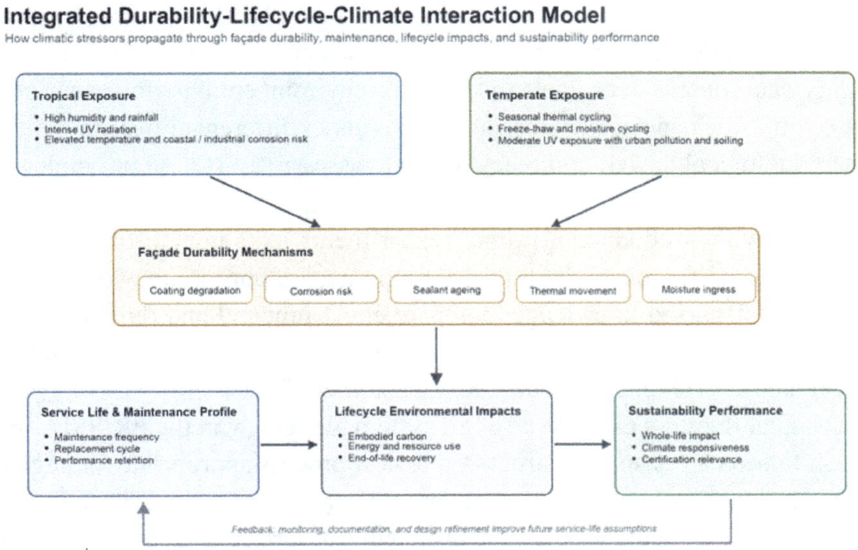

Source: Developed by the author based on Cabeza et al. (2014), Pomponi and Moncaster (2016, 2017), Straube and Burnett (2005), and Kumar and Imam (2013)

Within this framework, climate functions as a moderating variable that shapes façade degradation rates and maintenance requirements. Durability characteristics, in turn, affect lifecycle environmental outcomes by altering replacement intervals, repair cycles, and long-term operational performance. By synthesizing lifecycle assessment data with durability evidence across climatic contexts, the chapter develops a more realistic and performance-oriented basis for sustainability evaluation (Cabeza et al., 2014; Struhala et al., 2018).

This integrated approach forms the conceptual basis for the comparative analysis presented in subsequent sections of this chapter and is consistent with recent façade engineering research emphasizing multi-criteria and performance-oriented decision-making in material selection and system design (Bianchi et al., 2024; Naseer et al., 2024). It enables a balanced and context-sensitive evaluation of ACP façade sustainability, incorporating both environmental impact indicators and real-world performance conditions. As a result, the framework provides practical insights for façade engineers, designers, and sustainability practitioners seeking to optimize material selection and design strategies across diverse climatic environments.

3. METHODOLOGY

3.1 Research Design

This chapter adopts a structured literature-based comparative assessment using secondary data sources to evaluate the lifecycle environmental performance of aluminium composite panel (ACP) façade systems under different climatic conditions. The methodological design combines lifecycle assessment (LCA) principles with durability-informed interpretation to develop an integrated understanding of façade sustainability without conducting primary experiments, laboratory testing, or original simulation-based lifecycle modelling. This approach is appropriate when the aim is to synthesize dispersed knowledge, compare environmental and durability trends across existing studies, and derive design-oriented insights rather than generate new empirical datasets (Grant & Booth, 2009; Cabeza et al., 2014).

The chapter does not claim to be a full systematic review in the PRISMA sense. Instead, it follows a transparent structured review process appropriate for engineering and sustainability scholarship, drawing evidence from peer-reviewed studies, technical documents, standards, and environmental reporting sources. This approach is suitable for façade sustainability research because relevant evidence is distributed across lifecycle studies, durability investigations, façade case studies, building science literature, and product documentation rather than concentrated in a single standardized dataset (Pomponi & Moncaster, 2016; Han et al., 2015).

The study focuses specifically on tropical and temperate climatic contexts because these represent materially different environmental exposure conditions affecting façade durability, maintenance cycles, and lifecycle sustainability outcomes. By integrating published lifecycle environmental evidence with durability-related findings reported in the literature, the research seeks to provide a practice-oriented sustainability interpretation relevant to façade engineering decision-making (Struhala et al., 2018; Kumar & Imam, 2013).

3.2 Literature Search Strategy

A structured literature search was undertaken to identify academic publications, technical reports, environmental product declarations (EPDs), and standards documents related to ACP façade sustainability, lifecycle assessment, and durability performance. The search covered major academic databases, including Scopus, Web of Science, ScienceDirect, SpringerLink, and Google Scholar, because these sources collectively provide broad coverage of engineering, materials, sustainability, and construction research (Grant & Booth, 2009; Cabeza et al., 2014).

The search primarily targeted literature published during the last 10 to 15 years, while also retaining selected earlier references where they were considered foundational to lifecycle assessment, durability science, or building envelope methodology. Only sources published in English were included to maintain consistency of interpretation. The search terms were developed around the central themes of ACP façades, lifecycle sustainability, durability, and climatic exposure. Representative search strings included "aluminium composite panel façade lifecycle assessment," "façade embodied carbon sustainability," "ACP durability climate exposure," "building envelope lifecycle assessment," and "façade environmental performance." In addition to database searching, backward reference screening was used to identify key foundational studies and relevant frequently cited sources that may not have been captured through keyword searches alone (Grant & Booth, 2009; Pomponi & Moncaster, 2016).

To maintain relevance and methodological rigor, inclusion criteria were defined in advance. Sources were included if they comprised peer-reviewed journal articles, conference papers, standards, technical publications, or credible environmental reporting documents; if they addressed façade materials, lifecycle assessment, environmental durability, service life, or climate-sensitive performance; and if they provided either quantifiable environmental indicators or interpretable durability-related evidence. Studies were excluded if they were promotional in nature, lacked methodological transparency, focused solely on architectural aesthetics, or did not contain information relevant to sustainability or durability assessment. This screening process ensured that the final evidence base was sufficiently robust for comparative analysis while remaining aligned with the chapter's scope (Han et al., 2015; Struhala et al., 2018).

3.3 Secondary Data Extraction Framework

Data extracted from the selected literature were organized into three primary analytical categories: environmental impact indicators, durability performance evidence, and climate exposure characteristics. This framework was developed to support cross-study comparison while recognizing that the literature varies in scope, system boundary definition, and reporting depth. Organizing the evidence into these categories is consistent with the conceptual foundation of this chapter, which treats façade sustainability as the outcome of interactions among initial environmental burden, long-term durability, and environmental exposure context (Cabeza et al., 2014; Pomponi & Moncaster, 2017).

The first category, environmental impact indicators, included global warming potential (CO_2-equivalent emissions), embodied energy, resource depletion concerns, recyclability, and end-of-life implications. These indicators were selected because they are among the most commonly reported metrics in building and material LCA research and are especially relevant to aluminium-based façade systems, whose manufacturing and disposal stages can substantially influence overall environmental performance (Pomponi & Moncaster, 2016; Struhala et al., 2018).

The second category, durability-related evidence, included reported service life estimates for ACP systems or comparable façade materials, coating degradation observations, maintenance frequency recommendations, and documented failure modes associated with environmental exposure. Where exact numerical values were unavailable, qualitative performance evidence from façade studies and building science literature was incorporated to support interpretation. This category was essential because service life variation can alter maintenance intensity, replacement frequency, and cumulative lifecycle burden (Straube & Burnett, 2005; Kumar & Imam, 2013).

The third category, climate exposure characteristics, included humidity, ultraviolet radiation intensity, rainfall exposure, temperature variability, atmospheric pollution, and freeze–thaw conditions. Where source studies did not explicitly classify climatic zones, climatic conditions were interpreted and grouped into broader tropical or temperate categories using established building science reasoning. This enabled consistent comparison while acknowledging the diversity of terminology and reporting formats across the literature (Kubba, 2012; Kumar & Imam, 2013).

This structured extraction process supports a pattern-based synthesis rather than strict numerical pooling. It allows environmental data, durability evidence, and exposure conditions to be interpreted together in order to derive context-sensitive sustainability insights relevant to façade engineering practice. Table 3.1 summarizes the secondary data extraction structure and the analytical categories used to support the comparative sustainability assessment in this chapter.

Table 1. Secondary data extraction and comparative assessment framework

Analytical category	Variables / indicators extracted	Purpose in the study
Environmental impact indicators	Global warming potential (GWP); embodied energy; resource depletion; recyclability; end-of-life implications	To evaluate the lifecycle environmental burden associated with ACP façade systems and identify the principal sustainability indicators reported across studies
Durability performance evidence	Service life estimates; coating degradation observations; corrosion-related evidence; sealant performance; maintenance frequency; reported failure modes	To assess how long-term façade performance influences sustainability outcomes, particularly through maintenance intensity and replacement cycles
Climate exposure characteristics	Humidity; ultraviolet radiation exposure; rainfall intensity; temperature variability; atmospheric pollution; freeze–thaw conditions	To classify the environmental stress conditions affecting façade durability and to distinguish differences between tropical and temperate contexts
Comparative interpretation	Tropical versus temperate exposure profiles; durability-adjusted lifecycle implications; maintenance and replacement effects; context-sensitive sustainability trends	To synthesize climate-responsive sustainability insights relevant to façade material selection, design decisions, and long-term performance evaluation

3.4 Comparative Analytical Approach

The comparative analysis focuses on how climatic conditions influence lifecycle sustainability outcomes for ACP façade systems. Rather than conducting original lifecycle modelling, the study interprets recurring trends reported in published LCA studies, façade material comparisons, durability studies, and building science sources. This form of structured comparative synthesis is appropriate where the available literature provides complementary but methodologically varied evidence across environmental, durability, and climate dimensions (Grant & Booth, 2009; Pomponi & Moncaster, 2016).

For analytical clarity, two broad climatic categories were used. Tropical climates were defined as environments characterized by high humidity, elevated temperatures, intense solar radiation, and frequent rainfall. These conditions are associated with increased risks of coating degradation, corrosion, sealant ageing, and biological growth. Temperate climates were defined as environments characterized by moderate humidity but marked seasonal variation, where thermal expansion and contraction, freeze-thaw effects, and pollution-related surface deterioration may become significant (Kumar & Imam, 2013; Straube & Burnett, 2005).

Durability adjustments were incorporated conceptually by comparing reported or inferred service life ranges and maintenance demands across these climatic conditions. The underlying assumption is that more aggressive environmental exposure may reduce effective service life and increase maintenance or replacement frequency, thereby raising cumulative lifecycle environmental impacts. This logic is consistent with LCA principles that emphasize whole-life performance rather than initial production impacts alone (International Organization for Standardization, 2006a; Cabeza et al., 2014).

Accordingly, the comparative framework integrates three interacting dimensions: environmental impact indicators, durability evidence, and climatic exposure characteristics. The purpose is not to generate a universal pooled numerical value for ACP sustainability, but to develop a context-sensitive interpretation of how climate and durability influence lifecycle outcomes in façade systems. This makes the methodology especially relevant for design guidance, material selection, and sustainability-informed façade engineering decisions.

3.5 Reference Standards and Sustainability Frameworks

To ensure methodological consistency, the comparative assessment was interpreted with reference to internationally recognized lifecycle, durability, and building sustainability frameworks. Lifecycle principles were aligned primarily with ISO 14040 and ISO 14044, which establish the general framework, requirements, and guidance for life cycle assessment studies. EN 15978 was used as an important reference for environmental performance assessment at the building level, while EN 15804+A2 was used to contextualize environmental product declarations for construction products (International Organization for Standardization, 2006a; International Organization for Standardization, 2006b; American Institute of Steel Construction, 2011; American Institute of Steel Construction, 2019).

Durability-related interpretation was informed by ISO 15686, which emphasizes service life planning through the consideration of environmental exposure, maintenance conditions, and material performance characteristics. This standard is especially relevant to façade engineering because envelope systems are highly sensitive to climatic stressors, detailing quality, and maintenance practice, all of which directly influence effective service life and long-term sustainability outcomes (International Organization for Standardization, 2011; Straube & Burnett, 2005).

In addition to these standards, major green building certification systems such as LEED and BREEAM were reviewed at a conceptual level to understand how lifecycle performance, material selection, and durability considerations are reflected in professional sustainability evaluation. Although this chapter does not conduct a formal certification scoring analysis, these frameworks provide useful practical context for interpreting how durability-informed lifecycle assessment may support façade-related sustainability decision-making in actual projects (Kubba, 2012; Pomponi & Moncaster, 2016).

Taken together, these standards and frameworks provide a structured basis for interpreting the secondary evidence used in this study and help ensure that the comparative assessment remains aligned with internationally recognized sustainability evaluation practices.

3.6 Methodological Limitations

Several limitations of the methodology should be acknowledged. First, the reviewed studies differ in system boundaries, functional units, life cycle stages considered, data sources, and impact assessment methods. As a result, strict numerical comparability across all publications is limited, and this chapter emphasizes comparative pattern interpretation rather than direct statistical aggregation (Grant & Booth, 2009; Cabeza et al., 2014).

Second, the climatic categories used in this chapter are interpretive and are based on reported exposure characteristics rather than a single universal climate classification protocol. While this approach is appropriate for comparative façade analysis, regional variation within tropical and temperate zones may still influence actual durability and maintenance outcomes. Third, not all reviewed sources provide complete ACP-specific service life, maintenance, or end-of-life data, which means that some interpretations are informed by comparable façade material studies and established building science literature (Kumar & Imam, 2013; Straube & Burnett, 2005).

A further limitation is that the chapter does not include primary experimental validation, full simulation-based modelling, or statistical meta-analysis. Accordingly, the conclusions should be interpreted as evidence-based sustainability trends and design insights rather than as precise predictive outputs for all ACP façade applications. Nevertheless, the structured literature-based comparative approach remains valuable for synthesizing current knowledge, identifying recurring sustainability patterns, and informing façade engineering decision-making in a standards-informed manner (Pomponi & Moncaster, 2016; Grant & Booth, 2009).

4. COMPARATIVE LIFECYCLE SUSTAINABILITY ANALYSIS

4.1 Environmental Impact Comparison of ACP Façades Across Climatic Contexts

Aluminium composite panel (ACP) façade systems exhibit lifecycle sustainability profiles that are shaped not only by material composition but also by the climatic conditions under which they operate. Across building envelope and façade-related lifecycle assessment studies, the production stage consistently emerges as the dominant contributor to embodied environmental impact because aluminium manufacturing is highly energy-intensive, particularly when primary aluminium is used rather than recycled feedstock (Pomponi & Moncaster, 2016; Efthymiou et al., 2010). However, this initial embodied burden does not fully explain long-term sustainability performance, since climatic exposure influences subsequent maintenance, durability, and replacement requirements over the service life of the façade.

In tropical climates, high humidity, elevated temperatures, intense ultraviolet radiation, and frequent rainfall can accelerate degradation processes affecting façade performance. Although ACP systems are generally valued for their light weight, corrosion resistance, and architectural versatility, their long-term behaviour remains sensitive to coating durability, joint performance, drainage effectiveness, and maintenance quality. Under tropical exposure, intensified weathering may shorten the effective service life of façade coatings and increase maintenance frequency, thereby raising cumulative lifecycle impacts over time even when the initial material specification remains unchanged (Kumar & Imam, 2013; Riahinezhad et al., 2021).

By contrast, temperate climates often expose façade systems to seasonal thermal variation, moderate humidity, freeze–thaw risk in colder regions, and pollution-related ageing. These conditions also affect lifecycle performance, but the degradation pattern is often different from that observed in tropical environments. In many cases, lower humidity and reduced year-round UV intensity may slow certain forms of surface deterioration, although thermal cycling may still create movement-related stresses in joints, fixings, and sealants. As a result, ACP façades in temperate contexts may achieve longer effective service life under well-detailed conditions, thereby reducing replacement-related environmental burdens over the full life cycle (Straube & Burnett, 2005; Riahinezhad et al., 2021).

Lifecycle interpretation therefore requires attention not only to initial embodied carbon but also to exposure-sensitive durability outcomes. Studies that focus exclusively on production-stage emissions risk understating the long-term environmental significance of climate-driven maintenance and replacement effects. From a façade engineering perspective, climate-responsive service life behaviour is a central determinant of whether the initial embodied burden of ACP systems is either amplified or moderated over time (Cabeza et al., 2014; Pomponi & Moncaster, 2017).

4.2 Lifecycle Stage Contribution Analysis

The environmental performance of ACP façade systems can be interpreted across four life cycle stages: material production, transportation and installation, operational maintenance, and end-of-life management. Among these, production remains the dominant source of embodied impact because aluminium manufacturing is energy-intensive, particularly when primary rather than secondary aluminium is used (Pomponi & Moncaster, 2016; Efthymiou et al., 2010). For façade assemblies, however, use-stage assumptions and service-life scenarios can materially alter comparative outcomes, especially where maintenance or replacement intervals differ (Búryová & Sedlák, 2021; Park et al., 2018).

Transportation and installation generally contribute less than production, although ACP panels benefit from low weight and easier handling compared with heavier cladding systems. At the same time, comparative façade studies indicate that façade type can also affect broader performance outcomes, meaning that lifecycle interpretation should not be reduced to transport efficiency alone (Han et al., 2015; Tokbolat et al., 2020). The greater sustainability issue at installation is workmanship quality, since inadequate fixing, sealing, or drainage can shorten service life and increase future environmental burdens (Han et al., 2015; Straube & Burnett, 2005).

The maintenance stage is more climate-sensitive. Tropical exposure tends to intensify coating weathering, moisture-related deterioration, and sealant ageing, increasing the likelihood of more frequent inspection, repair, or replacement. Temperate climates impose different demands, especially thermal movement and freeze–thaw-related stress, but may permit longer maintenance intervals where detailing is well resolved (Kumar & Imam, 2013; Riahinezhad et al., 2021).

End-of-life interpretation remains uncertain because aluminium is recyclable in principle, but ACP panels are composite systems and may not be easy to separate efficiently in practice. Sustainability assessment should therefore distinguish between theoretical recyclability and realistic recovery under local waste-management, disassembly, and secondary-material market conditions (European Commission, 2020; Efthymiou et al., 2010; Trubina et al., 2024).

Table 2. Comparative lifecycle sustainability characteristics of ACP façades in tropical and temperate climates

Aspect	Tropical climate context	Temperate climate context	Sustainability implication
Dominant exposure factors	High humidity, intense UV, heavy rainfall, elevated temperature	Thermal cycling, freeze–thaw risk, moderate UV, pollution exposure	Climate affects degradation pathways differently
Main durability risks	Coating degradation, sealant ageing, corrosion susceptibility, biological growth	Joint movement, sealant fatigue, moisture ingress, freeze-thaw-related distress	Durability response shapes service life
Maintenance tendency	More frequent inspection, cleaning, repair, recoating	Moderate but detail-sensitive maintenance	Maintenance frequency affects cumulative impacts

Aspect	Tropical climate context	Temperate climate context	Sustainability implication
Likely service life trend	Potentially shorter without high-performance detailing and maintenance	Potentially longer under well-managed conditions	Service life influences replacement-related embodied carbon
End-of-life implication	Depends on local waste management and recycling infrastructure	Depends on disassembly and recycling feasibility	Theoretical recyclability may differ from actual recovery

Figure 3 translates the comparative interpretation from Table 2 into a design-oriented climate-durability-lifecycle matrix. It provides a practical bridge between the literature synthesis and the durability-adjusted interpretation that follows.

Figure 3. Climate-specific durability and lifecycle implication matrix for ACP façade systems

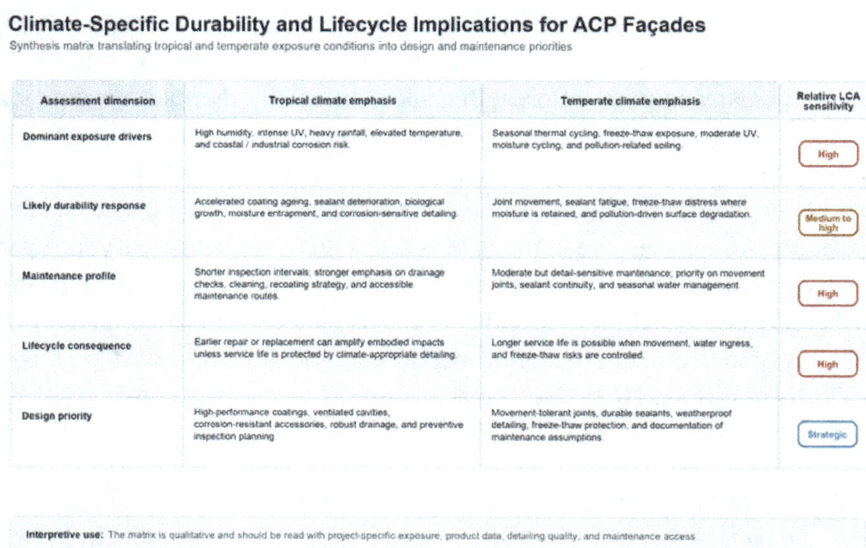

Climate-Specific Durability and Lifecycle Implications for ACP Façades

Synthesis matrix translating tropical and temperate exposure conditions into design and maintenance priorities

Assessment dimension	Tropical climate emphasis	Temperate climate emphasis	Relative LCA sensitivity
Dominant exposure drivers	High humidity, intense UV, heavy rainfall, elevated temperature, and coastal / industrial corrosion risk.	Seasonal thermal cycling, freeze-thaw exposure, moderate UV, moisture cycling, and pollution-related soiling.	High
Likely durability response	Accelerated coating ageing, sealant deterioration, biological growth, moisture entrapment, and corrosion-sensitive detailing.	Joint movement, sealant fatigue, freeze-thaw distress where moisture is retained, and pollution-driven surface degradation.	Medium to high
Maintenance profile	Shorter inspection intervals; stronger emphasis on drainage checks, cleaning, recoating strategy, and accessible maintenance routes.	Moderate but detail-sensitive maintenance; priority on movement joints, sealant continuity, and seasonal water management.	High
Lifecycle consequence	Earlier repair or replacement can amplify embodied impacts unless service life is protected by climate-appropriate detailing.	Longer service life is possible when movement, water ingress, and freeze-thaw risks are controlled.	High
Design priority	High-performance coatings, ventilated cavities, corrosion-resistant accessories, robust drainage, and preventive inspection planning.	Movement-tolerant joints, durable sealants, weatherproof detailing, freeze-thaw protection, and documentation of maintenance assumptions.	Strategic

Interpretive use: The matrix is qualitative and should be read with project-specific exposure, product data, detailing quality, and maintenance access.

Source: Developed by the author based on the comparative synthesis in this chapter and supporting literature on façade LCA, durability, and circular design.

4.3 Durability-Adjusted Lifecycle Interpretation

Durability is a critical determinant of façade sustainability because service life directly affects replacement frequency and cumulative environmental impact. A façade material with a relatively high initial embodied burden may still produce favourable whole-life performance if it remains functional for a long period with limited maintenance. Conversely, materials or systems that degrade more quickly may generate higher cumulative impacts through repeated repair and replacement cycles. For this reason, durability-adjusted lifecycle interpretation is essential when assessing ACP façades in different climatic regions (International Organization for Standardization, 2011; Pomponi & Moncaster, 2017).

In tropical climates, effective service life may be reduced by a combination of UV exposure, persistent humidity, heavy rainfall, and in certain locations salt-laden or polluted air. These conditions can accelerate coating deterioration, compromise sealants and joints, and increase the probability of moisture-related system distress. Under such circumstances, service life may be shorter unless high-performance protective coatings, appropriate detailing, and regular preventive maintenance are implemented (Kumar & Imam, 2013; Riahinezhad et al., 2021).

Temperate climates may support longer service life in cases where UV intensity and long-duration humidity exposure are lower. However, longer apparent service life should not be assumed automatically, since thermal cycling, joint movement, and freeze-thaw action can also create durability challenges. In these settings, the sustainability performance of ACP façades depends heavily on how well the system accommodates movement, protects against water ingress, and maintains seal integrity over time (Straube & Burnett, 2005; International Organization for Standardization, 2011).

A durability-adjusted reading of lifecycle sustainability therefore suggests that the environmental value of ACP systems cannot be judged from production-stage data alone. The decisive issue is whether the initial embodied burden is distributed across a sufficiently long, reliable, and well-managed service life. From this perspective, façade sustainability becomes inseparable from detailing quality, maintenance planning, climate-responsive durability strategy, and transparent assumptions about replacement intervals (Búryová & Sedlák, 2021; Riahinezhad et al., 2021).

4.4 Maintenance and Service Life Considerations

Maintenance strategy has a direct influence on the lifecycle sustainability of ACP façades because it affects both service continuity and the timing of material replacement. Preventive maintenance practices such as routine cleaning, sealant inspection, drainage checks, and coating condition monitoring can extend façade service life and reduce the environmental burdens associated with premature failure. Accordingly, maintenance should not be treated as a secondary operational issue but as a core sustainability variable in façade engineering (Straube & Burnett, 2005; International Organization for Standardization, 2011).

In tropical environments, the need for maintenance is often intensified by moisture-related exposure and faster surface ageing. Drainage detailing, cavity ventilation, corrosion-resistant accessories, and inspection accessibility become especially important under these conditions. Inadequate maintenance planning in such environments can lead to compounded deterioration effects, including staining, moisture entrapment, coating breakdown, joint leakage, and localized corrosion, all of which shorten the effective lifespan of the façade (Kumar & Imam, 2013; Kubba, 2012).

Temperate climates generally impose different service-life priorities. In these contexts, thermal movement accommodation, sealant elasticity, weather-sealing continuity, and moisture management during cold periods are especially significant. If these are poorly handled, façade deterioration may occur even where humidity levels are lower than in tropical settings. This indicates that service life is not simply a function of "harsh" or "mild" climate, but of how system design and maintenance strategy respond to the specific environmental stresses of each climatic zone (Riahinezhad et al., 2021; Straube & Burnett, 2005).

From a lifecycle perspective, maintenance planning affects environmental performance through three pathways: it delays premature replacement, reduces the scale of intervention required, and helps preserve the functional value of the original embodied material investment. Consequently, sustainability evaluation should incorporate maintenance frequency, accessibility, and durability management as integral components of whole-life façade performance.

4.5 Synthesis of Climate-Durability-Environment Interactions

The comparative evidence synthesized in this chapter indicates that climate functions as a moderating factor linking façade durability and lifecycle environmental performance. Tropical climates tend to intensify weathering-related degradation and increase maintenance demand, while temperate climates often impose movement-related and moisture-management challenges associated with seasonal change. In both cases, climate affects sustainability not only through direct material deterioration but also through its influence on service life, intervention frequency, and end-of-life timing (Kumar & Imam, 2013; International Organization for Standardization, 2011).

These interactions demonstrate that ACP façade sustainability cannot be adequately evaluated using initial embodied carbon alone. Instead, a meaningful assessment must consider the interaction of material composition, protective coating systems, detailing quality, maintenance planning, and regional environmental exposure. This confirms the conceptual framework developed in earlier sections of the chapter, where lifecycle indicators, durability science, and climatic context were treated as interdependent dimensions rather than isolated variables (Cabeza et al., 2014; Pomponi & Moncaster, 2017).

The synthesis also has practical implications for façade design and specification. Climate-responsive material selection, improved joint and drainage detailing, maintenance planning, and realistic end-of-life strategies all contribute to better sustainability outcomes. Moreover, as green building frameworks increasingly emphasize lifecycle thinking, durability and service life planning are likely to become more important in the environmental justification of façade systems used in different regions (European Commission, 2020; Kubba, 2012).

Overall, the analysis suggests that the environmental performance of ACP façades is dynamic rather than fixed. It is shaped by the interaction between initial material impacts and the real-world conditions that govern durability, maintenance, and recovery potential across the building lifespan. A climate-sensitive, durability-informed lifecycle perspective therefore provides a more accurate basis for sustainable façade engineering than production-stage assessment alone.

5. DISCUSSION: ENGINEERING INTERPRETATION AND DESIGN IMPLICATIONS

The analysis indicates that ACP sustainability is governed not only by embodied carbon but by the interaction of climatic exposure, detailing quality, maintenance strategy, and service life. In engineering terms, this means that façade sustainability is performance-dependent rather than material-dependent alone (Pomponi & Moncaster, 2016, 2017).

5.1 Climate-Responsive Façade Engineering Insights

A key finding is that climatic exposure changes the sustainability profile of ACP systems. In tropical climates, high humidity, rainfall, and solar radiation intensify degradation risk and increase the importance of moisture management, corrosion resistance, and coating durability. In temperate climates, thermal movement, freeze–thaw conditions, and joint behaviour become more decisive. Accordingly, façade engineering should respond to dominant local exposure conditions rather than applying uniform detailing assumptions across regions (Kumar & Imam, 2013; Riahinezhad et al., 2021).

5.2 Sustainable ACP Specification Strategies

Sustainable ACP specification should include not only cost and appearance but also coating durability, maintainability, disassembly potential, and end-of-life feasibility. Higher-performance coatings and better detailing may increase initial impact or cost, but they can improve whole-life outcomes by extending service life and reducing replacement frequency. Fasteners, joints, drainage paths, and substructure compatibility should therefore be treated as sustainability-critical components of the façade assembly, particularly where future disassembly or material recovery is expected (Straube & Burnett, 2005; Efthymiou et al., 2010; Trubina et al., 2024).

5.3 Lifecycle Sustainability Trade-Offs

The chapter highlights several trade-offs. Aluminium has a relatively high initial embodied burden, yet its durability, light weight, and recyclability can improve performance over time. Similarly, preventive maintenance requires periodic resource input, but it is usually preferable to early failure and major replacement. These examples show that whole-life interpretation is essential and that low initial impact does not necessarily mean low lifecycle impact (Pomponi & Moncaster, 2016, 2017).

5.4 Implications for Façade Engineers and Designers

For practitioners, the findings suggest four priorities: climate-responsive detailing, durability-informed lifecycle assessment, early interdisciplinary collaboration, and better post-installation performance documentation. Together, these measures can improve both environmental interpretation and practical façade reliability. In this sense, sustainable façade engineering depends as much on the design process and maintenance planning as on product selection alone (Attia et al., 2018; Bianchi et al., 2024).

The discussion also points to the value of stronger post-occupancy and post-installation performance documentation. Maintenance records, inspection findings, and durability observations from real projects would improve future façade sustainability assessments and reduce uncertainty in service-life assumptions. The advancement of façade sustainability therefore depends not only on better materials, but also on better feedback loops between practice and research.

6. IMPLICATIONS FOR GREEN BUILDING CERTIFICATION

6.1 Relevance to LEED Certification Framework

Green building certification systems increasingly recognize lifecycle-based material assessment, environmental transparency, and resource efficiency as essential components of sustainable building design. In this context, ACP façade systems can influence certification outcomes not only through their material composition but also through the quality of environmental documentation, maintenance planning, and end-of-life considerations. For projects pursuing LEED, the most relevant implications arise primarily within the Materials and Resources category, where environmental product declarations (EPDs), responsible material selection, construction waste reduction, and lifecycle impact reduction are emphasized (U.S. Green Building Council [USGBC], 2023; Kubba, 2012).

For ACP systems, verified EPDs are particularly valuable because they provide transparent environmental data that can support evidence-based material comparison during the design stage. This is especially important for aluminium-based façade products, since the production phase may carry substantial embodied impacts. In practice, ACP products accompanied by reliable environmental documentation place project teams in a stronger position to justify specification decisions within a lifecycle framework rather than on cost or aesthetic criteria alone (Pomponi & Moncaster, 2016; American Institute of Steel Construction, 2019; USGBC, 2023).

Durability also affects LEED relevance, even if it is not always represented as a single explicit credit line. Materials that maintain performance over longer service periods and require fewer replacements align more closely with lifecycle impact reduction goals. The findings of this chapter suggest that this issue is especially important in aggressive climatic conditions, where inadequate coating specification, poor detailing, or weak maintenance planning may reduce effective service life and undermine anticipated sustainability benefits. Therefore, ACP systems that combine documented environmental transparency with climate-appropriate durability strategies are more likely to support credible LEED-oriented design outcomes (Cabeza et al., 2014; Pomponi & Moncaster, 2017).

ACP systems may also contribute positively through recyclability and waste-management pathways. However, this contribution should be interpreted with caution because the recyclability value of ACPs depends not only on the aluminium itself but also on whether practical separation, recovery, and reuse systems are available locally. LEED-related sustainability claims should therefore distinguish between nominal recyclability and actual end-of-life feasibility in the project context (European Commission, 2020; Trubina et al., 2024).

6.2 Integration with BREEAM Sustainability Assessment

The implications for BREEAM are similarly important, and in some respects more directly connected to durability and long-term performance. BREEAM places strong emphasis on materials assessment, responsible sourcing, management quality, waste, and lifecycle-oriented design. For façade systems, this means that ACP sustainability is judged not only by material impact data but also by the way the assembly is specified, maintained, and integrated into a long-term building performance strategy (Building Research Establishment [BRE], 2018; Kubba, 2012).

One significant point of alignment is the use of lifecycle-based building assessment frameworks supported by standards such as EN 15978. Because this chapter evaluates ACP sustainability through the interaction of lifecycle impacts, durability, and climate-sensitive service life, its findings are especially relevant to BREEAM's whole-life orientation. ACP systems supported by credible lifecycle data, transparent sourcing information, and realistic maintenance expectations are therefore better positioned to contribute positively within a BREEAM-aligned project strategy (European Committee for Standardization, 2011; BRE, 2018).

BREEAM also places practical weight on management and long-term performance planning, which is especially relevant to façade engineering. This chapter has shown that climate-sensitive durability influences maintenance demand, intervention frequency, and overall environmental burden. Measures such as higher-performance coating specification, inspection accessibility, movement-tolerant detailing, and responsible maintenance planning therefore support not only technical performance but also certification-relevant sustainability outcomes (Straube & Burnett, 2005; Riahinezhad et al., 2021).

Another relevant issue is responsible sourcing. ACP manufacturers that provide stronger environmental transparency, better quality assurance, and clearer evidence of sustainable production and recovery practices are more likely to support favourable certification results. In this sense, the contribution of ACP systems to BREEAM depends not only on the panel product itself, but also on documentation quality, procurement strategy, and long-term management logic.

6.3 Opportunities for Enhanced Certification Integration

Although both LEED and BREEAM increasingly incorporate lifecycle thinking, the findings of this chapter suggest that there is still room to strengthen the integration of climate-responsive durability within certification systems. Current frameworks reward environmental transparency, reduced embodied impact, and responsible material use, but they are often less explicit in evaluating how local climatic exposure affects actual service life and long-term façade performance. For

ACP systems, this is important because materials with similar initial environmental profiles may perform very differently in tropical and temperate settings (Pomponi & Moncaster, 2017; Riahinezhad et al., 2021).

A stronger certification approach would therefore place greater emphasis on durability-informed lifecycle interpretation. In practical terms, this could include more explicit recognition of exposure-specific detailing, service-life-sensitive specification, maintenance planning, and realistic replacement assumptions. Such changes would improve the alignment between certification logic and actual façade performance in service.

There is also an opportunity to strengthen the role of performance evidence in certification. Wider use of verified EPDs, durability-related technical documentation, maintenance records, and post-installation monitoring data would make façade sustainability claims more defensible. Over time, integration between digital design tools, product databases, and lifecycle documentation could support more transparent and context-sensitive certification assessments, especially for complex façade assemblies (Bianchi et al., 2024; BRE, 2018).

Overall, ACP façade systems can contribute positively to green building certification, but the strength of that contribution depends on how lifecycle environmental data are interpreted together with durability, climate exposure, and practical end-of-life pathways. Certification systems already provide a useful incentive structure for more sustainable façade practice, yet they would become more effective if they gave stronger recognition to climate-sensitive service-life performance.

7. PRACTICAL RECOMMENDATIONS FOR INDUSTRY

7.1 Climate-Responsive Design and Maintenance Recommendations

The findings of this chapter suggest that industry practice should move toward a more climate-responsive and durability-informed approach when specifying ACP façade systems. Although ACPs are used widely across different regions, their long-term sustainability depends heavily on how well the façade assembly responds to local environmental exposure. This means that specification decisions should be based not only on initial cost, appearance, or nominal material properties, but also on expected weathering conditions, maintenance requirements, and service-life performance (Riahinezhad et al., 2021; Bianchi et al., 2024).

In tropical climates, façade design should prioritize resistance to humidity, solar radiation, rainfall, and corrosion risk. Higher-performance coatings, reliable drainage detailing, ventilated cavities, and corrosion-resistant accessories are particularly important under these conditions. Moisture management is essential because persistent wetting can accelerate deterioration of joints, coatings, and subcomponents, especially where installation quality is inconsistent. Industry practice in such climates should therefore emphasize protective detailing, careful product compatibility, and routine inspection planning from the beginning of the design process (Kumar & Imam, 2013; Straube & Burnett, 2005).

In temperate climates, the dominant technical concern shifts toward thermal movement, sealant fatigue, weather sealing, and moisture ingress associated with seasonal variation. ACP façade design in these settings should ensure adequate movement joints, flexible and durable sealant systems, and robust weatherproofing details. Freeze–thaw exposure in colder regions further reinforces the need for water-resistant detailing and joint integrity. These measures are important not only for performance but also for lifecycle sustainability, because premature failure increases maintenance demand and replacement-related environmental burdens (Riahinezhad et al., 2021; International Organization for Standardization, 2011).

Preventive maintenance should be treated as part of sustainability strategy rather than only an operational obligation. Routine inspection of coatings, sealants, fixings, and drainage paths can help detect local failures early and reduce the need for major interventions. Maintenance planning should therefore be considered during design, including access provision, inspection intervals, and documentation procedures. In practice, a façade that is easier to inspect and maintain is often more sustainable over its full life than one that appears efficient only at the specification stage (Pomponi & Moncaster, 2017; Attia et al., 2018).

7.2 Policy, Standards, and Industry Practice Implications

The study also points to several broader implications for standards, regulation, and professional practice. First, industry guidance on façade materials would benefit from stronger integration of lifecycle and durability considerations. Structural safety and fire performance will rightly remain essential regulatory concerns, but environmental performance and service-life planning should increasingly be incorporated into façade decision-making frameworks, especially as whole-life carbon assessment becomes more common in construction practice (Pomponi & Moncaster, 2016; Bianchi et al., 2024).

Second, climate-responsive façade principles should be more explicitly reflected in professional standards and sustainability-oriented specifications. Recent façade selection studies also show that material choice in practice depends on the integration of environmental, technical, and economic criteria at the early design stage, reinforcing the need for more systematic sustainability-oriented guidance in façade engineering (Naseer et al., 2024; Bianchi et al., 2024). The comparative analysis in this chapter shows that similar ACP systems may perform differently in tropical and temperate climates even when the base material is the same. This suggests that environmental exposure context should be more clearly recognized in material recommendations, durability expectations, and maintenance guidance.

Third, industry collaboration is essential. Manufacturers, façade engineers, contractors, architects, and sustainability consultants should work more closely to align environmental documentation, detailing strategy, durability expectations, and end-of-life planning. Product transparency, especially through environmental product declarations and clearer technical durability information, would improve specification quality and support more defensible sustainability claims (European Commission, 2020; USGBC, 2023).

Finally, professional training remains important. As sustainability requirements become more demanding, façade professionals will need stronger knowledge of lifecycle thinking, climate-sensitive detailing, maintainability, and material recovery pathways. In this sense, sustainable façade engineering is not only a technical challenge but also a capability-building challenge for the wider construction industry.

8. CONCLUSION AND FUTURE RESEARCH DIRECTIONS

Drawing on the lifecycle assessment literature reviewed by Cabeza et al. (2014) and the embodied carbon perspective advanced by Pomponi and Moncaster (2017), this chapter developed a durability-informed sustainability assessment of ACP façade systems through a comparative literature-based analysis of tropical and temperate climatic contexts. The chapter has shown that the sustainability of ACP façades cannot be evaluated on the basis of initial embodied carbon alone; whole-life performance is shaped by the interaction of material production impacts, climatic exposure, maintenance requirements, service-life variation, and end-of-life feasibility.

A central conclusion is that climate-sensitive durability is one of the most decisive variables in determining long-term environmental performance. As discussed by Kumar and Imam (2013) and supported by the durability-oriented review of Riahinezhad et al. (2021), tropical conditions tend to intensify coating weathering, moisture-related deterioration, and maintenance demand, while temperate conditions more often raise issues of thermal movement, sealant fatigue, and moisture management under seasonal change. In both cases, service life and detailing quality strongly influence whether the initial embodied burden of ACP systems is moderated or amplified over time. This confirms that façade sustainability must be interpreted as a dynamic performance outcome rather than as a fixed material attribute.

From an engineering and professional practice perspective, the chapter indicates that more sustainable ACP application depends on climate-appropriate specification, preventive maintenance planning, realistic recycling assumptions, and stronger integration of durability into lifecycle assessment and certification logic. The discussions on LEED and BREEAM further suggest that green building certification systems can support improved façade sustainability practice, but their effectiveness would be strengthened by greater recognition of climate-sensitive service life and long-term performance behaviour. In this respect, sustainable façade engineering requires not only better materials, but also better design judgement, documentation quality, and maintenance strategy.

Several future research directions emerge from this work. One priority is the integration of lifecycle assessment with long-term field performance evidence, since many existing evaluations still rely on generalized assumptions about service life and maintenance intervals. More empirical monitoring of ACP systems across different climatic regions would improve the accuracy of durability-adjusted sustainability interpretation. A second priority concerns ACP end-of-life recovery and circularity. Although aluminium has strong recycling potential, composite panel construction still complicates disassembly and material separation, making this a critical area for future innovation. Additional research is also needed on the relationship between fire safety, durability, and environmental performance, particularly in the development of ACP systems that can achieve safer and lower-impact whole-life outcomes (Khan et al., 2021; Schulz et al., 2020). Finally, digital tools such as BIM-linked lifecycle databases, sensor-based façade monitoring, and predictive maintenance systems offer promising opportunities for more evidence-based and adaptive sustainability assessment in façade engineering practice.

Overall, this chapter argues that a lifecycle-informed, climate-responsive, and durability-sensitive perspective provides a more accurate and professionally relevant basis for evaluating ACP façade sustainability than production-stage assessment alone. As environmental expectations continue to rise, façade engineering practice will increasingly need to integrate material science, building physics, lifecycle assessment, and service-life planning to deliver envelope systems that are technically reliable, environmentally responsible, and resilient over the long term.

REFERENCES

Antypa, D., Vlysidis, A., Gkika, A., et al. (2022). Life cycle assessment of advanced building materials towards NZEBs. E3S Web of Conferences, 349, 04001. DOI: 10.1051/e3sconf/202234904001

Asadi, E., da Silva, M. G., Antunes, C. H., & Dias, L. (2012). Multi-objective optimization for building retrofit strategies: A model and an application. *Energy and Building*, *44*, 81–87. DOI: 10.1016/j.enbuild.2011.10.016

Attia, S., Bilir, S., Safy, T., Struck, C., Loonen, R., & Goia, F. (2018). Current trends and future challenges in the performance assessment of adaptive façade systems. *Energy and Building*, *179*, 165–182. DOI: 10.1016/j.enbuild.2018.09.017

Bianchi, S., Andriotis, C. P., Klein, T., & Overend, M. (2024). Multi-criteria design methods in façade engineering: State-of-the-art and future trends. *Building and Environment*, *250*, 111184. DOI: 10.1016/j.buildenv.2024.111184

Building Research Establishment. (2018). *BREEAM UK new construction 2018 technical manual*. BRE.

Búryová, D., & Sedlák, P. (2021). Life cycle assessment of coated and thermally modified wood façades. *Coatings*, *11*(12), 1487. DOI: 10.3390/coatings11121487

Cabeza, L. F., Rincón, L., Vilariño, V., Pérez, G., & Castell, A. (2014). Life cycle assessment (LCA) and life cycle energy analysis (LCEA) of buildings and the building sector: A review. *Renewable & Sustainable Energy Reviews*, *29*, 394–416. DOI: 10.1016/j.rser.2013.08.037

Dixit, M. K., Fernández-Solís, J. L., Lavy, S., & Culp, C. H. (2010). Identification of parameters for embodied energy measurement: A literature review. *Energy and Building*, *42*(8), 1238–1247. DOI: 10.1016/j.enbuild.2010.02.016

Efthymiou, E., Cöcen, Ö. N., & Russo Ermolli, S. (2010). Sustainable aluminium systems. *Sustainability (Basel)*, *2*(9), 3100–3109. DOI: 10.3390/su2093100

European Commission. (2020). A new circular economy action plan: For a cleaner and more competitive Europe. https://eur-lex.europa.eu/legal-content/EN/TXT/?uri=CELEX:52020DC0098

European Committee for Standardization. (2011). *EN 15978:2011 Sustainability of construction works: Assessment of environmental performance of buildings: Calculation method*. CEN.

Grant, M. J., & Booth, A. (2009). A typology of reviews: An analysis of 14 review types and associated methodologies. *Health Information and Libraries Journal, 26*(2), 91–108. DOI: 10.1111/j.1471-1842.2009.00848.x PMID: 19490148

Han, B., Wang, R., Liang, Y., Dai, Y., Zhang, L., & Shen, J. (2015). Life cycle assessment of ceramic façade material and its comparative analysis with three other common façade materials. *Journal of Cleaner Production, 99*, 86–93. DOI: 10.1016/j.jclepro.2015.03.032

Huedo, P., Mulet, E., & López-Mesa, B. (2016). A model for the sustainable selection of building envelope assemblies. *Environmental Impact Assessment Review, 57*, 63–77. DOI: 10.1016/j.eiar.2015.11.005

International Organization for Standardization. (2006a). *ISO 14040:2006 Environmental management: Life cycle assessment: Principles and framework*. ISO.

International Organization for Standardization. (2006b). *ISO 14044:2006 Environmental management: Life cycle assessment: Requirements and guidelines*. ISO.

International Organization for Standardization. (2011). *ISO 15686-1:2011 Buildings and constructed assets: Service life planning: Part 1: General principles and framework*. ISO.

Jelle, B. P. (2011). Traditional, state-of-the-art and future thermal building insulation materials and solutions: Properties, requirements and possibilities. *Energy and Building, 43*(10), 2549–2563. DOI: 10.1016/j.enbuild.2011.05.015

Keles, Ç., & Yazıcıoğlu, F. (2023). Analyzing the environmental sustainability of primary schools' facades within the scope of life cycle assessment in Turkey and recommendations for improvement. *Smart and Sustainable Built Environment, 12*(2), 298–324. DOI: 10.1108/SASBE-04-2021-0072

Khan, A. A., Lin, S., Huang, X., & Usmani, A. (2021). Facade fire hazards of bench-scale aluminum composite panel with flame-retardant core. *Fire Technology, 59*(1), 5–28. DOI: 10.1007/s10694-020-01089-4

Kubba, S. (2012). *Handbook of green building design and construction: LEED, BREEAM, and Green Globes* (2nd ed.). Butterworth-Heinemann.

Kumar, P., & Imam, B. (2013). Footprints of air pollution and changing environment on the sustainability of built infrastructure. *The Science of the Total Environment, 444*, 85–101. DOI: 10.1016/j.scitotenv.2012.11.056 PMID: 23262326

Naseer, A. N., Al-Gahtani, K. S., Altuwaim, A., Alsanabani, N. M., & Almohsen, A. S. (2024). Selecting building façade materials by integrating stepwise weight assessment ratio analysis and weighted aggregated sum product assessment into value engineering. *Sustainability (Basel), 16*(11), 4611. DOI: 10.3390/su16114611

Park, J., Hengevoss, D., & Wittkopf, S. (2018). Industrial data-based life cycle assessment of architecturally integrated glass-glass photovoltaics. *Buildings (Basel, Switzerland)*, *9*(1), 8. DOI: 10.3390/buildings9010008

Pomponi, F., & Moncaster, A. (2016). Embodied carbon mitigation and reduction in the built environment: What does the evidence say? *Journal of Environmental Management*, *181*, 687–700. DOI: 10.1016/j.jenvman.2016.08.036 PMID: 27558830

Pomponi, F., & Moncaster, A. (2017). Scrutinising embodied carbon in buildings: The next performance gap made manifest. *Renewable & Sustainable Energy Reviews*, *81*, 2431–2442. DOI: 10.1016/j.rser.2017.06.049

Rabajczyk, A., Zielecka, M., Klapsa, W., & Dziechciarz, A. (2021). Self-cleaning coatings and surfaces of modern building materials for the removal of some air pollutants. *Materials (Basel)*, *14*(9), 2161. DOI: 10.3390/ma14092161 PMID: 33922766

Riahinezhad, M., Hallman, M., & Masson, J.-F. (2021). Critical review of polymeric building envelope materials: Degradation, durability and service life prediction. *Buildings (Basel, Switzerland)*, *11*(7), 299. DOI: 10.3390/buildings11070299

Schulz, J., Kent, D., Crimi, T., Glockling, J. L. D., & Hull, T. R. (2020). A critical appraisal of the UK's regulatory regime for combustible façades. *Fire Technology*, *57*(1), 261–290. DOI: 10.1007/s10694-020-00993-z

Straube, J. F., & Burnett, E. F. P. (2005). *Building science for building enclosures*. Building Science Press.

Struhala, K., Čekon, M., Slávik, R., & Kalousek, M. (2018). Life cycle assessment of solar façade concepts based on transparent insulation materials. *Sustainability (Basel)*, *10*(11), 4212. DOI: 10.3390/su10114212

Tokbolat, S., Naizabekov, Y., & Mariani, S. (2020). The impacts of different façade types on energy use in residential buildings. E3S Web of Conferences, 172, 24010. DOI: 10.1051/e3sconf/202017224010

Trubina, N., Lupíšek, A., Pešta, J., & Vlasatá, B. (2024). Demountable structures: Evaluation of a circular design approach through life cycle assessment. *IOP Conference Series. Earth and Environmental Science*, *1402*(1), 012040. DOI: 10.1088/1755-1315/1402/1/012040

U.S. Green Building Council. (2023). *LEED v4.1 building design and construction*. USGBC.

Chapter 6
Energy Performance Modelling and Optimization of Nanofluid Cooling Systems in Renewable Energy Applications

Dhirendra Patel
http://orcid.org/0000-0003-3308-168X
Amity University, Greater Noida, India

M. L. Kumar
Amity University, Greater Noida, India

Ankesh Kumar
Amity University, Greater Noida, India

ABSTRACT

This chapter presents a comprehensive treatment of energy performance modelling and optimization for nanofluid-based cooling systems used in renewable energy applications. It covers the physical principles of nanofluids, effective-property models, convective heat transfer correlations, numerical and reduced-order modelling approaches, experimental considerations, and optimization strategies (design of experiments, response surface methodology, evolutionary algorithms, and multi-objective techniques). Case studies — including solar photovoltaic (PV) cooling, concentrated solar power (CSP) receivers, and thermal energy storage augmen-

DOI: 10.4018/979-8-3373-6023-2.ch006

tation. practical modelling workflows, key results, and trade-offs. Figures, tables, and sample graphs are included to guide researchers and engineers in developing, validating, and optimizing nanofluid cooling solutions.

1. INTRODUCTION

Nanofluids, suspensions of nanoparticles in conventional fluids, have emerged as advanced heat transfer media in renewable energy applications. Their use improves thermal conductivity, heat transfer coefficients, and energy efficiency in solar collectors, PV panels, and thermal energy storage systems. Energy demand is rapidly increasing worldwide due to industrialization, urbanization, and improved living standards. At the same time, the depletion of fossil fuel reserves and environmental concerns such as greenhouse gas emissions have accelerated the global transition toward renewable energy technologies. Solar, wind, biomass, and geothermal systems are playing a vital role in achieving sustainable energy targets (Choi, 1995). However, renewable systems often face efficiency and reliability challenges.

Worldwide energy systems are undergoing a rapid transformation driven by the twin imperatives of climate mitigation and rising energy demand. Renewable technologies — especially solar photovoltaics (PV) and concentrated solar power (CSP), are central to that transition, but their real-world performance is limited by thermal management issues. For instance, PV modules lose electrical conversion efficiency as cell temperature rises, and CSP systems require heat transfer fluids that remain stable and efficient at high operating temperatures. Improving heat-transfer performance in these systems therefore directly increases energy yield, reduces operating costs, and prolongs component lifetime. The development of advanced working fluids is a promising route to meeting these needs (Xuan & Li, 2000).

Renewable energy systems (solar thermal collectors, photovoltaic panels, concentrated solar power, and thermal storage systems) often suffer performance limitations from elevated temperatures and suboptimal heat transfer. Advanced cooling strategies improve energy conversion efficiency and system longevity. Nanofluids—colloidal suspensions of nanoparticles (metals, metal oxides, carbon-based) in base fluids (water, ethylene glycol, propylene glycol, oils) offer enhanced thermal properties compared to conventional fluids (Das et al., 2003; Chen et al., 2005). This chapter focuses on modeling and optimizing nanofluid cooling systems to maximize energy performance in renewable energy applications as shown in figure 1.1 Growth of nanofluid publications in renewable energy applications and Growth of nanofluid publications in renewable energy applications.

Figure 1.

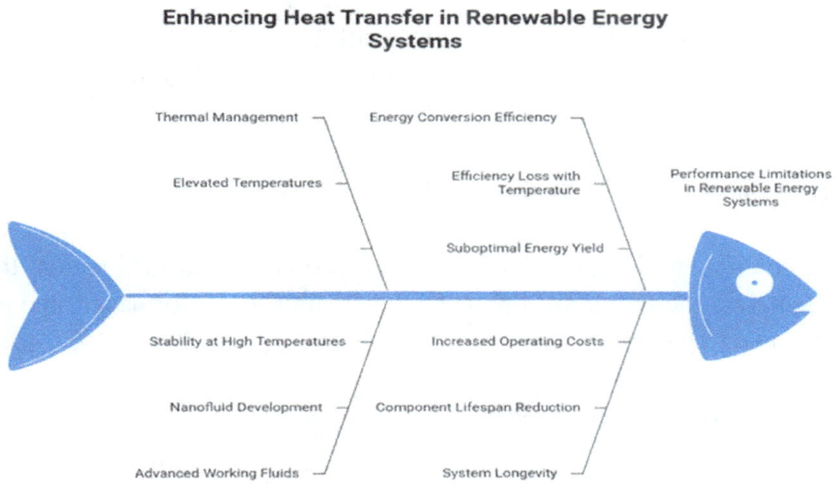

2. NANOFLUID FUNDAMENTALS

Nanofluids are engineered suspensions of nano-sized particles (typically 1–100 nm) dispersed in conventional base fluids such as water, ethylene glycol, oils, or molten salts. The nanoparticles can be metallic (Cu, Al), metal oxides (Al_2O_3, CuO, TiO_2), ceramic, or carbon-based (graphene, carbon nanotubes). When properly dispersed and stabilized, even small volume fractions (often < 1–5%) can measurably modify bulk thermo-physical properties: thermal conductivity, specific heat, viscosity, density, and optical absorption. The practical appeal of nanofluids lies in their potential to enhance convective heat transfer without extensive redesigning of existing heat-exchange hardware (Chen et al., 2005). The first widely cited experimental demonstrations of unusually large thermal conductivity increases in suspensions of nanoparticles date to the early 2000s and were pivotal in launching the field. Nanofluids: base fluid + dispersed nanoparticles (1–100 nm typically) (Keblinski et al., 2005; Eapen et al., 2007; Wang et al., 2019; Yasmin et al., 2023a; Gupta et al., 2014).

2.1 Definition and Properties

Detailed explanation of density, viscosity, specific heat, and thermal conductivity with derivations and comparative analysis between base fluids and nanoparticle combinations.

2.2 Preparation and Stability

The preparation and long-term stability of nanofluids are critical factors that directly affect their thermophysical properties, heat transfer performance, and applicability in renewable energy systems.

- **Two-step vs one-step methods**

Two-step method: Nanoparticles are first synthesized in dry powder form using chemical or physical methods (e.g., sol–gel, chemical vapor deposition). The powder is then dispersed into a base fluid with mechanical stirring, ultrasonication, or the addition of dispersants. This method is cost-effective and widely used but may suffer from agglomeration and sedimentation issues.

One-step method: Nanoparticles are simultaneously synthesized and dispersed into the base fluid during preparation (e.g., chemical precipitation, direct evaporation–condensation) (Jha et al., 2023). This reduces agglomeration and ensures better stability but is often more expensive and less scalable for large volumes.

- **Use of surfactants and ultrasonication**

Surfactants: Surface-active agents are added to improve nanoparticle dispersion by modifying the surface charge of particles and preventing agglomeration. Common surfactants include SDS (sodium dodecyl sulfate), CTAB (cetyltrimethylammonium bromide), and various polymers. The downside is that surfactants may alter fluid viscosity or break down under high temperatures.

Ultrasonication: Application of ultrasonic waves breaks apart nanoparticle clusters through cavitation effects, leading to uniform dispersion. It is a widely used technique in both lab-scale and industrial preparation. Parameters such as sonication time, frequency, and power strongly influence stability (Balani et al., 2021).

- **Stability assessment techniques (zeta potential, sedimentation tests).**

Zeta potential: A measure of the surface charge of nanoparticles in suspension, typically in millivolts (mV). A high absolute zeta potential ($>\pm30$ mV) indicates strong electrostatic repulsion and good stability. Sedimentation tests is Visual monitoring or gravimetric measurement of sediment formation over time. Stable nanofluids exhibit minimal sedimentation spectroscopy is Used to monitor light absorption characteristics over time, where changes indicate aggregation or precipitation[. Dynami (Duffie & Beckman, 2013; Demirbas, 2004; Saidur et al., 2011). Light Scattering (DLS)is used to Measures particle size distribution and helps detect agglomeration.as shown in Table 1.

Table 1. Typical nanofluids with base fluids, particle size, and stability range

Nanoparticle Material	Base Fluid(s)	Typical Particle Size (nm)	Stability Range / Remarks
Al_2O_3 (Alumina)	Water, EG, Water–EG, PG	20–50 nm	Good stability; zeta potential >30 mV with surfactants (SDS, CTAB). Stable for weeks to months.
CuO (Copper Oxide)	Water, EG	30–80 nm	Moderate stability; sedimentation reduced with pH control and PVP.
TiO_2 (Titanium Dioxide)	Water, EG, PG	15–40 nm	High stability in acidic pH; tends to agglomerate at neutral pH.
ZnO (Zinc Oxide)	Water, EG	20–70 nm	Good UV stability; requires surfactants for long-term suspension.
Fe_3O_4 (Magnetite)	Water, Kerosene, EG	10–40 nm	Magnetic nanofluid; stable under magnetic fields; stable with oleic acid coating.
SiO_2 (Silica)	Water, EG, PG	10–20 nm	Excellent stability due to hydrophilic surface; stable for months.
CNT(Carbon Nanotubes)	Water, EG	10–30 nm diameter	Requires strong ultrasonication; stability dependent on surfactants (SDS, SDBS).
Graphene/Graphene Oxide	Water, EG	5–20 nm thickness	High stability with functionalization (GO); tends to re-stack without additives.
Ag (Silver)	Water	10–40 nm	Good antibacterial stability; may require citrate or PVP stabilizer.
Cu (Copper)	Water, Oil	20–60 nm	Prone to oxidation; stability improved by oleic acid coating.

2.3 Mechanisms of Heat Transfer Enhancement in Nanofluids

Nanofluids—engineered colloidal suspensions of nanoparticles in base fluids—exhibit significantly enhanced heat transfer properties compared to conventional fluids like water, ethylene glycol, or propylene glycol. The improvement arises from multiple nanoscale mechanisms that act simultaneously (Wole-Osho et al., 2020; Kusure et al., 2014; Kamel et al., 2016; Elangovan et al., 2022; Okeke et al.,

2011). These mechanisms collectively increase thermal conductivity, convective heat transfer coefficient, and energy transport efficiency. As shown in Figure 2 Mechanisms of Heat Transfer Enhancement in Nanofluids

Figure 2. Mechanisms of heat transfer enhancement in nanofluids

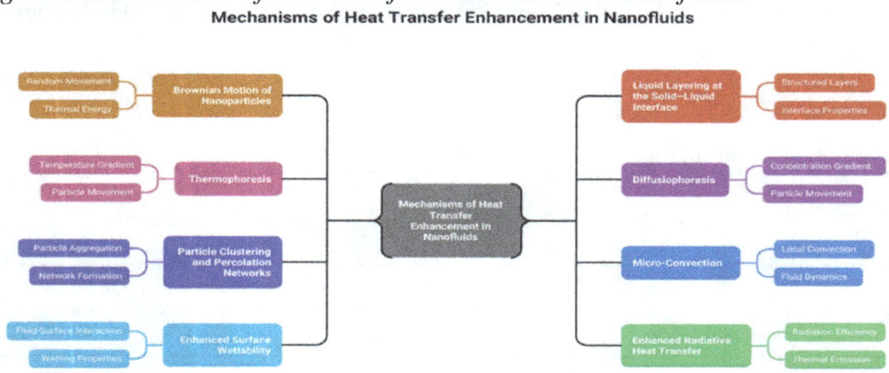

2.4 Brownian Motion of Nanoparticles

Nanoparticles undergo rapid random motion due to continuous collisions with fluid molecules. Random motion of nanoparticles increases microscopic mixing. Enhances energy exchange between particles and fluid molecules. More pronounced at low viscosity and high temperatures.

2.5 Liquid Layering at the Solid–Liquid Interface

One of the most widely accepted mechanisms behind the enhanced thermal conductivity of nanofluids is the formation of a liquid layering (also called the *interfacial nanolayer*) around nanoparticles. When nanoparticles are dispersed in a base fluid, the molecules of the fluid near the particle surface become more ordered and densely packed compared to the bulk fluid. This structured layer behaves like a solid-like or semi-crystalline layer, and it exhibits higher thermal conductivity than the surrounding fluid. Nanoparticles develop a high-density liquid layer (2–5 nm thick). This acts as a thermal bridge, increasing effective thermal conductivity (Patel et al., 2021; Ahmed & Bhargav, 2020; Bozorgan & Shafahi, 2015; Chaudhari & Walke, 2014; Bocanegra et al., 2025).

2.6 Thermophoresis and Diffusionphoresis

Nanofluids exhibit enhanced heat transfer primarily due to the dynamic behavior of nanoparticles dispersed in the base fluid. Among these dynamic mechanisms, thermophoresis and diffusiophoresis are two important particle migration phenomena that contribute to improved thermal transport, especially in systems with strong temperature and concentration gradients. Temperature gradient causes particle migration from hot to cold regions. Enhances convective heat transfer and reduces boundary layer resistance. hermophoresis (also called the Soret effect or thermal diffusion) refers to the movement of nanoparticles from a high-temperature region to a low-temperature region due to a temperature gradient in the fluid (Patel et al., 2022; Mahamude et al., 2022; Mousavi et al., 2025).

Temperature Gradient ⇒ Nanoparticle Migration

2.7 Particle Clustering and Network Formation

Particle clustering and network (percolation) formation is one of the most influential mechanisms contributing to the enhanced thermal conductivity of nanofluids. Although nanofluids are expected to behave as stable single-particle suspensions, nanoparticles in reality tend to interact, attach, and form aggregates or cluster networks. Surprisingly, controlled clustering can increase heat transfer, despite the general belief that agglomeration is undesirable. Formation of thermally conductive pathways through clusters. Improves conduction but excessive clustering may cause sedimentation. Depending on the size and structure of clusters, they may enhance or reduce heat transfer performance (Patel et al., 2022; Hussain et al., 2024; Nagar & Sreenivasa, 2025).

2.8 Micro-convection Effect

Micro-convection is one of the most widely discussed mechanisms proposed to explain enhanced thermal conductivity and convective heat transfer in nanofluids. It is based on the idea that random motion of nanoparticles at the molecular scale produces microscopic eddies in the surrounding fluid, increasing effective heat transport. Although debated in literature, micro-convection plays a supporting role in nanofluid thermal performance, especially at optimum particle concentrations and for low-viscosity base fluids. Nanoparticles create disturbances in the local velocity field. Increases convective heat-transfer coefficient. Localized fluid motion is induced by the random (Brownian) movement of nanoparticles. The nanoparticles continuously collide with fluid molecules, causing disturbances or eddies at the microscale. As shown in Figure 3 Mechanisms of Micro-convection Effect (Nagar & Sreenivasa, 2025; Naveed, 2021; Hannoura et al., 2025; Gopalsamy et al., n.d.).

Figure 3. Mechanisms of micro-convection effect

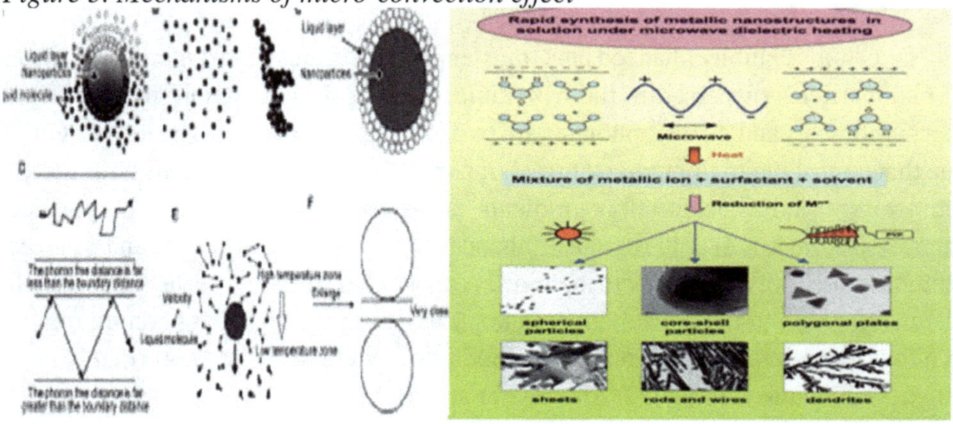

2.9 Enhanced Radiative Heat Transfer

Radiative heat transfer enhancement is an important mechanism in nanofluids, especially in solar thermal applications, high-temperature cooling, and optical–thermal energy systems.

Unlike conduction or convection—which depend on particle motion and fluid dynamics—radiative enhancement arises from interaction of nanoparticles with electromagnetic radiation, leading to improved absorption, scattering, and conversion of radiation to thermal energy. Nanoparticles act as radiation absorbers, spectrally selective media, and photothermal converters, increasing overall heat transfer performance. Metal-oxide nanoparticles absorb and re-radiate thermal energy. Relevant in high-temperature systems like CSP. As shown in Figure 4 Enhanced Radiative Heat Transfer.

Figure 4. Enhanced radiative heat transfer

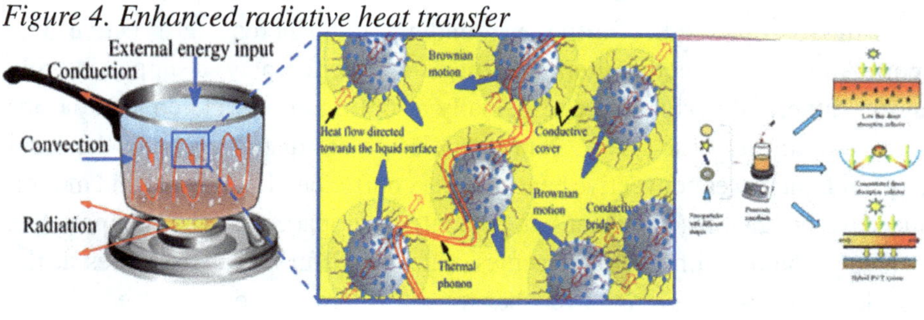

3. EFFECTIVE PROPERTY MODELLING

Nanofluids exhibit enhanced thermal performance compared to conventional fluids. Accurate prediction of their properties, particularly thermal conductivity and viscosity, is crucial for designing renewable energy systems such as PV cooling, CSP, and thermal energy storage. This section explains widely used models, the physical principles behind them, and their comparison with experimental observations (Kumar & Kumar, 2023; Patel et al., 2022; Patel et al., 2024; Rashidi & Zinatloo-Ajabshir, 2024).

3.1 Thermal Conductivity Models

The effective thermal conductivity k_{nf} of a nanofluid depends on the base fluid, particle material, volume fraction, particle shape, and interactions between particles

3.1.1 Maxwell Model

One of the earliest models for dilute suspensions. Assumes spherical particles uniformly dispersed in the fluid and negligible particle-particle interactions. Predicts modest increase in conductivity for small ϕ. Works well for low particle concentrations (< 5%) and spherical particles. The following formula use finds the thermal conductivity of nanofluid.

Where:

- k_b = thermal conductivity of base fluid
- kp = thermal conductivity of particles
- ϕ = volume fraction of nanoparticles
- Comparison with experimental data.

3.2 Viscosity Models

Viscosity is one of the most important thermophysical properties of nanofluids because it directly influences pumping power, heat transfer performance, and flow behaviour.

The addition of nanoparticles changes the rheology of the base fluid, and several empirical and theoretical models are used to predict the effective viscosity of nanofluid (Dehghan et al., 2024; Patel, 2023; Pereira et al., 2023; Patel, 2025; Panduro et al., 2022).

3.2.1. Classical Viscosity Models

Einstein's Model (for Dilute Suspensions)

The earliest and simplest model for low particle concentration (≤ 2%):

Where:

- : base fluid viscosity
- : particle volume fraction

Brinkman's Model

Extends Einstein's model to moderately concentrated systems:

Batchelor Model

Includes hydrodynamic interactions between particles:

3.3.2 Semi-empirical Models for Nanofluids

Chen et al. Model

Accounts for particle aggregation:

Applicable for:

- Water-based oxide nanofluids
- Aggregation-dominant systems

Nguyen et al. Temperature-Dependent Model

Where:

- : empirical constants
- : temperature

Important:

- Captures temperature sensitivity.
- Suitable for Al_2O_3 and CuO nanofluids.

Masoumi et al. Model (Brownian Motion Contribution)

Incorporates micro-convection due to Brownian motion:

Where:

- : particle diameter
- : Brownian velocity

3.3.3 Advanced Viscosity Models

Krieger–Dougherty Model

Widely used for concentrated suspensions:

Where:

- : maximum packing fraction (≈ 0.63 for spheres)
- : intrinsic viscosity (≈ 2.5 for spheres)

Useful for:

- clustering
- non-Newtonian behavior at high φ

Wang et al. Model

Specifically developed for nanofluids:

Where:

- : fit parameters

Advantages:

- Simple
- Effective for metal oxide nanofluids

Non-Newtonian Models

Some nanofluids exhibit shear-thinning or shear-thickening behaviour:
Power-law model

Where:

[1]. : shear-thinning
[2]. : Newtonian
[3]. : shear-thickening

Used for:

- CNT nanofluids
- Graphene nanofluids
- High concentration oxide nanofluids

3.3.4 Temperature Influence on Viscosity

Most nanofluid viscosity models incorporate temperature effects:

Table 2. Summary of viscosity model applicability

Model	Best for	Features
Einstein	Very dilute	Linear φ dependence
Brinkman	Moderate φ	Accounts for crowding
Batchelor	Slightly higher φ	Includes particle interactions
Chen	Real nanofluids with aggregation	Strong φ^2 term
Nguyen	Temperature-dependent	Suitable for Al_2O_3, CuO
Masoumi	Brownian motion effects	Includes micro-convection
Krieger–Dougherty	High φ	Concentrated suspensions
Power-law	CNT/graphene	Non-Newtonian fluids

Table 3. Comparison of theoretical and experimental thermal conductivity enhancement

Nanofluid Type	Base Fluid	Particle Size (nm)	Vol. Fraction (φ%)	Theoretical Enhancement (Maxwell / EMT)	Experimental Enhancement (Reported)	Remarks / Reasons for Deviation
Al_2O_3–Water	Water	20–50	1.0	2–3%	8–12%	Micro-convection, liquid layering, aggregation
CuO–Water	Water	25–40	1.0	3–4%	10–18%	Brownian motion + particle clustering
TiO_2–Water	Water	20–40	2.0	4–5%	10–14%	Strong interfacial layer formation
ZnO–Water	Water	30–60	1.0	3–4%	9–13%	Higher phonon transport through oxide
Fe_3O_4–Water	Water	10–30	1.0	2–3%	8–15%	Magnetic clustering + Brownian motion
SiO_2–EG/ Water	EG–Water (60:40)	10–20	1.0	1–2%	4–7%	Stable dispersion → higher measured Keff
CNT–Water	Water	Diameter 10–20	0.1	5–7%	20–35%	Very high aspect ratio enhances conduction
Graphene–Water	Water	1–5 layers	0.1	6–8%	25–40%	High intrinsic thermal conductivity
Ag–Water	Water	10–15	0.3	3–5%	10–20%	Plasmonic heating contributes
Cu-EG	Ethylene Glycol	40–80	1.0	4–5%	12–22%	Enhanced phonon transport + micro-convection
Al_2O_3–PG/ Water	Propylene Glycol/Water	20–50	1.0	1–2%	3–6%	Higher viscosity suppresses Brownian motion
Hybrid (Al_2O_3–CuO) Water	Water	20–40	1.0	3–4%	10–16%	Synergistic effect of mixed particles

4. CONVECTIVE HEAT TRANSFER AND FLUID DYNAMICS

Convective heat transfer plays a fundamental role in determining the thermal performance of nanofluids in engineering systems. When nanoparticles are dispersed in base fluids such as water, ethylene glycol, propylene glycol, or oils, the resulting suspension exhibits modified thermophysical properties—thermal conductivity, viscosity, specific heat, density, and radiative characteristics all of which influence convective behavior. Convective heat transfer in nanofluids occurs under two major categories: natural convection driven by buoyancy forces, and forced convection driven by external flow mechanisms such as pumps or fans. In both cases, nanoparticles influence the flow behavior, heat transfer coefficient, boundary layer structure, and fluid dynamics, often resulting in enhanced thermal performance compared to conventional heat transfer fluids (Dehghan et al., 2024; Patel, 2023; Pereira et al., 2023; Patel, 2025; Panduro et al., 2022; Nassan et al., 2023).

5. APPLICATIONS IN RENEWABLE ENERGY TECHNOLOGIES

Nanofluid-enhanced cooling systems have emerged as a highly effective solution for improving the thermal performance, energy efficiency, and reliability of renewable energy technologies. Renewable systems such as solar PV modules, solar thermal collectors, wind turbine generators, geothermal heat exchangers, and biomass reactors suffer from thermal losses that significantly reduce their efficiency. Integrating nanofluid-based cooling can enhance heat absorption, accelerate heat removal, and minimize energy losses.

Energy performance modelling provides a quantitative prediction of how nanofluids improve system efficiency under various operating scenarios. Optimization strategies help identify the best combination of nanoparticle type, concentration, flow rate, geometry, pumping power, and operating conditions (Nassan et al., 2023; Patel, 2024; Saraswat & Johari, n.d.; Porgar et al., 2023).

Solar PV–Thermal (PVT) Systems

A Solar Photovoltaic–Thermal (PVT) System is a hybrid solar technology that simultaneously generates electricity and captures useful thermal energy from the same surface area. Unlike conventional solar panels, which convert only 15–20% of solar radiation into electrical energy and lose the rest as heat, PVT systems recover this heat using a thermal collector placed behind the PV module.Integrating nanofluids as the cooling/working fluid significantly improves system performance by enhancing heat removal, reducing PV temperature, and increasing thermal output. Nanofluid removes heat increases both thermal and electrical efficiency.CNT/water and Al_2O_3/water are commonly used (Kalsi et al., 2025; Patel et al., 2022; Kaggwa & Carson, 2019; Hussein et al., 2024; Nassan et al., 2023; Ahmed & Podder, 2024).

Figure 5. Solar PV–Thermal (PVT) systems

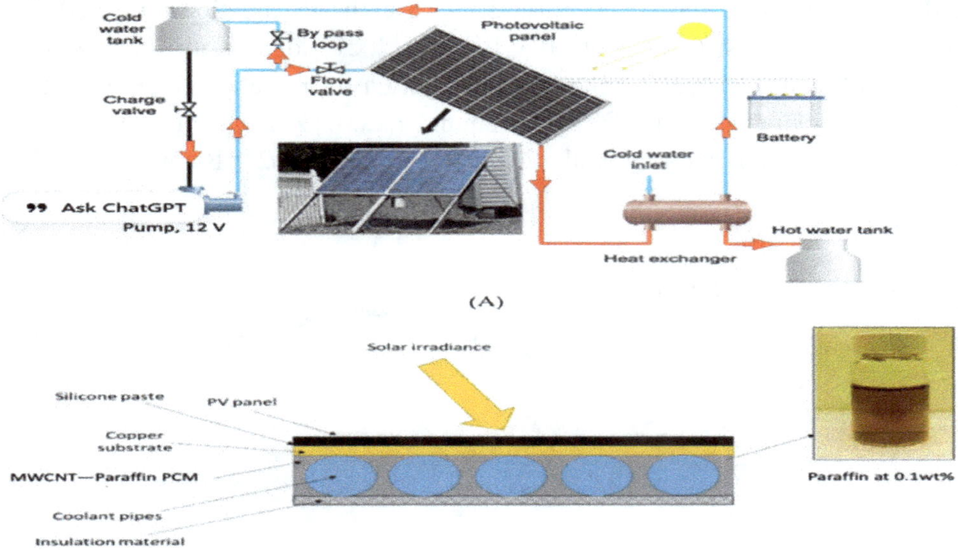

Solar Thermal Collectors

Solar thermal collectors are devices that absorb solar radiation and convert it into useful thermal energy for applications such as water heating, space heating, industrial process heat, desalination, and power generation. They represent one of the most efficient methods of utilizing solar energy because they directly convert sunlight into heat without intermediate conversion losses. Solar collectors are widely classified into low-temperature, medium-temperature, and high-temperature systems depending on their working temperature and optical configuration. Nanofluid acts as an advanced working fluid.10–30% thermal enhancement observed (Chaudhari & Walke, 2014; Rajput et al., 2023; Okeke et al., 2011; Ahmed & Bhargav, 2020).

Geothermal Heat Exchangers

Geothermal heat exchangers are systems that transfer heat to and from the ground using buried piping networks. Because the earth's subsurface maintains a relatively stable temperature throughout the year (18–25°C depending on location), geothermal heat exchangers provide highly efficient heating and cooling for residential, commercial, and industrial applications. These systems are a core component of Ground Source Heat Pumps (GSHP), Enhanced Geothermal Systems (EGS), and Low-temperature geothermal energy extraction. Nanofluids used in geothermal loops can significantly improve heat extraction by enhancing thermal conductivity and convective heat transfer. Nanofluid improves heat extraction rate. Reduces borehole thermal resistance (Ahmed & Bhargav, 2021; Soltan et al., 2024; Kumar et al., 2013; Jaafar et al., 2024; Gupta et al., 2014; Eastman et al., 2001; Keblinski et al., 2002; Saidur et al., 2011; Wang & Majumdar, 2007; Zhang et al., 2006; Yadav & Sanserwal, 2023; Nabwey et al., 2023; Porgar et al., 2023; Barai et al., 2023; Kalsi et al., 2025).

Figure 6. Geothermal heat exchangers

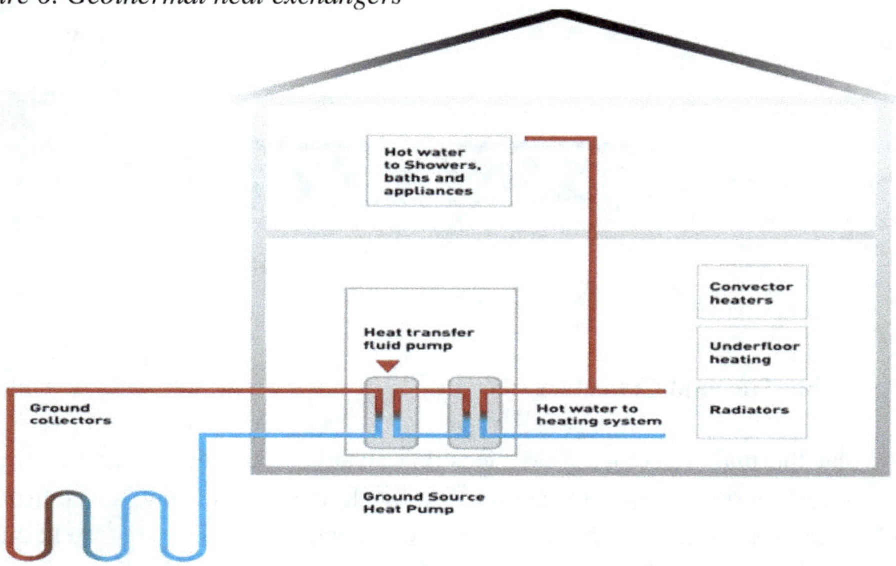

Wind Turbine Generator Cooling

Wind turbines convert kinetic energy from wind into electrical energy using rotor blades, a gearbox, and a generator housed in the nacelle. Modern direct-drive or geared generators produce significant heat due to Efficient cooling is crucial to maintain generator efficiency, prevent overheating, extend lifespan, and reduce maintenance (Yang et al., 2022; Nassan et al., 2023; Tanuja et al., 2023; Adio et al., 2025).

- Copper winding losses (I^2R losses)
- Core losses (hysteresis and eddy currents)
- Frictional and mechanical losses in bearings and gearbox
- Enhances cooling of permanent-magnet generators.

- Maintains torque efficiency.

Figure 7. Wind turbine generator cooling

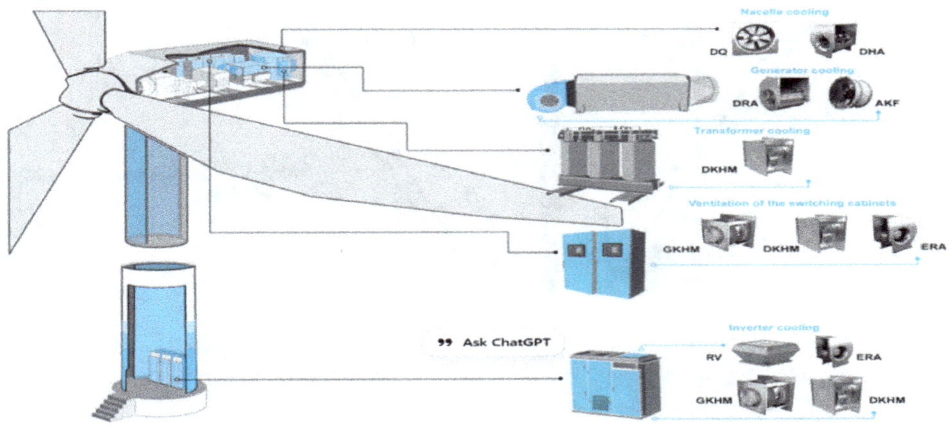

6. BIOMASS & BIOGAS REACTORS

Biomass and biogas reactors are engineered systems designed to convert organic matter into energy through biological, chemical, or thermal processes. Biomass can include agricultural residues, food waste, animal manure, and industrial organic waste. Biogas reactors primarily focus on anaerobic digestion, producing methane-rich biogas for heating, electricity generation, or fuel (Jaafar et al., 2024; Kalsi et al., 2023; Hussein et al., 2024; Panduro et al., 2022; Manimaran et al., 2025; Saraf et al., 2025).

The integration of nanofluids in heat exchange loops of these reactors can enhance thermal management, improve digestion efficiency, and stabilize reactor temperature, leading to higher biogas yield.Nanofluid loop helps maintain optimal digestion temperatures (Ahmed et al., 2022; Elmnefi & Al-Khazraji, 2023; Farhan et al., 2025; Yasmin et al., 2023b).

Figure 8. Biomass & biogas reactors

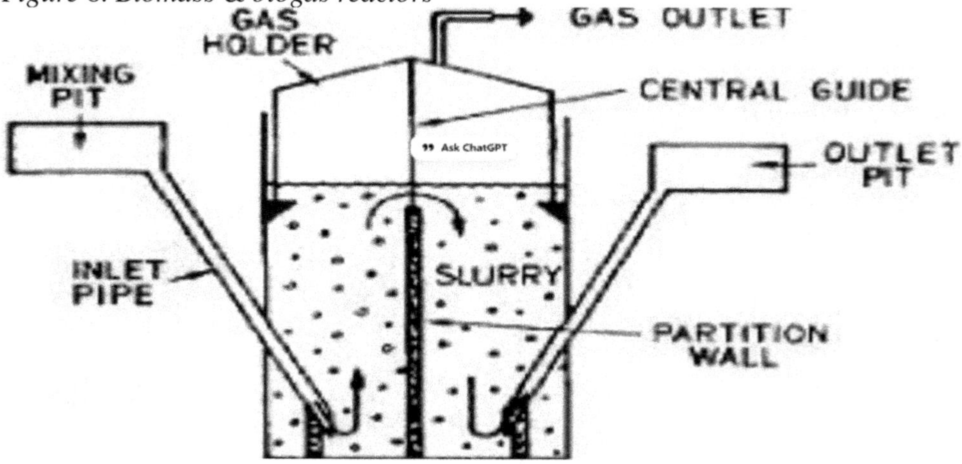

CONCLUSION

Nanofluid-based cooling systems hold strong potential to improve renewable energy technologies. Modelling and optimization remain crucial for real-world implementation. Particle clustering and network formation play a major role in explaining the large, often unexpected increases in thermal conductivity observed in nanofluids. When appropriately controlled, these structures. This mechanism is especially important in advanced nanofluids (graphene, CNT, hybrid nanofluids) and contributes significantly to their superior heat transfer performance.

Nanofluid-based cooling systems have emerged as a highly effective solution for enhancing the thermal performance and energy efficiency of renewable energy technologies such as solar PV–thermal systems, solar thermal collectors, geothermal heat exchangers, wind turbine generators, and biomass/biogas reactors. In summary, nanofluid cooling systems, when properly modeled and optimized, offer significant energy and efficiency advantages across renewable energy applications, contributing to more sustainable and high-performance energy systems.

FUTURE DIRECTIONS

The field of nanofluid cooling for renewable energy applications continues to evolve, with several promising research and development directions:

Advanced Nanofluid Formulations

- o Development of hybrid nanofluids, magnetic nanofluids, and functionalized nanoparticles to further enhance thermal conductivity, stability, and optical absorption.
- o Exploration of low-cost, eco-friendly nanoparticles to improve economic feasibility.

AI-Driven Energy Optimization

- o Integration of artificial intelligence (ANN, ML, XGBoost) for predictive modeling and real-time control of nanofluid flow, temperature, and energy output.
- o Dynamic optimization of multi-input renewable energy systems based on varying environmental conditions.

Hybrid Renewable Systems

- o Combining solar, wind, geothermal, and biomass systems with nanofluid cooling loops to improve overall system efficiency.
- o Optimization for multi-energy storage and hybrid energy harvesting.

Thermal Energy Storage Integration

- o Integration with phase-change materials (PCM) or thermal storage tanks for load leveling and improved energy availability.
- o Nanofluid-PCM systems offer enhanced heat transfer for both charging and discharging phases.

Long-Term Stability and Lifecycle Assessment

- o Studies on nanofluid durability, erosion/corrosion of pipes, and operational maintenance.
- o Environmental impact and sustainability assessments of nanofluid use in large-scale systems.

Experimental Validation at Scale

- o Pilot-scale and field tests for solar farms, geothermal plants, and industrial biogas reactors.
- o Optimization of system layout, flow channels, and pump selection based on real-world performance data.

Cost Reduction and Commercialization

- o Development of low-cost nanofluids and modular cooling systems.
- o Market-driven strategies to enhance adoption in residential, industrial, and utility-scale renewable energy systems.

REFERENCES

Adio, S. A., Atofarati, E. O., Muritala, A. O., Huan, Z., & Veeredhi, V. R. (2025). Nanofluids flow boiling and convective heat transfer in microchannels: A systematic review. *Journal of Thermal Analysis and Calorimetry, 150*(12), 8879–8911. DOI: 10.1007/s10973-025-14265-x

Ahmed, H., & Podder, C. (2024). Mixed convection heat transfer and flow of Al_2O_3–water nanofluid in a square enclosure with heated obstacles and varied boundary conditions. *arXiv preprint arXiv:2401.05497.*

Ahmed, M., Meteab, M. M., Salih, Q. O., Mohammed, H. A., & Alawi, O. A. (2022). Experimental investigation of aqueous hybrid nanofluid in flat plate solar collectors. *Energies, 15*(22), 8541. DOI: 10.3390/en15228541

Ahmed, Z., & Bhargav, A. (2020). Thermal interfacial resistance and nanolayer effect on the thermal conductivity of Al_2O_3–CO_2 nanofluid: a molecular dynamics approach. arXiv preprint arXiv:2006.12805.

Ahmed, Z., & Bhargav, A. (2021). Effect of aggregation morphology on thermal conductivity and viscosity of Al_2O_3–CO_2 nanofluid: a molecular dynamics approach. *Nanoscience and Technology: An International Journal, 12*(1).

Balani, R. S., Patel, D., Barthwal, S., & Arun, J. (2021). Fabrication of Air Conditioning System Using the Engine Exhaust Gas. In *Advances in Interdisciplinary Research in Engineering and Business Management* (pp. 369–378). Springer Nature Singapore. DOI: 10.1007/978-981-16-0037-1_29

Barai, R., Kumar, D., & Wankhade, A. (2023). Heat transfer performance of nanofluids in heat exchangers: A review. *Journal of Thermal Engineering, 9*(1), 86–106. DOI: 10.18186/thermal.1243398

Bocanegra, J. A., Marchitto, A., & Misale, M. (2025). Nanofluids in solar collectors: A comprehensive review focused on its sedimentation. *Clean Technologies and Environmental Policy, 27*(4), 1753–1784. DOI: 10.1007/s10098-024-02964-2

Bozorgan, N., & Shafahi, M. (2015). Performance evaluation of nanofluids in solar energy: A review of the recent literature. *Micro and Nano Systems Letters, 3*(1), 5. DOI: 10.1186/s40486-015-0014-2

Chaudhari, K. S., & Walke, P. V. (2014). Applications of nanofluid in solar energy—A review. *International Journal of Engine Research, 3*(3).

Chen, Y., Groll, M., Mertz, R., Maydanik, Y. F., & Vershinin, S. V. (2005, January). Steady-state and transient performance of a miniature loop heat pipe. In *International Conference on Nanochannels, Microchannels, and Minichannels* (Vol. 41855, pp. 183–189). DOI: 10.1115/ICMM2005-75122

Choi, S. U. (1995, November). Enhancing thermal conductivity of fluids with nanoparticles. In *ASME International Mechanical Engineering Congress and Exposition* (Vol. 17421, pp. 99–105). American Society of Mechanical Engineers. DOI: 10.1115/IMECE1995-0926

Das, S. K., Putra, N. S. D., Thiesen, P., & Roetzel, W. (2003). Temperature dependence of thermal conductivity enhancement for nanofluids. *Journal of Heat Transfer*, *125*(4), 567–574. DOI: 10.1115/1.1571080

Dehghan, M., Akbari, G., Montazerin, N., & Maroufi, A. (2024). Thermoeconomic prioritization of heat transfer fluids in solar-driven tri-generation systems using MADM methods. *Energy Equipment & Systems, 12*(2).

Demirbas, A. (2004). Combustion characteristics of different biomass fuels. *Progress in Energy and Combustion Science*, *30*(2), 219–230. DOI: 10.1016/j.pecs.2003.10.004

Duffie, J. A., & Beckman, W. A. (2013). *Solar Engineering of Thermal Processes*. John Wiley & Sons. DOI: 10.1002/9781118671603

Eapen, J., Li, J., & Yip, S. (2007). Beyond the Maxwell limit: Thermal conduction in nanofluids with percolating fluid structures. *Physical Review. E*, *76*(6), 062501. DOI: 10.1103/PhysRevE.76.062501 PMID: 18233882

Eastman, J. A., Choi, S. U. S., Li, S., Yu, W., & Thompson, L. J. (2001). Anomalously increased effective thermal conductivities of ethylene glycol-based nanofluids containing copper nanoparticles. *Applied Physics Letters*, *78*(6), 718–720. DOI: 10.1063/1.1341218

Elangovan, M., Srimanickam, B., Čep, R., Saranya, A., & Ramachandran, M. (2022). Experimental study of a hybrid solar collector using TiO_2/water nanofluids. *Energies*, *15*(12), 4425. DOI: 10.3390/en15124425

Elmnefi, M., & Al-Khazraji, W. (2023). Thermal performance enhancement of parabolic trough collectors using CuO nanofluid. *International Journal of Numerical Methods for Heat & Fluid Flow*, *33*(9), 3124–3163. DOI: 10.1108/HFF-11-2022-0659

Farhan, M., Hassan, F., Zubair, M., Jamil, F., Shahid, I., Khiadani, M., & Ali, H. M. (2025). Experimental performance of auto-tracking parabolic trough solar collector with nanofluids. *Journal of Thermal Analysis and Calorimetry*. Advance online publication. DOI: 10.1007/s10973-025-14001-5

Gopalsamy, V., Rajasekaran, K., & Baccoli, R. Experimental and analytical study of nanofluid-based parabolic trough collectors.

Gupta, M., Arora, N., Kumar, R., Kumar, S., & Dilbaghi, N. (2014). A comprehensive review of experimental investigations of forced convective heat transfer characteristics for various nanofluids. *International Journal of Mechanical and Materials Engineering*, *9*(1), 11. DOI: 10.1186/s40712-014-0011-x

Hannoura, A. A.. (2025). Thermal properties of heat transfer fluids in solar collectors. *Journal of Thermal Analysis and Calorimetry*. Advance online publication. DOI: 10.1007/s10973-025-14912-3

Hussain, M.. (2024). Direct absorption solar collectors working with nanofluids: A review. *Journal of Thermal Analysis and Calorimetry*, *149*(15), 7919–7946. DOI: 10.1007/s10973-024-13304-3

Husseın, A. M., Awad, A. T., & Alı, H. H. M. (2024). Evaluation of the thermal efficiency of nanofluid flows in flat plate solar collector. *Journal of Thermal Engineering*, *10*(2), 299–307. DOI: 10.18186/thermal.1448578

Jaafar, A. D., Al-Timimi, D. A. H., Alatabe, M. J. A., Shorbaz, M., & Kadhim Abed, B. (2024). Thermal conductivity of nanofluid: a mini review.

Jha, P., Das, B., Gupta, R., Mondol, J. D., & Ehyaei, M. A. (2023). Review of recent research on photovoltaic thermal solar collectors. *Solar Energy*, *257*, 164–195. DOI: 10.1016/j.solener.2023.04.004

Kaggwa, A., & Carson, J. K. (2019). Developments and future insights of using nanofluids for heat transfer enhancements in thermal systems: A review of recent literature. *International Nano Letters*, *9*(4), 277–288. DOI: 10.1007/s40089-019-00281-x

Kalsi, S., Kumar, S., Kumar, A., Alam, T., & Dobrotă, D. (2023). Thermophysical properties of nanofluids and applications in heat transfer. *Arabian Journal of Chemistry*, *16*(11), 105272. DOI: 10.1016/j.arabjc.2023.105272

Kalsi, S., Kumar, S., Kumar, A., Alam, T., Sharma, A., & Yadav, A. S. (2025). A review on hybrid nanofluids for heat transfer: Advancements, synthesis, challenges and applications. *Discover Applied Sciences*, *7*(7), 698. DOI: 10.1007/s42452-025-07141-8

Kamel, M. S., Syeal, R. A., & Abdulhussein, A. A. (2016). Heat transfer enhancement using nanofluids: A review of the recent literature. *American Journal of Nano Research and Applications*, *4*(1), 1–5.

Keblinski, P., Eastman, J. A., & Cahill, D. G. (2005). Nanofluids for thermal transport. *Materials Today*, *8*(6), 36–44. DOI: 10.1016/S1369-7021(05)70936-6

Keblinski, P., Phillpot, S. R., Choi, S. U. S., & Eastman, J. A. (2002). Mechanisms of heat flow in nanofluids. *International Journal of Heat and Mass Transfer*, *45*(4), 855–863. DOI: 10.1016/S0017-9310(01)00175-2

Kumar, A., Patel, C., Maurya, S. K., & Patel, D. (2013). *Characteristics of incompatible and strong oxidizing agents in e-glass random fibre reinforced composites*. MANFEX.

Kumar, S., & Kumar, A. (2023). *Review of nanofluids in parabolic trough solar collectors*. Energy Sources Part A.

Kusure, B. D., Warkhedkar, R. M., Harde, P. R., & Shirke, P. K. (2014). Heat transfer enhancement by using nanofluids in heat pipe—A review. *Applied Mechanics and Materials*, *592*, 932–938. DOI: 10.4028/www.scientific.net/AMM.592-594.932

Mahamude, A. S. F., Kamarulzaman, M., Harun, W., Kadirgama, K., Ramasamy, D., Farhana, K., Bakar, R., Yusaf, T., Subramanion, S., & Yousif, B. (2022). A comprehensive review on efficiency enhancement of solar collectors using hybrid nanofluids. *Energies*, *15*(4), 1391. DOI: 10.3390/en15041391

Manimaran, M., Norizan, M. N., Kassim, M. H. M., Adam, M. R., Abdullah, N., & Norrrahim, M. N. F. (2025). Stability and thermal conductivity of water-based hybrid nanofluids. *RSC Advances*, *15*(18), 14088–14125. DOI: 10.1039/D5RA00844A PMID: 40313320

Mousavi, S. M.. (2025). Surface engineered nanofluids in geothermal systems: Experimental and numerical investigation. *Journal of Molecular Liquids*.

Nabwey, H. A., Armaghani, T., Azizimehr, B., Rashad, A. M., & Chamkha, A. J. (2023). Review of nanofluid heat transfer in porous media. *Nanomaterials (Basel, Switzerland)*, *13*(5), 937. DOI: 10.3390/nano13050937 PMID: 36903815

Nagar, S., & Sreenivasa, S. (2025). Thermal performance investigations using nanofluids and PCM in parabolic trough collectors. *Proceedings of the IMechE Part A*.

Nassan, T. H., Heris, S. Z., & Bagheban, S. H. N. (2023). Experimental investigation of nanofluid heat transfer in a square cross-sectional duct. *arXiv preprint arXiv:2311.07447*.

Naveed, A. (2021). Thermal Performance Analysis of Flat-Plate Solar Collector Using Nanofluids (Doctoral dissertation, University of Malaya).

Okeke, G., Witharana, S., Antony, S. J., & Ding, Y. (2011). Computational analysis of factors influencing thermal conductivity of nanofluids. *Journal of Nanoparticle Research, 13*(12), 6365–6375. DOI: 10.1007/s11051-011-0389-9

Panduro, E. A. C., Finotti, F., Largiller, G., & Lervåg, K. Y. (2022). A review of the use of nanofluids as heat-transfer fluids in parabolic-trough collectors. *Applied Thermal Engineering, 211*, 118346. DOI: 10.1016/j.applthermaleng.2022.118346

Patel, D. (2023). Exploring the frontiers of microfluidics: challenges and future prospects. In *Advances in MEMS and Microfluidic Systems* (pp. 11–31). DOI: 10.4018/978-1-6684-6952-1.ch002

Patel, D. (2024). Sustainable renewable energy sources and emerging technologies. In *Optimization Techniques for Hybrid Power Systems* (pp. 343–361). Renewable Energy, Electric Vehicles, and Smart Grid. DOI: 10.4018/979-8-3693-0492-1.ch015

Patel, D. (2025). Emerging sustainable nanomaterials and their applications and future scope. In *Advances in Sustainable Materials* (pp. 107–135). Elsevier. DOI: 10.1016/B978-0-443-13849-2.00005-3

Patel, D., Pandey, G., Mourya, V. K., & Kumar, R. (2024). Sustainable base isolation: A review. *Sadhana, 49*(2), 173. DOI: 10.1007/s12046-024-02511-1

Patel, D., Sharma, A., & Mishra, A. (2021, November). Study of convective heat transfer characteristics of nanofluids in a circular tube. In *2021 International Conference on Technological Advancements and Innovations (ICTAI)* (pp. 264–267). IEEE. DOI: 10.1109/ICTAI53825.2021.9673432

Patel, D., Singh, R. P., Rajput, R. S., & Tiwari, P. (2022, December). Thermophysical properties and applications of nanofluids—A review. In 2022 5th International Conference on Contemporary Computing and Informatics (IC3I) (pp. 1324–1328). IEEE.

Patel, D., Varma, I. P., & Khan, F. A. (2022). A review advanced vehicle with automatic pneumatic bumper system using two cylinder. *International Research Journal of Engineering and Technology, 9*, 264–267.

Patel, H. R.. (2022). Review on design aspects of parabolic trough collectors. *International Journal of Renewable Energy Research, 12*(3), 1566–1588.

Pereira, J., Souza, R., Moreira, A., & Moita, A. (2023). A review on the nanofluids–PCMs integrated solutions for solar thermal heat transfer enhancement purposes. *Technologies, 11*(6), 166. DOI: 10.3390/technologies11060166

Porgar, S., Oztop, H. F., & Salehfekr, S. (2023). A comprehensive review on thermal conductivity and viscosity of nanofluids and their application in heat exchangers. *Journal of Molecular Liquids*, *386*, 122213. DOI: 10.1016/j.molliq.2023.122213

Rajput, R. S., Patel, D., & Singh, R. P. (2023). Applications and feasibility of large-scale solar-powered Peltier refrigeration systems.

Rashidi, M. M., & Zinatloo-Ajabshir, S. (Eds.). (2024). *Nanofluids: Preparation.* Applications and Simulation Methods.

Saidur, R., Leong, K. Y., & Mohammed, H. A. (2011). A review on applications and challenges of nanofluids. *Renewable & Sustainable Energy Reviews*, *15*(3), 1646–1668. DOI: 10.1016/j.rser.2010.11.035

Saraf, A. K., Goyal, R., & Yusop, A. F. (2025). Application of nanofluids under laminar flow in heat exchangers. *Discover Applied Sciences*, *7*(10), 1154. DOI: 10.1007/s42452-025-07726-3

Saraswat, S., & Johari, R. Review of advancements in heat transfer enhancement through nanofluids.

Soltan, Y. I., Nasser, M. S., Almomani, F., Mahmoud, K. A., & Onaizi, S. A. (2024). Thermal conductivity of different materials nanofluids including MXenes and metal-organic frameworks. *Journal of Materials Research and Technology*, *31*, 2723–2761. DOI: 10.1016/j.jmrt.2024.07.023

Tanuja, T. N.. (2023). Flow and heat transfer analysis on micropolar fluid through porous media. *Frontiers in Materials*, *10*, 1216757. DOI: 10.3389/fmats.2023.1216757

Wang, M., Wang, C. C., Cai, H. Q., Li, Y. Y., Zhang, Q. W., & Yi, H. B. (2019). Molecular dynamics simulation study on distinctive hydration characteristics of highly coordinated calcium chloride complexes. *Journal of Molecular Liquids*, *274*, 261–269. DOI: 10.1016/j.molliq.2018.10.136

Wang, X. Q., & Mujumdar, A. S. (2007). Heat transfer characteristics of nanofluids: A review. *International Journal of Thermal Sciences*, *46*(1), 1–19. DOI: 10.1016/j.ijthermalsci.2006.06.010

Wole-Osho, I., Okonkwo, E. C., Abbasoglu, S., & Kavaz, D. (2020). Nanofluids in solar thermal collectors: Review and limitations. *International Journal of Thermophysics*, *41*(11), 157. DOI: 10.1007/s10765-020-02737-1

Xuan, Y., & Li, Q. (2000). Heat transfer enhancement of nanofluids. *International Journal of Heat and Fluid Flow*, *21*(1), 58–64. DOI: 10.1016/S0142-727X(99)00067-3

Yadav, D., & Sanserwal, M. (2023). Effects of factors on thermal conductivity and rheology of CNT nanofluids. *Journal of Thermal Analysis and Calorimetry*, *148*(5), 1723–1763. DOI: 10.1007/s10973-022-11821-7

Yang, J., Yang, X., Wang, J., Chin, H. H., & Sundén, B. (2022). Review on thermal performance of nanofluids with/without magnetic fields. *Frontiers in Energy Research*, *10*, 822776. DOI: 10.3389/fenrg.2022.822776

Yasmin, H., Giwa, S. O., Noor, S., & Sharifpur, M. (2023a). Thermal conductivity enhancement of metal oxide nanofluids: A critical review. *Nanomaterials (Basel, Switzerland)*, *13*(3), 597. DOI: 10.3390/nano13030597 PMID: 36770558

Yasmin, H., Giwa, S. O., Noor, S., & Sharifpur, M. (2023b). Experimental exploration of hybrid nanofluids for solar and thermal energy storage. *Nanomaterials (Basel, Switzerland)*, *13*(2), 278. DOI: 10.3390/nano13020278 PMID: 36678031

Zhang, X., Gu, H., & Fujii, M. (2006). Effective thermal conductivity and diffusivity of nanofluids. *Journal of Applied Physics*, *100*(4). Advance online publication. DOI: 10.1063/1.2259789

Chapter 7
AI and Machine Learning in Façade Performance Prediction

Ansuman Samal
http://orcid.org/0000-0002-4213-5988
Siksha O Anusandhan (Deemed to be) University, India

R. Rukmathan
http://orcid.org/0000-0002-4989-3407
Dr. NGP Arts and Science College, India

V. C. Vandana
http://orcid.org/0009-0006-1851-0975
Government First Grade College, Malur, India

Vijay Uprikar
http://orcid.org/0000-0003-2546-8307
Datta Meghe Institute of Management Studies, Nagpur, India

P. Selvakumar
http://orcid.org/0000-0002-3650-4548
Department of Science and Humanities, Nehru Institute of Technology, Coimbatore, India

Ninad Rithe
http://orcid.org/0009-0004-5729-6081
Datta Meghe Institute of Management Studies, Nagpur, India

ABSTRACT

The integration of Artificial Intelligence (AI) and Machine Learning (ML) in the built environment is transforming how buildings are designed, constructed, operated, and maintained, with far-reaching implications for sustainability, occupant comfort, operational efficiency, and cost-effectiveness. In the context of building performance, AI and ML act as pivotal technologies that enable data-driven decision-making and proactive control strategies. As buildings become more complex and demand greater energy efficiency and user responsiveness, traditional rule-based systems often fall short in handling the volume, velocity, and variety of data generated by modern sensors and control devices. AI and ML offer powerful alternatives, with

DOI: 10.4018/979-8-3373-6023-2.ch007

the ability to learn from historical and real-time data, recognize patterns, forecast performance, and autonomously adapt to changing conditions. One of the most critical areas where AI and ML contribute to building performance is energy management and optimization.

THE ROLE OF AI AND MACHINE LEARNING IN BUILDING PERFORMANCE

The integration of Artificial Intelligence (AI) and Machine Learning (ML) in the built environment is transforming how buildings are designed, constructed, operated, and maintained, with far-reaching implications for sustainability, occupant comfort, operational efficiency, and cost-effectiveness. In the context of building performance, AI and ML act as pivotal technologies that enable data-driven decision-making and proactive control strategies (Arowoiya et al., 2024; Yang et al., 2022). As buildings become more complex and demand greater energy efficiency and user responsiveness, traditional rule-based systems often fall short in handling the volume, velocity, and variety of data generated by modern sensors and control devices. AI and ML offer powerful alternatives, with the ability to learn from historical and real-time data, recognize patterns, forecast performance, and autonomously adapt to changing conditions (Arowoiya et al., 2024; Zhang et al., 2022; Fakhabi et al., 2024). One of the most critical areas where AI and ML contribute to building performance is energy management and optimization. Smart buildings generate vast amounts of energy data through IoT-enabled systems, including HVAC, lighting, water systems, and renewable energy sources. AI algorithms can analyze this data to detect inefficiencies, predict peak loads, and dynamically adjust settings to reduce energy consumption without compromising occupant comfort. Machine learning models such as neural networks, reinforcement learning, and decision trees can forecast energy demand, optimize the operation of building systems, and identify abnormal usage patterns indicative of faults or inefficiencies. For example, predictive analytics powered by AI can anticipate temperature fluctuations based on weather data and adjust HVAC systems accordingly, achieving significant energy savings and lowering carbon emissions. Another significant contribution of AI is in predictive maintenance and fault detection. Building systems often degrade over time or develop faults that go unnoticed until they cause failures or increase operational costs. AI-powered diagnostics can continuously monitor sensor data and detect anomalies that may indicate equipment failure or underperformance. By applying supervised and unsupervised ML models, facility managers can receive early alerts about potential issues and schedule maintenance before critical failures occur, reducing downtime and repair costs. This proactive approach not only improves equipment lifespan but also enhances the overall reliability and performance of building systems.

AI also plays a crucial role in design optimization and performance simulation during the early stages of building planning. Traditionally, design iterations for energy modeling, daylighting analysis, and ventilation assessments are time-consuming and resource-intensive. AI accelerates this process by enabling generative design, where algorithms propose optimal configurations based on predefined objectives such as energy efficiency, material usage, or thermal comfort (Abediniangerabi et al., 2021; El-Mowafy et al., 2022; Yu et al., 2022). ML models trained on historical building performance data can guide architects and engineers toward design choices that meet or exceed performance targets. This integration of AI in the design phase ensures more sustainable and high-performing buildings from the outset, reducing the need for costly retrofits and adjustments later. Taken together, these developments show that AI and ML are reshaping the building industry by enhancing performance across the entire lifecycle, from design and construction to operation and end-of-life. By leveraging vast datasets and advanced algorithms, these technologies enable more intelligent, adaptive, and sustainable buildings. Their impact spans energy efficiency, occupant well-being, maintenance efficiency, and long-term resilience. As buildings continue to evolve into complex, data-rich ecosystems, AI and ML will be indispensable tools in meeting the growing demands of environmental stewardship, economic efficiency, and human-centric design. Investing in these technologies today ensures that the buildings of tomorrow will not only perform better but also contribute meaningfully to a smarter, more sustainable built environment.

FAÇADE PERFORMANCE PREDICTION: CHALLENGES AND OPPORTUNITIES

Façade performance prediction plays a pivotal role in the architectural, environmental, and energy performance of modern buildings. As façades serve as the primary interface between indoor and outdoor environments, they are crucial for controlling solar heat gain, daylighting, ventilation, acoustic performance, and thermal insulation. Predicting how a façade will perform under real-world conditions is essential for achieving sustainable design goals, enhancing occupant comfort, reducing operational energy demand, and meeting increasingly stringent environmental regulations. However, despite the growing sophistication of modeling tools and analytical methods, façade performance prediction remains a complex and multifaceted challenge (Zhang et al., 2022; Gonçalves et al., 2024). This is primarily due to the dynamic, non-linear, and context-dependent behavior of building envelopes, combined with limitations in current simulation practices, material characterization, climate responsiveness, and data availability. As the built environment faces mounting pressures to reduce carbon emissions and adapt to climate change, the accurate prediction of façade performance has never been more critical, and simultaneously, never more difficult. One of the central challenges in façade performance prediction lies in the complexity of environmental interactions. Façades are exposed to variable and often unpredictable external conditions, including solar radiation, wind, temperature fluctuations, humidity, and pollution (Bueno et al., 2021; Gonçalves et al., 2024). These factors can have compound and interactive effects on the thermal and optical behavior of façade materials and systems. Predictive models must therefore account for highly dynamic boundary conditions, which can vary not only seasonally but also hourly or even minute-by-minute. Most current simulation tools, however, are built on deterministic or static assumptions, often relying on historical weather files or standardized climate data that do not capture real-time or future climatic variability. As a result, predictions may not accurately reflect actual building performance, especially in the context of climate change or extreme weather events. This discrepancy poses risks for both energy performance guarantees and occupant well-being. Data quality, availability, and granularity remain persistent issues. Accurate façade performance prediction depends on access to detailed climate data, occupancy schedules, and operational patterns (Arowoiya et al., 2024).

On the technological front, computational limitations can hinder the implementation of high-fidelity, time-resolved simulations required for detailed façade analysis. While tools like EnergyPlus, TRNSYS, Radiance, or CFD packages offer powerful simulation capabilities, they often demand high computational effort, specialized expertise, and extended simulation times. This makes it difficult to apply these tools in early design stages, where rapid feedback is essential for design iteration (Yang et al., 2022; Nazari & Matusiak, 2024). Additionally, many façade designers and architects lack access to or training in using these advanced tools, creating a knowledge gap between design intent and performance simulation. Despite these challenges, numerous opportunities are emerging to enhance façade performance prediction and narrow the gap between design and reality. The integration of AI and machine learning is transforming how façade systems are analyzed and optimized. Data-driven approaches can learn from historical performance data, detect patterns, and predict outcomes based on a multitude of variables, including occupancy, climate, and control strategies. These models can complement physics-based simulations, offering rapid performance feedback and scenario analysis capabilities (Abediniangerabi et al., 2021; Joe et al., 2022; Adegoke et al., 2022; Takhmasib et al., 2023). Additionally, real-time sensor networks and IoT-enabled façades now allow for continuous monitoring of performance, providing a rich dataset for model calibration, performance verification, and adaptive control. Overall, while façade performance prediction remains a technically and contextually challenging domain, it is also a field ripe with innovation and transformative potential. Accurately predicting how a façade will perform over time, across seasons, occupant behaviors, and technological changes, is vital for designing buildings that are not only energy-efficient but also resilient, adaptive, and human-centered. Addressing the multifactorial challenges outlined, from environmental variability and material uncertainty to occupant behavior and integration gaps, requires both technological advancement and methodological innovation.

PREDICTING ENERGY EFFICIENCY AND THERMAL PERFORMANCE

Predicting energy efficiency and thermal performance in buildings is a cornerstone of sustainable architecture, environmental responsibility, and economic optimization. As buildings account for a substantial portion of global energy consumption and greenhouse gas emissions, accurately forecasting their energy behavior has become a fundamental strategy for reducing environmental impacts, minimizing operational costs, and improving occupant comfort. The ability to predict how a building will consume, conserve, and manage energy, and how effectively it will maintain thermal comfort under varying climatic and operational conditions, offers a wide array of benefits across the entire lifecycle of the building. From early-stage design and material selection to system integration, operational strategies, and post-occupancy evaluation, energy and thermal performance prediction enables data-driven decisions that lead to better-performing, healthier, and more resilient buildings (Yang et al., 2022; Yu et al., 2022; Joe et al., 2022; Arowoiya et al., 2024). One of the most significant benefits of predicting energy efficiency is enhanced environmental sustainability. By leveraging simulation tools and analytical models, designers and engineers can assess the potential energy demand of different design alternatives and optimize building geometry, orientation, envelope properties, HVAC systems, and renewable energy integration before construction begins. This proactive approach helps reduce the building's operational carbon footprint and supports compliance with energy codes, sustainability standards (such as LEED, BREEAM, and WELL), and national or regional decarbonization targets. Predictive modeling allows stakeholders to evaluate the long-term implications of design decisions, encouraging the adoption of low-energy systems, passive design strategies, and high-performance materials that reduce energy demand while maintaining or enhancing thermal comfort. Another key advantage is the optimization of building systems for operational efficiency (Abediniangerabi et al., 2021; El-Mowafy et al., 2022; Yu et al., 2022; Adegoke et al., 2022). Predictive models simulate how different components, such as heating and cooling systems, ventilation, lighting, and glazing, interact under varying loads, occupancy patterns, and environmental conditions. This enables system sizing and specification to be based on realistic performance expectations rather than static assumptions, avoiding issues of over- or under-sizing that can lead to energy waste or discomfort. For example, thermal simulation tools can identify peak heat gains and losses, allowing for more precise design of insulation, shading devices, and HVAC capacities. By understanding and modeling thermal behavior in different zones and timeframes, designers can develop systems that are not only more efficient but also more responsive to actual building use, improving comfort and reducing reliance on mechanical conditioning (Yang et al., 2022). A related benefit is regulatory compliance and performance benchmarking.

Furthermore, energy and thermal prediction plays a vital role in retrofitting and renovation strategies. For existing buildings, simulation models calibrated with operational data can identify inefficiencies and opportunities for targeted interventions, such as façade upgrades, equipment replacements, or insulation improvements. Predictive modeling helps quantify the potential energy savings and thermal comfort improvements associated with each intervention, enabling cost-effective and impactful retrofits. It also allows prioritization of investments based on predicted performance gains, helping to build a strong business case for energy upgrades. Finally, predictive modeling of energy and thermal performance contributes to the larger goal of decarbonizing the built environment (Yang et al., 2022; Yu et al., 2022; Sheng et al., 2022; Arowoiya et al., 2024). Accurate forecasting is essential for achieving net-zero energy buildings (NZEBs), passive house standards, and carbon-neutral portfolios. It enables stakeholders to assess the impact of electrification, on-site renewable generation, storage integration, and grid-responsive design on both individual building and district-wide scales. Predictive insights also inform urban planning strategies, helping cities and regions plan for future energy demands, manage peak loads, and reduce dependency on fossil fuels. In summary, predicting energy efficiency and thermal performance offers a multifaceted set of benefits that extend far beyond energy savings alone. From enabling climate-adaptive design and cost-effective operations to enhancing occupant health and supporting decarbonization goals, predictive modeling is a powerful enabler of better-performing, more resilient, and more intelligent buildings (Arowoiya et al., 2024; Zhang et al., 2022). As modeling tools become more sophisticated and data-rich, especially with the integration of AI, IoT, and digital twins, the role of prediction will continue to grow as a foundational component of building design and operation. Embracing predictive strategies today is essential for shaping a built environment that meets the challenges of tomorrow with efficiency, adaptability, and sustainability at its core.

SIMULATING DAYLIGHT AND GLARE ANALYSIS: IMPORTANCE IN SUSTAINABLE BUILDING DESIGN

Simulating daylight and glare analysis has become a foundational aspect of sustainable and occupant-centric building design. As daylight is a natural and renewable source of illumination, its thoughtful integration into architectural spaces offers immense benefits that span energy conservation, occupant well-being, visual comfort, and aesthetic quality. The accurate simulation of daylight distribution and glare potential provides architects, engineers, and lighting designers with critical insights that enable them to make informed design decisions early in the process, optimize building orientation, façade design, window placement, material selection, and control strategies (Wang et al., 2022; Akimov et al., 2023; Nazari & Matusiak, 2024). These simulations ensure that buildings are not only energy-efficient and environmentally responsible but also conducive to human comfort and productivity. In an era increasingly defined by climate responsiveness and wellness-centered environments, the importance of daylight and glare analysis through advanced simulation tools cannot be overstated. One of the foremost advantages of daylight simulation is its ability to significantly reduce reliance on artificial lighting, which typically constitutes a substantial portion of a building's total energy consumption. By accurately modeling how natural light penetrates and distributes within interior spaces throughout the day and across seasons, designers can maximize daylight autonomy and minimize the use of electric lighting during daytime hours (Wang et al., 2022; Akimov et al., 2023; Nazari & Matusiak, 2024). Simulation tools such as Radiance, Daysim, DIVA, and Honeybee allow for quantitative assessments of daylight metrics such as Spatial Daylight Autonomy (sDA), Annual Sunlight Exposure (ASE), and Daylight Factor (DF). These metrics provide detailed insights into how well spaces are naturally lit and help identify areas that may be underlit or overexposed. Optimizing these parameters through simulation reduces lighting energy demand, thereby contributing directly to building energy efficiency and helping achieve sustainability certifications like LEED, WELL, or BREEAM. Through parametric analyses and iterative design feedback, simulations support the creation of spaces that are both functionally effective and psychologically uplifting. A crucial component of daylight analysis is glare simulation, which addresses visual discomfort caused by excessive brightness or high luminance contrasts within a visual field. Glare, whether direct (from sunlight entering through glazing) or reflected (from surfaces such as screens or white walls), can severely impair visual tasks and reduce occupant satisfaction (Nazari & Matusiak, 2024). Predicting and mitigating glare through simulation ensures that the visual environment remains comfortable under all lighting conditions. Tools like Radiance-based glare analysis (using metrics such as Daylight Glare Probability, or DGP) help identify problematic zones where occupants might be exposed to intolerable brightness.

Daylight and glare simulations also play an instrumental role in supporting passive design strategies. Passive solar heating, cooling, and lighting strategies depend heavily on the predictable behavior of sunlight within the building envelope. Through simulation, designers can assess how different window sizes, positions, orientations, and shading systems impact solar heat gain and interior light levels throughout the year. This information is essential for minimizing the building's dependence on mechanical systems, thus reducing overall energy use and improving thermal comfort. For example, daylight analysis can help identify opportunities for winter solar gain while preventing overheating during summer, all while maintaining a high level of natural illumination (Xie & Sawyer, 2021). These simulations thus become vital tools for climate-responsive architecture and bioclimatic design, especially in regions with high solar exposure or fluctuating seasonal conditions. From an operational perspective, daylight simulation informs the design of automated control systems that regulate lighting and shading based on actual daylight conditions. By predicting how natural light fluctuates in specific zones at different times, control strategies for electric lighting and blinds can be programmed to maintain consistent light levels, minimize glare, and conserve energy. These systems not only enhance energy efficiency but also improve occupant satisfaction by adapting to changing environmental conditions. Advanced simulation-driven controls are particularly important in smart buildings, where integration with IoT devices and Building Management Systems (BMS) allows for dynamic, responsive operation tailored to real-time conditions and user preferences (Takhmasib et al., 2023; Isaia et al., 2021; Luna-Navarro et al., 2022; Xie & Sawyer, 2021). Overall, simulating daylight and glare analysis is a vital practice that transcends technical optimization to embrace human-centered, sustainable, and resilient design. The ability to visualize and quantify how natural light interacts with architecture empowers designers to create spaces that are environmentally responsible, energy-efficient, visually comfortable, and emotionally engaging. As tools and methodologies continue to evolve, incorporating AI, real-time data, and immersive visualization, the role of daylight simulation will only grow in importance.

ASSESSING STRUCTURAL INTEGRITY AND DURABILITY: BENEFITS FOR LONG-TERM BUILDING PERFORMANCE

The assessment of structural integrity and durability is a critical pillar in the design, construction, and maintenance of the built environment. Structural integrity refers to the ability of a structure to withstand intended loads without experiencing failure due to deformation, fatigue, or degradation, while durability relates to the capacity of materials and components to perform reliably over time despite exposure to environmental, chemical, or mechanical stressors. The proactive assessment of these characteristics offers numerous benefits, ensuring occupant safety, optimizing material usage, reducing lifecycle costs, enhancing sustainability, and improving resilience against climate extremes and natural disasters. As buildings and infrastructure continue to age and confront increasingly complex performance demands, the importance of structural integrity and durability assessment becomes more pronounced, driving innovations in materials science, modeling, inspection, and monitoring technologies (Falorca & Lanzinha, 2020; Mathur et al., 2024). First and foremost, the most fundamental benefit of assessing structural integrity is ensuring the safety and reliability of buildings. Structural failures can result in catastrophic consequences, ranging from financial loss and functional disruption to injury or loss of life. By conducting detailed structural assessments during both the design and operational phases, engineers can ensure that critical components such as beams, columns, joints, and connections are designed and maintained to resist anticipated loads, including dead loads, live loads, wind, seismic forces, and dynamic impacts. These evaluations are based on established design codes and standards (e.g., Eurocode, ACI, ASCE), which are supported by simulations, laboratory testing, and real-world performance data. Assessing integrity not only confirms compliance with safety regulations but also provides a safety margin to accommodate unforeseen conditions, material defects, or load changes over time. In addition to safety, assessing structural integrity and durability offers substantial economic benefits through lifecycle cost reduction. Buildings and infrastructure that are evaluated and maintained for long-term performance typically experience fewer failures, lower maintenance costs, and extended service lives. Early detection of structural vulnerabilities or material degradation, such as corrosion in steel, spalling in concrete, fatigue in metal components, or delamination in composites, allows for timely and cost-effective interventions before more severe damage occurs. Predictive maintenance strategies, supported by non-destructive testing (NDT), structural health monitoring (SHM), and digital modeling, can forecast when and where repairs are needed, reducing the reliance on costly emergency repairs or complete replacements. This approach supports asset managers in making informed investment decisions and optimizing the allocation of resources across large building portfolios. Digital twin technology and structural health monitoring (SHM) further enhance the ongoing assessment of structural integrity and durability (Arowoiya et al., 2024; Elghaish et al., 2021). By embedding sensors in key structural components, such as strain gauges, accelerometers, and corrosion sensors, engineers can monitor real-time performance and degradation. These data are fed into digital twin models that replicate the physical structure virtually, enabling predictive analytics and performance simulations throughout the building's lifecycle. SHM systems detect anomalies early, identify patterns of deterioration, and allow for remote diagnostics and decision-making. The integration of AI and machine learning with SHM enables even more precise failure forecasting and condition-based maintenance, reducing downtime and maximizing safety. These technologies are particularly valuable in high-risk or mission-critical facilities such as hospitals, transportation hubs, industrial plants, and heritage structures.

In the context of heritage conservation and adaptive reuse, structural integrity assessments allow architects and engineers to repurpose old buildings without compromising safety. Many historical structures were built without modern codes or knowledge of long-term material behavior. Through techniques like ground-penetrating radar (GPR), ultrasonic testing, and finite element modeling (FEM), experts can assess hidden damage, load-bearing capacity, and remaining service life. This allows for safe renovations, conversions, or modernizations while preserving architectural character and cultural value. The accurate assessment of existing structures also supports sustainability goals by reducing demolition waste and maintaining the embodied energy invested in the original construction. Risk management and insurance are also strengthened by structural assessment practices. Insurers and investors rely on verified performance data to evaluate liability and set premiums for buildings and infrastructure. Properties with well-documented integrity assessments, maintenance records, and durability testing are seen as lower-risk assets, which may qualify for better financing and insurance terms. In turn, building owners benefit from reduced exposure to structural risk, better capital planning, and increased asset value. Within façade engineering specifically, AI-assisted inspection methods are becoming more useful because they can process façade imagery, thermal data, and drone-based capture workflows more consistently than purely manual screening, especially when integrated with BIM or asset records (Dias et al., 2021; Falorca & Lanzinha, 2020; Faltynova et al., 2016; Zhang et al., 2023; Yang, Liu, et al., 2024; De Filippo et al., 2023; Mathur et al., 2024). Overall, assessing structural integrity and durability offers far-reaching benefits that extend across safety, sustainability, economy, and performance. It empowers designers, engineers, and facility managers to anticipate risks, optimize material usage, prolong building lifespan, and respond effectively to changing environmental conditions. As buildings evolve into complex systems with increasing demands for resilience and intelligence, structural assessment will continue to play a foundational role in shaping the future of the built environment. By integrating advanced technologies, data analytics, and interdisciplinary expertise, the assessment of structural performance is not merely a technical necessity, it is a strategic imperative for creating buildings that are safe, enduring, adaptable, and future-ready.

SUPERVISED LEARNING FOR FAÇADE PERFORMANCE PREDICTION: IMPORTANCE AND APPLICATIONS

Supervised learning, a subset of machine learning (ML), has emerged as a powerful and increasingly essential tool in the prediction and optimization of façade performance. In the context of the built environment, where building envelopes play a critical role in thermal regulation, energy efficiency, daylighting, ventilation, and overall occupant comfort, accurate façade performance prediction is indispensable. Supervised learning techniques, which involve training algorithms on labeled datasets to learn relationships between input features and target outputs, offer significant advantages over traditional rule-based or purely physics-based approaches (Tarkhan et al., 2024; Gonçalves et al., 2024). By harnessing historical performance data, environmental inputs, material properties, and system behaviors, supervised learning models can uncover complex patterns, enable rapid simulations, and generate highly accurate predictions about how façades perform under various conditions. The importance of supervised learning lies not only in its predictive accuracy and computational efficiency but also in its adaptability, scalability, and potential to inform sustainable, data-driven design decisions throughout a building's lifecycle. One of the primary reasons supervised learning is so impactful in façade performance prediction is its ability to model complex, nonlinear relationships that are often too intricate for traditional analytical or deterministic models to capture effectively. Façade systems interact dynamically with multiple variables, including solar radiation, wind, temperature, humidity, building orientation, occupancy schedules, material properties, and control mechanisms like shading devices or operable windows (Takhmasib et al., 2023; Arowoiya et al., 2024). These interactions are rarely linear and often vary across time and spatial dimensions. Supervised learning algorithms such as support vector machines (SVM), random forests, gradient boosting machines, decision trees, and artificial neural networks (ANNs) excel at identifying these hidden correlations.

Once trained on historical data, these models can predict façade performance outcomes, such as thermal transmittance (U-values), solar heat gain coefficients (SHGC), cooling loads, daylight autonomy, or glare probability, with a high degree of accuracy. Another important benefit of supervised learning in this domain is its capacity for real-time and data-driven prediction, which supports adaptive and intelligent façade systems. In smart buildings, sensor data collected from various parts of the façade, such as temperature sensors, light meters, occupancy detectors, and weather stations, can be used as input for supervised learning models trained to forecast performance metrics in real-time. For example, a model might predict interior thermal comfort levels or daylight availability based on current environmental conditions, and then inform the building management system (BMS) to adjust shading or HVAC operation accordingly (Abediniangerabi et al., 2021). This level of automation and responsiveness improves building efficiency, reduces energy consumption, and enhances occupant comfort. Supervised learning models can be continuously refined as more data is collected, making them increasingly accurate over time and uniquely suited for dynamic environments. Supervised learning is particularly valuable in the early design phases, where rapid feedback is essential for evaluating multiple design alternatives. Traditional simulation tools for façade performance, such as EnergyPlus, Radiance, or TRNSYS, can be computationally expensive and time-consuming, especially when modeling high-resolution temporal data or detailed geometries. By contrast, once a supervised learning model is trained on simulation outputs or empirical data, it can serve as a fast, surrogate predictor that delivers near-instantaneous performance feedback. This enables architects and engineers to explore a wide range of façade configurations, materials, glazing types, and orientations without running time-intensive simulations each time. As a result, supervised learning supports faster, more informed design iterations, leading to optimized façade systems that are both high-performing and cost-effective.

Figure 1. System architecture for machine learning-based TC prediction

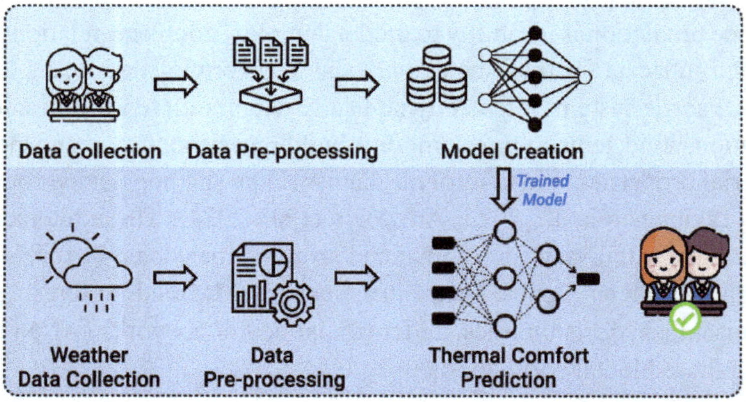

Another major area where supervised learning demonstrates importance is in predictive maintenance and fault detection related to façade components. Over time, façades may develop performance issues due to material aging, mechanical failures (e.g., in operable windows or shading devices), or unexpected environmental stressors. By analyzing historical sensor data, labeled with known fault occurrences, supervised learning models can be trained to recognize patterns that precede performance degradation or failure. These models can alert facility managers before failures become critical, allowing for proactive maintenance and reducing costs associated with unscheduled downtime or occupant discomfort. This predictive capability is especially useful for complex or kinetic façades that incorporate moving parts, embedded sensors, and smart materials. Supervised learning also contributes to climate-responsive and location-specific design, which is essential in the context of global climate variability and localized urban microclimates (El-Mowafy et al., 2022; Arowoiya et al., 2024). By training models on region-specific data, including local solar radiation patterns, ambient temperatures, humidity levels, and wind speeds, supervised learning can support the design of façades tailored to specific environmental contexts. For example, a façade optimized for a tropical climate might require different thermal and solar management strategies than one designed for a temperate or arid region. Supervised learning models can adapt to such conditions, helping designers select materials, shading strategies, and geometries that respond appropriately to local climatic forces while ensuring energy efficiency and comfort. Additionally, supervised learning facilitates post-occupancy performance evaluation and model calibration, helping to close the gap between predicted and actual building performance. During post-occupancy evaluations, performance data collected from building sensors can be compared against predictions generated by supervised learning models. Discrepancies can reveal where the design assumptions may have been flawed or where operational issues (such as occupant overrides or system malfunctions) may be impacting performance. Supervised learning can also be used to recalibrate existing physics-based models based on measured performance, thereby improving future design predictions and informing retrofit strategies (Seghezzi et al., 2021; Zheng et al., 2021; Zhang et al., 2022; Khan et al., 2024). This creates a feedback loop where real-world data continuously enhances predictive accuracy. Overall, supervised learning is a vital tool for enhancing the accuracy, efficiency, and scope of façade performance prediction in the built environment. Its ability to model complex relationships, integrate diverse data sources, support rapid design iteration, and facilitate intelligent, responsive control strategies makes it indispensable for modern architecture and engineering. As sustainability goals, occupant expectations, and regulatory demands evolve, supervised learning offers the flexibility and precision needed to optimize façade systems for energy efficiency, thermal comfort, daylighting, durability, and more. By embedding intelligence into the design and operation of façades, supervised learning not only advances the capabilities of performance prediction but also plays a crucial role in shaping buildings that are adaptive, efficient, resilient, and human-centered.

DEEP LEARNING FOR COMPLEX FAÇADE ANALYSIS: IMPORTANCE AND EMERGING APPLICATIONS

Deep learning, a subfield of machine learning that leverages artificial neural networks with multiple layers, has emerged as a transformative technology in the analysis and optimization of complex building systems, particularly façade systems. Façades, as the primary interface between the building and its environment, play a critical role in regulating thermal performance, energy use, acoustic insulation, daylight penetration, visual aesthetics, structural behavior, and occupant comfort. However, the interactions between these variables are often nonlinear, dynamic, and highly context-dependent (Klimkowska et al., 2022; Goncalves et al., 2024). Traditional analytical or physics-based simulation tools, while effective, can struggle with the complexity and computational intensity involved in modeling these multifactorial systems, especially when applied to innovative, adaptive, or high-performance façades. Deep learning offers a powerful alternative: by learning directly from data, be it images, simulation outputs, or sensor readings, it can model intricate façade behavior with unprecedented accuracy and efficiency. Its ability to capture hidden relationships, scale across large datasets, and support real-time analysis makes deep learning an indispensable tool in modern façade design, evaluation, and operation. The most fundamental advantage of deep learning in façade analysis lies in its ability to model high-dimensional, nonlinear relationships inherent in building envelope performance (Dias et al., 2021; Elghaish et al., 2021; Guo et al., 2020; De Filippo et al., 2023). Façade systems are influenced by a multitude of variables, including geometry, material properties, solar exposure, climatic conditions, thermal mass, occupancy behavior, and control strategies (e.g., shading or operable elements). Deep learning models, particularly deep neural networks (DNNs), convolutional neural networks (CNNs), and recurrent neural networks (RNNs), are designed to handle this complexity. They can automatically extract hierarchical features from raw data and learn patterns that might be missed by conventional statistical approaches or simplified regression models. For example, CNNs can analyze façade imagery or architectural renderings to classify material types, detect anomalies, or assess construction quality. Similarly, RNNs and long short-term memory (LSTM) networks can model time-dependent patterns in energy performance or daylight penetration, providing accurate temporal forecasts that inform control strategies and occupant behavior modelling (Chew & Gan, 2022; Guo et al., 2020). Deep learning also excels at image-based and spatial analysis, making it particularly valuable for tasks involving visual evaluation, computer vision, and pattern recognition within façades. For instance, CNNs can be used to analyze infrared or thermal images to detect heat leakage, thermal bridging, or moisture intrusion within façade assemblies. By training on labeled examples, these models can automate the identification of performance defects, assess the effectiveness of insulation systems, or monitor the condition of façade components over time. Similarly, deep learning can classify and segment façade types from satellite imagery or urban scans, supporting urban-scale analyses of envelope efficiency, aesthetic trends, or compliance with building codes. This capability is especially important for retrofitting initiatives, urban heat island mitigation, and heritage conservation, where large-scale, non-invasive assessment methods are needed.

In performance-driven design and simulation, deep learning supports rapid surrogate modeling and predictive analytics that significantly accelerate design workflows. Traditional simulation tools like EnergyPlus or Radiance, while accurate, can be computationally expensive and time-consuming, particularly when evaluating parametric variations or high-resolution time-series outputs. Deep learning models trained on these simulation outputs can act as high-speed surrogates, producing near-instantaneous predictions of thermal load, daylight metrics, glare risk, or visual comfort based on input design parameters. This enables real-time design feedback, multi-objective optimization, and generative design exploration in complex façade systems. Architects and engineers can use deep learning-powered tools to evaluate thousands of design permutations within minutes, identify optimal configurations, and visualize performance trade-offs early in the design process (Abediniangerabi et al., 2021; El-Mowafy et al., 2022; Sheng et al., 2022; Takhmasib et al., 2023). Moreover, deep learning supports the development of adaptive and intelligent façade systems that respond in real-time to environmental and operational conditions. Integrated with sensor networks and building management systems (BMS), deep learning models can analyze live data streams, such as solar irradiance, indoor temperatures, occupancy levels, or daylight availability, and make real-time decisions about the operation of dynamic façade elements like louvers, electrochromic glazing, or ventilation panels. Reinforcement learning, a subset of deep learning, further allows these systems to learn optimal control policies over time, improving energy efficiency and occupant comfort without human intervention. In this way, façades become not just static envelopes but intelligent, learning systems capable of adapting to their context and enhancing overall building performance. Structural analysis and defect detection also benefit greatly from deep learning, especially in façade systems composed of modular or prefabricated elements. Using deep learning techniques like object detection, segmentation, and generative modeling, engineers can automate quality control in manufacturing or construction (Elghaish et al., 2021; Zhang et al., 2023; Yang, Liu, et al., 2024). For example, deep learning models can process imagery from construction sites to verify the alignment and installation quality of panels, identify missing components, or assess adherence to design specifications. In structural monitoring, sensor data such as strain gauges, accelerometers, or vibration records can be analyzed using deep networks to detect early signs of fatigue, material degradation, or potential failure. These capabilities support predictive maintenance, extend façade lifespan, and reduce long-term operational costs. Overall, the importance of deep learning for complex façade analysis is substantial and multifaceted. From predictive modeling and real-time control to defect detection, design optimization, and personalized user experience, deep learning empowers architects, engineers, and building operators to navigate complexity with greater intelligence and agility. Its ability to learn from diverse data types, numerical, visual, temporal, and spatial, makes it uniquely suited to address the challenges of high-performance, sustainable, and adaptive façade systems in a rapidly evolving built environment. As computational power increases and data availability expands, deep learning will continue to redefine what is possible in façade design, transforming static surfaces into intelligent interfaces that enhance both human comfort and environmental stewardship.

REDUCED ENERGY CONSUMPTION AND COSTS: BENEFITS FOR SUSTAINABILITY, EFFICIENCY, AND LONG-TERM VALUE

Reducing energy consumption and operational costs in buildings has become one of the most critical objectives in modern architecture, engineering, and facility management. Buildings are responsible for a significant share of global energy use, accounting for roughly 30-40% of total energy consumption, and contribute substantially to greenhouse gas emissions. In an era marked by rising energy prices, climate change concerns, and stricter regulatory frameworks, energy-efficient building design and operation are no longer optional but essential. Strategies aimed at minimizing energy use, through passive design, high-efficiency systems, smart controls, renewable energy integration, and data-driven optimization, offer a range of far-reaching benefits. From reducing carbon emissions and conserving natural resources to lowering operational expenses and enhancing occupant comfort, reduced energy consumption drives sustainability, improves financial performance, and fosters innovation in the built environment. One of the most immediate and tangible benefits of reducing energy consumption in buildings is significant cost savings (Yang et al., 2022; Gonçalves et al., 2024). Operational energy, used for heating, cooling, ventilation, lighting, water heating, and equipment, constitutes a major portion of building maintenance expenses. By implementing energy-efficient technologies and strategies, such as high-performance insulation, LED lighting, energy recovery ventilators (ERVs), and variable refrigerant flow (VRF) HVAC systems, building owners and operators can cut utility costs by 20-50% or more depending on the building type and climate. Over the life of a building, these reductions accumulate into substantial financial savings, often surpassing initial investment premiums associated with sustainable design. Moreover, energy-efficient buildings are typically less affected by fluctuations in energy prices, offering increased budget predictability and resilience in volatile markets. Beyond direct financial benefits, reduced energy consumption contributes to environmental sustainability and climate change mitigation. Energy generation, particularly from fossil fuels, is a primary source of carbon dioxide emissions and air pollution. By reducing energy demand through efficient design and operations, buildings lower their environmental footprint and contribute to broader decarbonization goals. This supports international climate commitments, such as the Paris Agreement, and aligns with national targets for net-zero emissions (Yang et al., 2022; Arowoiya et al., 2024). Energy-efficient buildings reduce reliance on coal, oil, and gas, curbing not only CO_2 but also harmful pollutants like nitrogen oxides (NO_X) and particulate matter that negatively impact public health. In this way, every kilowatt-hour saved through better design translates into fewer emissions, cleaner air, and a healthier planet. Energy savings are closely tied to building performance and occupant comfort. High-performing, energy-efficient buildings are better equipped to maintain stable indoor conditions with lower energy input. For example, superior insulation and high-performance windows reduce unwanted heat loss in winter and heat gain in summer, creating more thermally stable environments. Advanced HVAC systems with demand-controlled ventilation ensure that heating and cooling are delivered only when and where needed, preventing overconditioning and ensuring personalized comfort. Similarly, daylight-responsive lighting systems dim or turn off electric lights when sufficient natural light is available, reducing energy use while enhancing visual comfort. These technologies not only cut energy waste but also improve the user experience, increasing satisfaction, productivity, and well-being in workplaces, schools, and homes.

Advanced energy monitoring and management technologies have made it easier than ever to identify and eliminate inefficiencies in real time. Smart building systems and Internet of Things (IoT) sensors collect detailed data on energy usage, temperature, occupancy, and equipment status. This data can be used to generate energy profiles, identify peak usage patterns, detect equipment malfunctions, and suggest conservation strategies. Machine learning algorithms can analyze this data to predict energy needs, optimize system performance, and adapt operations dynamically (Arowoiya et al., 2024; Zheng et al., 2021; Fakhabi et al., 2024; Kozlovska et al., 2023). By continuously tracking and managing energy performance, buildings can remain efficient over time, even as usage patterns or environmental conditions change. Predictive maintenance also becomes possible, reducing downtime and avoiding costly repairs. Lifecycle cost reduction and improved asset value are additional financial advantages associated with reduced energy consumption. Energy-efficient buildings typically have lower operating and maintenance costs over time, resulting in a more favorable total cost of ownership (TCO). Reduced energy bills, fewer system breakdowns, and lower equipment wear translate into long-term savings. In commercial real estate, energy-efficient buildings often command higher rents, attract premium tenants, and have shorter vacancy periods. Certification under green building rating systems, such as LEED, BREEAM, ENERGY STAR, or NABERS, adds value and credibility, signaling to investors and tenants that a building is well-managed and future-ready. In addition, buildings with strong energy performance are better positioned to comply with current and upcoming energy disclosure regulations and carbon reduction mandates, reducing legal and reputational risks. Overall, the benefits of reduced energy consumption and costs in buildings are broad, deep, and multifaceted. From financial savings and environmental stewardship to occupant well-being and regulatory compliance, energy-efficient design and operation serve as pillars of sustainable development. As energy prices rise, climate regulations tighten, and public expectations evolve, reducing energy use is no longer just a technical objective, it is a strategic imperative. Embracing this shift empowers architects, engineers, building owners, and occupants to create spaces that are not only efficient and affordable but also healthier, more resilient, and aligned with the global mission of a low-carbon, equitable future.

DATA QUALITY AND AVAILABILITY: IMPORTANCE IN BUILDING PERFORMANCE, AI INTEGRATION, AND SUSTAINABLE DEVELOPMENT

In the age of digital transformation, the role of data in the architecture, engineering, construction, and operations (AECO) industry has become increasingly critical. With the integration of Building Information Modeling (BIM), smart sensors, Internet of Things (IoT), artificial intelligence (AI), and machine learning (ML) tools, buildings have evolved into data-rich systems capable of generating, analyzing, and responding to vast volumes of information throughout their life cycle (Arowoiya et al., 2024; Kozlovska et al., 2023; Massafra et al., 2023). However, the true value of this transformation depends not merely on the quantity of data collected but on its quality, accuracy, accessibility, and contextual relevance. The quality and availability of data determine the effectiveness of performance evaluations, predictive models, operational efficiency, and sustainability assessments. Poor or incomplete data can lead to faulty analysis, misinformed decisions, energy inefficiencies, and even system failures, while high-quality, well-managed, and easily accessible data enables evidence-based planning, advanced analytics, and continuous improvement in building performance. At its core, data quality refers to the accuracy, completeness, consistency, reliability, and timeliness of data. In the context of buildings, this encompasses data from various sources such as energy meters, weather stations, occupancy sensors, HVAC systems, lighting controls, BIM models, utility bills, commissioning reports, and post-occupancy evaluations. Data must be accurate to reflect the true conditions of building systems and environments. Inaccurate sensor readings, misaligned timestamps, or incorrect units can significantly skew performance analysis or AI training outcomes. Completeness is equally vital, partial datasets or missing variables can lead to flawed conclusions and undermine optimization efforts. Consistency ensures that data collected across different systems and timeframes can be reliably compared and aggregated. Timeliness, meanwhile, enables real-time monitoring, fault detection, and adaptive control responses. Data availability involves having timely, accessible, and interoperable datasets that stakeholders can readily use for decision-making and analytics. In building performance analysis, readily available data supports the benchmarking of energy use, evaluation of indoor environmental quality, simulation of system behavior, and identification of operational inefficiencies. For AI and machine learning applications in façade design, energy prediction, thermal analysis, and occupant comfort modeling, large and high-quality datasets are the foundation. These models rely on diverse training data to detect patterns, predict outcomes, and learn adaptive strategies. Without access to comprehensive and representative data, AI systems cannot generalize effectively or provide accurate insights (Yang et al., 2022; Zhang et al., 2022; Khan et al., 2024). Building Information Modeling (BIM) serves as a key repository and structure for organizing and accessing building data. A well-implemented BIM model contains geometric information, material specifications, spatial relationships, system configurations, construction schedules, cost estimates, and maintenance requirements. High-quality BIM data supports lifecycle analysis, facility management, clash detection, and sustainability assessment. However, the full potential of BIM is realized only when the underlying data is accurate, up-to-date, and interoperable across platforms. Poor-quality BIM models with outdated specifications or missing metadata can hinder coordination, slow decision-making, and introduce costly errors during construction or operations (Htet et al., 2023).

AI and machine learning models used in building performance prediction, fault detection, and design optimization are especially sensitive to data quality and quantity. For supervised learning models, labeled datasets must be accurately annotated and representative of the problem domain. Biases, anomalies, or gaps in training data can lead to overfitting, underfitting, or erroneous predictions. For instance, an energy prediction model trained only on data from temperate climates may fail in tropical or arid regions. Similarly, façade performance analysis using deep learning requires high-resolution images, sensor readings, and simulation results with correct alignment and labeling. Without high-quality data, AI models cannot generalize, adapt, or provide reliable outputs. Furthermore, unsupervised learning and clustering techniques depend on the quality of raw data to identify meaningful patterns or anomalies. Poor data quality undermines their ability to extract useful insights or detect system faults. Data availability also underpins transparency, compliance, and accountability in sustainability reporting and building certification. Many green building standards, such as LEED, BREEAM, NABERS, and ENERGY STAR, require detailed documentation of energy consumption, water use, waste generation, indoor air quality, and system performance. This data must be collected accurately, verified, and maintained over time to support audits, recertification, and benchmarking. Public and private stakeholders increasingly demand transparent reporting of energy use intensity (EUI), greenhouse gas emissions, and environmental impacts. Digital twins and occupancy-oriented monitoring platforms further depend on calibration and validation workflows that screen outliers, sensor faults, and missing data before analytics are trusted in operation (Seghezzi et al., 2021; Fotopoulou et al., 2017; Kozlovska et al., 2023; Massafra et al., 2023). Without reliable and available data, organizations risk falling short of compliance requirements, losing certification, or facing reputational damage. Overall, the importance of data quality and availability in the building industry is foundational to achieving high-performance outcomes, enabling intelligent systems, complying with sustainability standards, and supporting continuous improvement. As buildings become increasingly dynamic, interconnected, and data-driven, the ability to collect, validate, manage, and share trustworthy data will define success in design, construction, and operations. High-quality data empowers innovation, enhances decision-making, reduces risk, and supports sustainability across the entire building life cycle. Investing in robust data infrastructure, standardization, security, and accessibility is no longer optional, it is essential for creating smarter, greener, and more resilient buildings for future generations.

COMPLEXITY OF FAÇADE SYSTEMS AND INTERACTIONS: CHALLENGES IN DESIGN, INTEGRATION, AND PERFORMANCE MANAGEMENT

Modern building façades have evolved far beyond their traditional role as static exterior envelopes. Today, façades are dynamic, high-performance systems that must fulfill a multitude of functions, providing thermal insulation, daylight transmission, acoustic control, structural stability, energy generation, weather resistance, ventilation, and aesthetic identity. This increasing functional diversity and the integration of advanced technologies have introduced significant complexity into façade systems. As buildings grow more sophisticated and performance-driven, the façade becomes a multidisciplinary nexus that intertwines architectural design, structural engineering, environmental science, building physics, materials technology, and digital control systems (Gonçalves et al., 2024). While this evolution has opened new possibilities for innovation and sustainability, it also presents a range of formidable challenges related to design coordination, performance prediction, system integration, maintenance, cost management, and regulatory compliance. One of the primary challenges arising from façade complexity is multifunctionality and performance trade-offs. A high-performance façade must simultaneously regulate heat transfer, facilitate daylighting, prevent glare, provide visual transparency, buffer sound, resist wind and rain, and allow for natural ventilation, all while contributing to the architectural expression of the building. Achieving this balance requires careful material selection, detailed modeling, and nuanced trade-off analysis (Gonçalves et al., 2024; Klimkowska et al., 2022). For example, increasing glazing to maximize daylight can reduce thermal insulation and increase cooling loads unless mitigated by advanced coatings, shading systems, or double-skin façades. Optimizing for one function may inadvertently compromise another, necessitating iterative, multi-objective optimization techniques and cross-disciplinary collaboration. Another layer of complexity emerges from the integration of active and responsive technologies into the building envelope. Contemporary façades may incorporate photovoltaic panels, dynamic glazing, kinetic shading devices, actuated vents, embedded sensors, and intelligent control systems. These active façade elements respond in real time to environmental conditions, occupant preferences, and energy management goals. While they enable adaptability and improved performance, they also introduce challenges related to interoperability, control logic, system tuning, and long-term reliability (Gonçalves et al., 2024; Takhmasib et al., 2023). The coordination of these active components with passive design strategies requires precise calibration and often relies on sophisticated algorithms, machine learning models, and real-time data streams, further intensifying system complexity. Thermal and structural interactions also contribute to façade system complexity. The façade must accommodate both static and dynamic loads, including dead weight, wind pressure, thermal expansion, and seismic forces, while maintaining air-tightness, water resistance, and visual integrity. Differential movement between materials, such as glass, aluminum, steel, and composites, can lead to fatigue, sealant failure, or thermal bridging if not properly addressed during the design phase. In double-skin or ventilated façades, thermal behavior is further complicated by convective airflows and temperature stratification. Predicting and controlling these interactions requires advanced simulation tools, such as Computational Fluid Dynamics (CFD) and Finite Element Analysis (FEA), which demand accurate inputs, deep expertise, and intensive computation.

Design coordination across disciplines is another major challenge due to the interdependencies of façade systems. Architects, façade engineers, mechanical engineers, structural consultants, energy modelers, and contractors must work together from early design stages to ensure that the façade meets all performance and aesthetic goals. However, siloed workflows, incompatible software tools, and communication gaps can lead to misalignment, late-stage design changes, or performance discrepancies. For instance, a design that meets aesthetic intentions may fail to satisfy ventilation or shading requirements without early input from mechanical and environmental consultants. Integrated design processes supported by BIM, collaborative platforms, and shared simulation environments are essential but can be difficult to implement consistently across all project teams (Tarkhan et al., 2024; Klimkowska et al., 2022). The construction and fabrication processes for complex façade systems introduce further obstacles. Prefabrication, modularization, and custom fabrication are often required to meet intricate design geometries or material combinations. However, ensuring precision, minimizing on-site assembly errors, and coordinating delivery schedules can be difficult, especially when working with global supply chains or tight timelines. Installation tolerances, weather constraints, and site logistics must be meticulously planned to avoid delays or defects. Moreover, quality assurance and control become more challenging when dealing with custom components or novel materials without well-established standards. Overall, the complexity of façade systems and their interactions presents a wide array of challenges, technical, operational, financial, regulatory, and organizational. These challenges stem from the growing multifunctionality of façades, integration of active technologies, dynamic environmental interactions, diverse materials, and the need for interdisciplinary coordination. While these complexities can drive innovation and improve building performance, they also demand rigorous planning, advanced tools, skilled professionals, and a commitment to lifecycle thinking. Navigating this complexity is essential to realizing the full potential of façades as intelligent, adaptive, and high-performance systems that contribute to sustainable and resilient buildings in the 21st century.

CONCLUSION

The integration of artificial intelligence (AI) and machine learning (ML) into façade performance prediction marks a pivotal transformation in the architecture, engineering, and construction (AEC) industry. As building envelopes become increasingly dynamic, multifunctional, and responsive to complex environmental and operational demands, traditional deterministic approaches are often insufficient to address the intricate interplay of variables involved in façade design and optimization. AI and ML have emerged as highly influential technologies that offer advanced capabilities for modeling, simulating, and managing these interactions with unprecedented accuracy, efficiency, and adaptability. From early-stage design analytics and predictive modeling to real-time control and lifecycle management, the application of intelligent algorithms is fundamentally reshaping the way façades are conceived, executed, and maintained. Looking ahead, the future of AI and ML in façade performance prediction is poised to be even more expansive and deeply integrated. One of the most transformative developments will be the growing use of predictive and prescriptive analytics, wherein AI models not only forecast how a façade will perform under varying conditions but also prescribe the optimal configurations or control strategies to enhance that performance. These models will increasingly leverage vast and diverse datasets from climate databases and material libraries to real-time sensor feeds and occupant feedback to deliver more contextualized and granular predictions. As the quality and availability of façade-related data improve, AI systems will evolve from being primarily analytical tools to becoming proactive, autonomous agents in the design and operation of high-performance envelopes. Furthermore, deep learning and reinforcement learning algorithms will play a greater role in enabling real-time decision-making and adaptive control. Façades will no longer be static barriers but intelligent interfaces that dynamically respond to solar gain, airflow, humidity, temperature, noise, and user presence. Reinforcement learning, in particular, holds promise for optimizing the behavior of operable components, such as shading devices, smart glazing, and ventilated panels, based on ongoing interaction with their environments. These self-learning systems will continuously fine-tune performance metrics such as daylight availability, glare control, and thermal comfort, thereby contributing to occupant well-being and energy efficiency in a more resilient and automated manner. Moreover, the emphasis on explainable AI (XAI) will grow in tandem with the increasing use of black-box models in design decision-making. As the stakes of façade performance, ranging from occupant health and energy consumption to fire safety and capital investment, are high, stakeholders must be able to interpret and trust AI recommendations. Explainability frameworks will provide clear visualizations, rationale, and confidence levels for AI outputs, enabling architects, consultants, regulators, and clients to make informed decisions. The incorporation of ethical and transparent AI practices will be crucial in aligning façade design processes with regulatory standards, user expectations, and professional accountability. In conclusion, the future of AI and ML in façade performance prediction is one of profound potential and multidisciplinary synergy. These technologies will continue to reshape the boundaries of what is possible in façade design, pushing toward buildings that are not only more efficient and intelligent but also more resilient, sustainable, and responsive to human needs. Through the convergence of predictive analytics, generative design, digital twins, explainable AI, and IoT integration, façades will evolve into living systems, intuitively attuned to their environments, users, and operational contexts. As the built environment becomes increasingly data-driven and performance-focused, AI and ML will serve as the catalysts for a smarter, greener, and more adaptive architectural future. To realize this potential, ongoing research, cross-disciplinary collaboration, investment in data infrastructure, and a commitment to transparency and ethics will be essential. With these foundations in place, AI-driven façade performance prediction can make a substantive contribution to future architectural performance and environmental stewardship.

REFERENCES

Abediniangerabi, B., Makhmalbaf, A., & Shahandashti, M. (2021). Deep learning for estimating energy savings of early-stage facade design decisions. *Energy and AI*, *5*, 100077. DOI: 10.1016/j.egyai.2021.100077

Adegoke, M., Alaka, H., & Ajayi, S. (2022). Application of multilayer extreme learning machine for efficient building energy prediction. *Energies*, *15*(24), 9512. DOI: 10.3390/en15249512

Akimov, L., Bezborodov, A. A., & Badenko, V. (2023). Example-based dynamic facade design using the facade daylighting performance improvement (FDPI) indicator. *Building Simulation*, *16*(12), 2261–2283. DOI: 10.1007/s12273-023-1073-6

Arowoiya, V. A., Moehler, R. C., & Fang, Y. (2024). Digital twin technology for thermal comfort and energy efficiency in buildings: A state-of-the-art and future directions. *Energy and Built Environment*, *5*(5), 641–656. DOI: 10.1016/j.enbenv.2023.05.004

Bueno, A. M., de Paula Xavier, A. A., & Broday, E. E. (2021). Evaluating the connection between thermal comfort and productivity in buildings: A systematic literature review. *Buildings (Basel, Switzerland)*, *11*(6), 244. DOI: 10.3390/buildings11060244

Chew, M. Y. L., & Gan, V. J. L. (2022). Long-standing themes and future prospects for the inspection and maintenance of facade falling objects from tall buildings. *Sensors (Basel)*, *22*(16), 6070. DOI: 10.3390/s22166070 PMID: 36015831

De Filippo, M., Asadiabadi, S., Kuang, J. S., Mishra, D. K., & Sun, H. (2023). AI-powered inspections of facades in reinforced concrete buildings. *HKIE Transactions*, *30*(1), 1–14. DOI: 10.33430/V30N1THIE-2020-0023

Dias, I. S., Flores-Colen, I., & Silva, A. (2021). Critical analysis about emerging technologies for building's facade inspection. *Buildings (Basel, Switzerland)*, *11*(2), 53. DOI: 10.3390/buildings11020053

El-Mowafy, B. N., Elmokadem, A., & Waseef, A. A. (2022). Evaluating adaptive facade performance in early building design stage: An integrated daylighting simulation and machine learning. In *Recent Advances in Advanced Machine Learning Approaches for Data Science* (pp. 211–223). Springer., DOI: 10.1007/978-3-031-03918-8_20

Elghaish, F., Matarneh, S. T., Talebi, S., Abu-Samra, S., Salimi, G., & Rausch, C. (2021). Deep learning for detecting distresses in buildings and pavements: A critical gap analysis. *Construction Innovation, 22*(3), 554–579. DOI: 10.1108/CI-09-2021-0171

Fakhabi, M. M., Hamidian, S. M., & Aliehyaei, M. (2024). Exploring the role of the Internet of Things in green buildings. *Energy Science & Engineering, 12*(9), 3779–3822. DOI: 10.1002/ese3.1840

Falorca, J. F., & Lanzinha, J. (2020). Facade inspections with drones: Theoretical analysis and exploratory tests. *International Journal of Building Pathology and Adaptation, 39*(2), 235–258. DOI: 10.1108/IJBPA-07-2019-0063

Faltynova, M., Matouskova, E., Sedina, J., & Pavelka, K. (2016). Building facade documentation using laser scanning and photogrammetry and data implementation into BIM. *The International Archives of the Photogrammetry, Remote Sensing and Spatial Information Sciences, B3*(B3), 215–220. DOI: 10.5194/isprs-archives-XLI-B3-215-2016

Fotopoulou, E., Zafeiropoulos, A., Terroso-Saenz, F., Şimşek, U., González-Vidal, A., Tsiolis, G., Gouvas, P., Liapis, P., Fensel, A., & Skarmeta, A. (2017). Providing personalized energy management and awareness services for energy efficiency in smart buildings. *Sensors (Basel), 17*(9), 2054. DOI: 10.3390/s17092054 PMID: 28880227

Goncalves, M., Figueiredo, A., & Almeida, R. M. S. F. (2024). Dynamic facades in buildings: A systematic review across thermal comfort, energy efficiency and daylight performance. *Renewable and Sustainable Energy Reviews, 199, 114474.* _Guo, J., Wang, Q., & Li, Y. (2020). Facade defects classification from imbalanced dataset using meta learning-based convolutional neural network. *Computer-Aided Civil and Infrastructure Engineering, 35(12), 1403-1418.* DOI: 10.1016/j.rser.2024.114474

Isaia, F., Fiorentini, M., Serra, V., & Capozzoli, A. (2021). Enhancing energy efficiency and comfort in buildings through model predictive control for dynamic facades with electrochromic glazing. *Journal of Building Engineering, 43*, 102535. DOI: 10.1016/j.jobe.2021.102535

Joe, J., Min, S., Oh, S.-H., Jung, B., Kim, Y. M., Kim, D. W., Lee, S. E., & Yi, D. H. (2022). Development of simplified building energy prediction model to support policymaking in South Korea: Case study for office buildings. *Sustainability (Basel), 14*(10), 6000. DOI: 10.3390/su14106000

Htet, A., Liana, S. R., Aung, T., & Bhaumik, A. (2023). The Intersection of Facade Engineering and Building Information Modeling: Opportunities and Challenges. *Journal of Technology Innovations and Energy*, *2*(3), 94–110. DOI: 10.56556/jtie.v2i3.643

Khan, I., Zedadra, O., Guerrieri, A., & Spezzano, G. (2024). Occupancy prediction in IoT-enabled smart buildings: Technologies, methods, and future directions. *Sensors (Basel)*, *24*(11), 3276. DOI: 10.3390/s24113276 PMID: 38894069

Klimkowska, A., Cavazzi, S., Leach, R., & Grebby, S. (2022). Detailed three-dimensional building facade reconstruction: A review on applications, data and technologies. *Remote Sensing (Basel)*, *14*(11), 2579. DOI: 10.3390/rs14112579

Kozlovska, M., Petkanic, S., Vranay, F., & Simko, M. (2023). Enhancing energy efficiency and building performance through BEMS-BIM integration. *Energies*, *16*(17), 6327. DOI: 10.3390/en16176327

Luna-Navarro, A., Hunt, G. R., & Overend, M. (2022). Dynamic facades: An exploratory campaign to assess occupant multi-domain environmental satisfaction and facade interaction. *Building and Environment*, *211*, 108703. DOI: 10.1016/j.buildenv.2021.108703

Massafra, A., Costantino, C., Predari, G., & Gulli, R. (2023). Building information modeling and building performance simulation-based decision support systems for improved built heritage operation. *Sustainability (Basel)*, *15*(14), 11240. DOI: 10.3390/su151411240

Mathur, P., Sharma, C., & Azeemuddin, S. (2024). Autonomous inspection of high-rise buildings for facade detection and 3D modeling using UAVs. *IEEE Access : Practical Innovations, Open Solutions*, *12*, 18251–18258. DOI: 10.1109/ACCESS.2024.3360209

Nazari, M., & Matusiak, B. (2024). Daylighting simulation and visualisation: Navigating challenges in accuracy and validation. *Energy and Building*, *312*, 114188. DOI: 10.1016/j.enbuild.2024.114188

Seghezzi, E., Locatelli, M., Pellegrini, L., Pattini, G., Di Giuda, G. M., Tagliabue, L. C., & Boella, G. (2021). Towards an occupancy-oriented digital twin for facility management: Test campaign and sensors assessment. *Applied Sciences (Basel, Switzerland)*, *11*(7), 3108. DOI: 10.3390/app11073108

Sheng, Y., Ward, W. O. C., Arbabi, H., Álvarez, M., & Mayfield, M. (2022). Deep multimodal learning for residential building energy prediction. *IOP Conference Series. Earth and Environmental Science*, *1078*(1), 012038. DOI: 10.1088/1755-1315/1078/1/012038

Takhmasib, M., Lee, H. J., & Yi, H. (2023). Machine-learned kinetic facade: Construction and artificial intelligence enabled predictive control for visual comfort. *Automation in Construction*, *156*, 105093. DOI: 10.1016/j.autcon.2023.105093

Tarkhan, N., Szczesniak, J. T., & Reinhart, C. (2024). Facade feature extraction for urban performance assessments: Evaluating algorithm applicability across diverse building morphologies. *Sustainable Cities and Society*, *105*, 105280. DOI: 10.1016/j.scs.2024.105280

Wang, T. H., Huang, Y., & Park, J. (2022). Development of daylight glare analysis method using an integrated parametric modelling approach: A comparative study of glare evaluation standards. *Buildings (Basel, Switzerland)*, *12*(11), 1810. DOI: 10.3390/buildings12111810

Xie, J., & Sawyer, A. O. (2021). Simulation-assisted data-driven method for glare control with automated shading systems in office buildings. *Building and Environment*, *196*, 107801. DOI: 10.1016/j.buildenv.2021.107801

Yang, L., Liu, K., Ou, R., Qian, P., Wu, Y., Tian, Z., Zhu, C., Feng, S., & Yang, F. (2024). Surface defect-extended BIM generation leveraging UAV images and deep learning. *Sensors (Basel)*, *24*(13), 4151. DOI: 10.3390/s24134151 PMID: 39000929

Yang, X., Liu, S., Zou, Y., Ji, W., Zhang, Q., Ahmed, A., Han, X., Shen, Y., & Zhang, S. (2022). Energy-saving potential prediction models for large-scale building: A state-of-the-art review. *Renewable & Sustainable Energy Reviews*, *156*, 111992. DOI: 10.1016/j.rser.2021.111992

Yu, J., Chang, W.-S., & Dong, Y. (2022). Building energy prediction models and related uncertainties: A review. *Buildings (Basel, Switzerland)*, *12*(8), 1284. DOI: 10.3390/buildings12081284

Zhang, C., Wang, F., Zou, Y., Dimyadi, J., Guo, B. H. W., & Hou, L. (2023). Automated UAV image-to-BIM registration for building facade inspection using improved generalised Hough transform. *Automation in Construction*, *153*, 104957. DOI: 10.1016/j.autcon.2023.104957

Zhang, W., Wu, Y., & Calautit, J. K. (2022). A review on occupancy prediction through machine learning for enhancing energy efficiency, air quality and thermal comfort in the built environment. *Renewable & Sustainable Energy Reviews*, *167*, 112704. DOI: 10.1016/j.rser.2022.112704

Zheng, L., Mueller, M., Luo, C., & Yan, X. (2021). A review of data-driven approaches for occupant's behaviour in building energy conservation. *Energy Procedia*. Advance online publication. DOI: 10.46855/energy-proceedings-7460

Chapter 8
The Role of Entrepreneurial Leadership in Advancing Façade Performance and Sustainability

Sriparna Guha
http://orcid.org/0000-0003-3092-3722
Amity University, Kolkata, India

Anirban Mandal
http://orcid.org/0000-0002-5130-5424
ICFAI Business School, Kolkata, India

Sandeep Poddar
http://orcid.org/0000-0001-9771-877X
Lincoln University College, Malaysia

ABSTRACT

The construction and building envelope industry is evolving as consumers worldwide desire to be greener, employ smart technologies, and conserve energy. The chapter emphasises the need for entrepreneurial leadership for façade engineering, particularly in performance and sustainability. In the value chain of façade solutions, entrepreneurial leaders innovate not just new materials and systems but also green practices, digital technologies, and market-orientated tactics. This chapter uses theory and case studies to examine how entrepreneurial leadership promotes innovation in smart and adaptable façades, teamwork across disciplines, and

DOI: 10.4018/979-8-3373-6023-2.ch008

technical uncertainty and market acceptability. It highlights how executives affect sustainability goals, resource gathering, and strategic decisions in new and current façade firms. The findings contribute to sustainable construction management and built environment technopreneurship discussions.

1. INTRODUCTION

The ecosystem is presently encountering significant economic, environmental, and sociological strains. The economic crisis, disparities in opportunities, unemployment, illnesses, wars, natural catastrophes, climate change, and poverty have been prioritised on the global agenda for sustainable development. Sustainable development denotes the satisfaction of the current generation's wants without jeopardising the capacity of future generations to satisfy their own needs (Nor-Aishah et al., 2020). Environmental concerns, urbanisation, and the need for climate-responsive design are all driving a major shift in the construction industry, especially in the area of façade engineering. Façade systems used to be mostly for looks and structure, but now they are a key part of how well a building works, affecting things like energy efficiency, thermal comfort, and carbon emissions. In this situation, the role of leadership has changed from just running projects to also include strategic thinking, managing innovation, and integrating sustainability.

There is a notable and increasing interest in the application of highly glazed facades in commercial structures. A significant amount of the façade, or in some cases the whole façade, is composed of high transmittance glass systems, sometimes accompanied by sun control mechanisms. The tendency, originating in Europe, is proliferating to other areas, including the United States(Lee et al., 2002; Poddar & Guha, 2021). A subset of these systems utilises a secondary layer to provide a double envelope system, facilitating enhanced venting and ventilation procedures. The articulated justification for employing these design methodologies differs, although it frequently is linked to occupant advantages and sustainable design related to daylighting and energy conservation (Valladares-Rendón et al., 2017; Fernando et al., 2023).

To comply with the advancing criteria of sustainable architectural design, façade materials and systems must conform to these requirements. A significant restriction is the requirement for specific experience and abilities in the design, installation, and maintenance of these systems. Moreover, façade technology may exhibit greater susceptibility to deterioration compared to conventional materials, hence escalating maintenance and replacement expenses over time (Prieto et al., 2019; Fernando et al., 2023). The integration of complicated systems, such as lighting and shading, can present issues associated with high construction costs and the use of expensive materials (Attia et al., 2020).

The significance of leaders has been recognised by previous scholars, including Hooi (2016). Sustainable entrepreneurship encourages leaders or business owners to integrate their operations with financial, environmental, and social objectives to establish a distinctive Triple Bottom Line (TBL). Hamann et al. (2017) assert that managers' environmental obligations are crucial for establishing a competitive advantage. The notion of entrepreneurial leadership (EL) has been insufficiently explored in the context of sustainability, necessitating additional study focus (He et al., 2017). Entrepreneurial leadership, which combines looking for opportunities with being proactive about innovation and taking risks, has become a crucial factor in moving façade technology towards solutions that are both sustainable and high-performing. Entrepreneurial leaders work in risky situations with few resources, where they have to find market gaps, encourage innovation, and generate value under pressure. This type of management is different from traditional leadership. Their importance in façade engineering is especially clear as the field faces worldwide problems, including the need for smart and adaptable façades, the shift to net-zero buildings, and the advent of digital construction methods (Nor-Aishah et al., 2020).

This chapter examines how entrepreneurial leadership fosters innovation in façade systems, including smart control systems, digital modelling tools, the development of sustainable materials, and modular designs by thematic analysis. It examines how leaders in startups and construction firms are redefining sustainability by proactively making decisions, engaging stakeholders, and experimenting with innovative approaches. The chapter starts by discussing the concepts of entrepreneurial leadership and their connection to sustainability in façade innovation. It provides case-based insights from practical implementations, identifies leadership issues, and proposes a conceptual framework linking entrepreneurial leadership to sustainable façade performance. The objective is to contribute to the growing body of knowledge at the intersection of leadership, sustainability, and façade technology, while providing valuable guidance to innovators, managers, and policymakers in the built environment.

2. LITERATURE REVIEW

The built environment has been under more and more pressure to deal with global sustainability issues, energy efficiency needs, and technological disruptions recently. In this complicated world, façades have changed from passive design elements to performance-driven systems that need new ideas and strong leadership. As buildings become smarter, more climate-resilient, and more compliant, those responsible for implementing these changes must also adapt.

The environmental and economic elements of sustainability in façade systems have been widely discussed, but the social aspects especially when it comes to occupant health and human-centered design have not received as much attention. The goal of social sustainability is to build places that make people more comfortable, healthy, safe, fair, and able to live there for a long time. Entrepreneurial leadership is very important for developing socially sustainable façade solutions that put these human-centred issues first (Popoola & Aluko, 2025). The performance of the façade has a direct effect on the quality of the inside environment, such as access to natural light, thermal comfort, soundproofing, and views of the outside world. All of these things have a big effect on user pleasure, productivity, and mental health (Lahjiri & Benhida, 2025). Leaders who incorporate biophilic ideas into façade systems like movable windows, shading components, or façades that include nature show that they know how design can improve mental and physical wellness. These design principles are especially crucial in cities, where the physical environment may make people feel alone or stressed. More and more, entrepreneurial leaders in the façade sector are making design choices that are in line with social performance criteria from worldwide certification frameworks like the WELL Building Standard, Living Building Challenge, and Fitwel (Faton & Hermans, 2025). These requirements focus on energy and material use, as well as how well people live in the building, its accessibility, and its inclusivity. Facade leaders go beyond just following the rules by following these criteria. They show that they are morally and strategically committed to holistic sustainability. Additionally, incorporating end-users in the design process via feedback loops, post-occupancy assessments, or participatory design techniques bolsters legitimacy and guarantees that façades genuinely cater to the individuals who engage with them every day. As cities work to become more open and stronger, entrepreneurial leadership in façade development has to place social fairness, user dignity, and community involvement at the centre of sustainable innovation (Baburajan, 2025).

Sustainable façade engineering has changing, cross-disciplinary, and innovation-focused needs that traditional models of construction leadership don't always meet. This trend has led both scholars and practitioners to look into how entrepreneurial leadership can help make the built environment more sustainable by encouraging innovation. The next sections look at research in five areas that are important to this study.

2.1 Entrepreneurial Leadership: An Emerging Construct

Entrepreneurial leadership is a new way of leading that combines seeing opportunities, coming up with new ideas, and taking strategic risks in changing situations (Renko et al., 2015). Entrepreneurial leadership is better than transactional or transformational leadership approaches when there is a lot of uncertainty, new ideas, and problems with long-term growth. It means being able to see and use resources to create new value, frequently outside of the normal boundaries of an organisation (Gupta et al., 2004). In the construction and built environment industry, which has always been cautious about risk and full of rules, this kind of leadership is being considered a way to bring about long-lasting and technical transformation. In the more volatile and competitive landscape that businesses encounter today, a certain sort of "entrepreneurial" leader, separate from other leadership styles, is essential.

The paper by Gupta et al. (2004) formulates the concept of entrepreneurial leadership, drawing upon existing literature on entrepreneurship and leadership. The proposal includes a way to measure entrepreneurial leadership based on the Global Leadership and Organisational Behaviour Effectiveness (GLOBE) research, which studied 62 different cultures and more than 15,000 middle managers. The results show that the idea of entrepreneurial leadership is attractive in many cultures and provide early understanding of how different social factors affect how effective people think entrepreneurial leadership is.

The objective of the paper by Bagheri & Harrison (2020) is to enhance comprehension of entrepreneurial leadership by creating a multi-dimensional metric for the construct and utilising a thorough approach that encompasses the skills, competencies, behaviours, and roles of entrepreneurial leaders. The study resulted in a comprehensive conceptualisation of entrepreneurial leadership, which subsequently laid the foundation for an empirical construct of this phenomenon and its assessment from a cross-cultural viewpoint. Their study specifically identifies the factors that most accurately characterise each facet of entrepreneurial leadership. This study defines the talents, competences, and distinctive behaviours of entrepreneurial leaders based on its results.

2.2 Leadership and Sustainability in the Built Environment

Leadership and sustainability have been studied together. Substantial information exists about sustainable practices, sustainability, and environmental leadership, both as distinct study domains and as successful leadership styles that promote the attainment of sustainability objectives. Although these study streams have evolved concurrently, there is less understanding of the similarities and distinctions across these three leadership approaches (Boeske, 2023). Social value in the built environment pertains to the social impact that any organisation, initiative, or program within that sector has on the lives of the stakeholders influenced by its actions. Social values serve as a practical mechanism at the national or organisational level for achieving the United Nations' Sustainable Development Goals (SDGs). This point is especially pertinent given the ongoing expansion of the built environment globally through building and infrastructure initiatives involving various stakeholders (Raiden & King, 2021).

In the past, construction industry leaders have focused on efficiency, managing budgets, and following the rules. But the advent of green buildings and climate-responsive design has made it necessary for leaders to focus on sustainability (Goh et al., 2013). Even with little time or money, leaders must consider the environment when planning projects. Façades are the main link between a structure and its surroundings. They have a big impact on how well a building keeps its heat, how much energy it uses, and how it looks. Darko and Chan (2017) say that strategic leadership choices made during the design and implementation stages have a big effect on a building's energy use and operating costs.

2.3 Innovation and Façade Performance

Innovation in building is increasingly essential globally as contemporary social and environmental concerns demand swift and extensive solutions. This dynamic landscape prompts inquiry into the role of the façade business in defining the future of architecture, engineering, and construction (Gasparri et al., 2022). The building industry, especially the façade sector, is markedly fragmented. The presence of several parties with varying goals and objectives is frequently recognised as a barrier to the dissemination of innovation within the market. This chapter aims to serve as a catalyst, fostering a multilayered discourse that connects universities with industry, applied research with product development, and design with building (Sangiorgio, 2022).

Façade systems have changed from passive design elements to building envelopes that can change based on what is happening outside (Htet et al., 2023). New technologies like double-skin façades, photovoltaic-integrated panels, and smart shading systems can now help cut down on energy use while making the indoors more comfortable (Poirazis, 2006). But how well these systems work often depends less on the availability of technology and more on how leaders feel about trying new things, working together across disciplines, and being open to new ideas (Bluyssen, 2010).

2.4 Entrepreneurial Leadership in Sustainable Innovation

In the contemporary business environment, possessing only leadership or entrepreneurial characteristics is inadequate for organisations' success. Effective managers must have a blend of leadership and entrepreneurial qualities that influence their employees' creativity and sustainable innovation outcomes. Entrepreneurial leadership plays a paramount role in shaping sustainable organisational success in a dynamic and uncertain business environment (Ibrahim et al., 2024). The findings indicated that entrepreneurial leadership directly influences workers' creativity and sustainable innovation performance, with employees' creativity serving as a mediator in the interaction between entrepreneurial leadership and organisational success in sustainable innovation.

In modern business contexts, possessing only leadership or entrepreneurial traits is inadequate to ensure a company's success. Effective managers must exhibit a blend of leadership and entrepreneurial attributes that influence their innovative work behaviours and sustainable innovation outcomes. Given the dynamic and uncertain nature of the organisational environment, entrepreneurial leadership plays a crucial role in achieving sustained organisational success (Al-Refaei et al., 2024).

Rae's (2017) paper examines the burgeoning role of leadership development in sustainable business. It tackles the necessity of cultivating research and effective methodologies, proposing strategies to overcome the challenges that organisations pursuing sustainability encounter in fostering enduring entrepreneurial leadership, nurturing an entrepreneurial culture, and enabling individuals to assume leadership positions that drive ongoing innovation, development, and growth. A model to improve sustainable entrepreneurial leadership is created by carefully studying relevant literature and real-life examples, focussing on four key areas: strategic direction, culture, community, and entrepreneurial innovation. They identify these as crucial factors in fostering leadership for the long-term viability of such groups and propose a future research trajectory.

Entrepreneurial leaders are in a unique position to change the way things are done and come up with new ways to think about sustainability in organisations. They can lead sustainable changes in complicated fields because they are proactive, willing to take risks, and able to frame strategic opportunities (Hargreaves, 2011). In fields like design and engineering, research reveals that these kinds of leaders are more likely to support green innovation, promote lifecycle thinking, and question the rules about how materials should be used and how well they should work (Dean & McMullen, 2007).

2.5 Digital Transformation and the Role of Leadership

Sainger's (2018) conceptual paper is based on a thorough analysis of numerous survey results concerning leadership and digital transformation. The presentation begins with a theoretical exposition on digitalisation and digital transformation, which are characterised as processes of change. The study elucidates that effective digital transformation hinges on leaders' use of digital technologies for organisational growth. Various survey reports and instances of diverse leaders and organisations have been utilised to substantiate that the outcome reflects the leader's vision, foresight, and their application of digital technologies to enhance organisational efficiency.

Innovation in building is increasingly essential globally as contemporary social and environmental concerns necessitate swift and extensive actions. This dynamic landscape prompts inquiry into the façade industry's role in defining the future of architecture, engineering, and construction. Shahruddin & Husain (2024) conducted a critical examination of the façade sector through the lens of innovation, highlighting significant technological advancements in façade design, analysing key obstacles to sectoral innovation, and exploring new opportunities and research directions for reimagining future facades. It outlined primary avenues for innovation in the areas of 'product' and 'process,' propelled by emerging sustainability frameworks and facilitated by digital transformation.

Figure 1. Summary of literature themes

Figure 1 shows the five main ideas that came out of the literature review: Entrepreneurial Leadership in Dynamic Contexts, Sustainability and Environmental Responsibility, Innovation in Façade Systems, Digital Transformation and BIM Integration, and Stakeholder Collaboration and Co-creation. This theme synthesis demonstrates the interconnectedness of leadership, innovation, and sustainability within the built environment. Entrepreneurial leadership literature (for example, Renko et al., 2015; Gupta et al., 2004) talks a lot about attributes like having a vision and being willing to take risks. These are crucial for leading high-stakes, creative initiatives like adaptable façade systems. These ideas add to what has already been said about sustainability (Dean & McMullen, 2007; Darko & Chan, 2017), which stresses the need for more leaders who take environmental factors into account when making strategic choices. The topic of "Innovation in Façade Systems" shows how more and more buildings are using innovative building envelopes, including double-skin façades, solar panels, and dynamic shading systems. But most studies focus on technical factors, leaving out the leadership behaviours that make it easier for people to accept new ideas. Digital transformation and BIM integration became important tools for leaders to talk about the advantages of façade performance. BIM, IoT sensors, and energy modelling help with this. Finally, stakeholder collaboration stresses that entrepreneurial leadership works best when people from different backgrounds work together to create something new, such as architects, engineers, regulators, and building owners. Figure 1 is both a map of what is already known and the basis for the conceptual framework that this study builds on. It shows how these principles come together to create the leadership styles needed to make facade innovation last.

Building Information Modelling (BIM), the Internet of Things (IoT), and AI-enabled energy management have all added a digital layer to the responsibilities of leaders in façade systems. Now, leaders have to deal with complicated technology, training their employees, and performance goals based on data. Leaders who are open to digital transformation are more likely to get people to use new technologies faster and make sure that new ideas are in line with what customers want (Lu & Weng, 2018). As a result, the merging of digital and green innovation is both a challenge and an opportunity for leaders in businesses that focus on façades.

Figure 2. Conceptual model

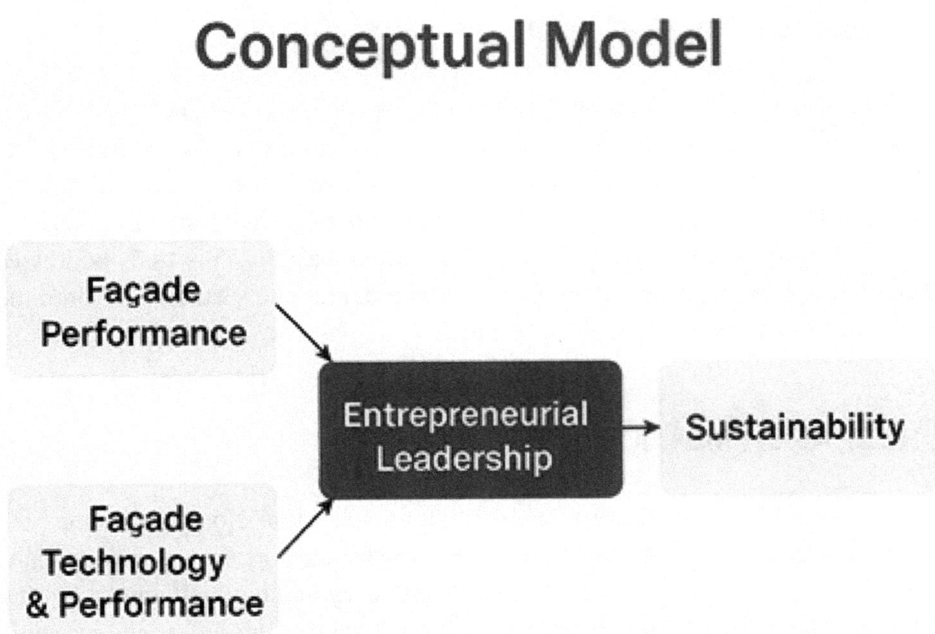

Figure 2 shows the conceptual model that brings together what we learnt from the literature review and what we found in our study. Entrepreneurial Leadership is the main force behind façade innovation, sustainability, and performance optimisation. The model shows that there are three main areas of leadership: visionary thinking, taking measured risks, and making decisions based on sustainability. These factors have a direct impact on the use of technical tools like Building Information Modelling (BIM), energy analytics, and IoT-based adaptable façades. Leaders may make decisions based on data and show stakeholders how much value they are getting from these technologies. The two case studies used here show that working together, with leadership actively including clients, regulators, and technical specialists, is necessary for the effective use of new façade technologies. This portion builds on conventional leadership ideas by focussing on participatory design and shared responsibility for sustainable results. The model's end goal is Enhanced Façade Performance, which is characterised in terms of both the environment (energy savings, a smaller carbon footprint) and the economy (lower lifespan costs, long-term value generation). So, entrepreneurial leadership becomes a strategic and unifying idea that brings together sustainability, innovation, and digital change in the physical environment. Figure 2 not only brings together different theoretical points of view, but it also provides a useful framework for future study and practice. It shows how leadership behaviours may lead to demonstrable gains in façade performance.

2.6 Research Gap

There is a lot of research on sustainable building and leadership theory on its own, but not much on the confluence of entrepreneurial leadership, façade innovation, and sustainability, especially in emerging countries. Most research on façade-related innovation looks at materials or performance indicators, but they generally don't look at the leadership systems that make such innovation feasible. This study addresses this gap by examining how entrepreneurial leaders integrate sustainability into the development of façade systems and business strategies.

3. RESEARCH METHODOLOGY

This research used a qualitative case study approach to look closely at the complex impact of entrepreneurial leadership in improving façade performance and sustainability. This is a subject that is both multidisciplinary and not well covered in the literature. The study's focus on understanding behaviours, attitudes, and strategic actions in real-world organisational settings makes this method effective. Semi-structured interviews gave us deep, detailed information about how entrepreneurial leaders make decisions, and more papers added depth and meaning to the data (Braun, V., & Clarke, V., 2006). Authors used thematic analysis to systematically find and explain commonalities across instances. The process helped us come up with theory-informed themes that show both how complicated the data is and how clear the ideas are. This strategy makes important contributions to the growing conversation about sustainable leadership and innovation in the built environment by putting depth ahead of breadth.

3.1 Research Design

This study used a qualitative exploratory research approach to look at how entrepreneurial leadership helps façade engineering stay innovative and sustainable. Because the issue is new and involves several fields, qualitative research allowed for the discovery of deep, context-specific insights from important players who are making changes in this area.

3.2 Research Approach

A multiple case study approach was adopted, with a focus on a small number of organisations that are actively working on sustainable façade innovation. These comprised new businesses, small and medium-sized enterprises (SMEs), and innovation-driven parts of major construction companies that work in the façade engineering industry. Using case studies made it easier to compare how leaders perform in different types of organisations and on projects of different sizes.

3.3 Sampling Strategy

Authors used purposive and snowball selection to choose participants to make sure that people who had direct experience managing façade innovation initiatives were included. Five business leaders were interviewed, including founders, directors, and project heads. All of them had proven experience in designing, developing, and putting into place façade systems that included sustainability and performance goals.

Authors chose two embedded case studies Green-Skin Façade Solutions and Aludecor Lamination Pvt Ltd to show two very different types of entrepreneurial leadership: a startup focused on sustainability and a small business that uses technology. Both companies have used new façade solutions and shown strategic leadership in promoting sustainability, getting stakeholders involved, and improving performance. These examples gave us a lot of information about the setting that helped us come up with themes and made the data more reliable.

3.4 Data Collection

The study relied on semi-structured interviews as the primary data source, allowing participants the flexibility to share experiences while ensuring thematic consistency across cases. Interviews lasted between 45-60 minutes and were conducted in-person or via video conferencing. To enrich the analysis, two case studies were also reviewed.

4. ANALYSIS AND DISCUSSION

The study's thematic analysis, backed up by two in-depth case studies, shows how entrepreneurial leadership can change the façade engineering industry by encouraging innovation and sustainability. The data showed five themes that were all connected: visionary leadership for green potential, measured risk-taking for innovation, collaborative stakeholder management, sustainability-driven value generation, and technology integration for performance. All of these characteristics help us better understand how entrepreneurial leaders start, manage, and make sustainable practices a part of the fabric of façade innovation.

Case Study 1: GreenSkin Façade Solutions (Startup in India)

Summary: GreenSkin Façade Solutions is a firm in Pune, India, that focusses on sustainability and makes modular, energy-efficient façade solutions for mid-rise commercial buildings. A group of architect-engineers started the firm in 2019. It used an entrepreneurial leadership strategy that focused on seeing early market opportunities, making experimental prototypes, and creating value that is good for the environment.

Relevance to Study: The creator showed clear visionary leadership by finding a niche in retrofittable façade solutions for Tier 2 cities. Even though not many people knew about the market, the team took a chance and made a hybrid bio-composite panel. They tested it for two years with a local university. The initiative also showed how to work together as a leader by getting city officials and real estate developers to help with the pilot testing phase.

Data Sources: Authors used an in-depth interview with the founder, product development reports, customer feedback forms, and marketing documents to find out about leadership methods, choices for sustainability, and ways to innovate.

Narration: GreenSkin Façade Solutions was started in 2019 because there was a growing need for eco-friendly retrofit solutions in India's Tier 2 cities. The creator, who was an architect before starting a business, saw a need for economical, modular façade retrofits that might increase energy performance without needing to change the structure.

"When we started, retrofitting was largely for looks. We thought it would be possible to create façades that minimise heat gain by 30% without changing the building's core". – Founder, GreenSkin.

Early pioneers concentrated on collaborative innovation by working with engineering universities to make and test eco-friendly panels composed of bamboo-reinforced bio-composites. The company invested significantly in educating customers by using computer simulations to demonstrate the long-term benefits of the product. Their leadership as entrepreneurs is shown in:

- First ideas about how to frame green retrofitting markets
- Taking risks with biomaterial research that are worth it
- Engaging stakeholders with cities and housing boards
- A value proposition that focuses on sustainability

The project's first big accomplishment was a makeover of a city library that cut energy use by 22% in its first year of operation.

Case Study 2: Aludecor Lamination Pvt Ltd (Small Business in India)

Summary: Aludecor Lamination is a mid-sized company that helps people with façades and provides solutions in India. The business has been in charge of a number of smart façade projects that use IoT sensors, dynamic shading systems, and Building information modelling (BIM) integration to keep an eye on performance and make changes in real time.

Relevance to Study: The CEO, who is a technopreneur with a background in mechanical engineering, showed risk-enabled innovation by investing in smart façades before these kinds of technologies were widely used in the area. He made digital transformation a significant competitive advantage by linking the use of technology to sustainability indicators like energy savings and thermal comfort. The company's business strategy, which was based on sustainability, focused on lifecycle value and meeting green building standards like LEED and GRIHA.

Data Sources: An interview with the CEO, internal project paperwork, IoT performance dashboards, and project reports from two high-rise commercial towers in Kolkata and Mumbai that are particular to the case.

Narration: Aludecor Lamination works in India, specialising in high-performance façades that leverage technology. The CEO of the company, who had experience with mechanical systems and automation, made the company a digital leader in façade systems. Their solutions include IoT sensors to change the shade in real time and BIM tools to coordinate the system and show how it will work over time.

"People don't simply want glass and panels; they want intelligence. Our façades react to sunshine, which is what sells."— CEO of Aludecor Lamination.

It was clear that the business had entrepreneurial leadership in:

- The company demonstrated technological foresight by investing in smart façades at an early stage.
- Taking risks with sensor integration when there isn't much demand
- There is significant emphasis on performance metrics and green certifications.
- Aligning tech and sustainability goals with leadership

Their installation of adaptable façades on a commercial skyscraper in Dubai cut cooling loads by 17% and became a case study for the Smart Building standards from the Dubai Municipality.

Table 1. Cross case comparison

Dimension	GreenSkin Façade Solutions	Aludecor Lamination Pvt. Ltd.
Leadership Orientation	Sustainability-driven, grassroots innovation	Technology-driven, performance-focused
Market Focus	Tier 2 cities, retrofit projects	Urban high-rise commercial buildings
Innovation Approach	Bio-composite materials, low-cost design	IoT sensors, AI-enabled smart façades
Key Risks Taken	New material experimentation	Investment in intelligent automation tools
Stakeholder Strategy	Community and public institutions	Corporate clients and international firms
Value Proposition	Lifecycle affordability + climate response	Real-time adaptation + energy savings
Key Performance Outcome	22% energy savings in retrofit project	17% drop in HVAC load in smart tower

Source: Authors' Compilation

4.1 Visionary Leadership for Green Opportunity

Both case studies showed that entrepreneurial leaders had visionary insight. The creator of GreenSkin Façade Solutions saw the untapped potential in cheap retrofit façades for communities that aren't in the metro area years before there was any regulatory or commercial impetus. This proactive approach to addressing environmental issues as strategic opportunities not only gave the company a competitive advantage, but also shaped its primary objective. Aludecor Lamination also showed that they were forward-thinking leaders by seeing how building intelligence and energy efficiency will come together. The CEO's idea to combine adaptable façade systems came before the rest of the industry became interested in smart building envelopes.

"We felt structures required smarter skins... We were already testing low-carbon façades before the green construction movement took off." Founder, GreenSkin

"Retrofit isn't sexy, but it's the future. We started small because we believed the regulations would catch up."— Founder, GreenSkin

"Everyone was busy chasing big cities, but we saw potential in mid-sized buildings in hot zones where thermal discomfort was the norm." CEO of Aludecor Lamination

This proactive way of thinking fits with what the research says about entrepreneurial leadership, which says that visionary framing is an important skill for taking advantage of new chances in uncertain situations (Renko et al., 2015).

4.2 Calculated Risk-Taking for Innovation

In both situations, the entrepreneurial leaders were prepared to take risks in order to come up with new ideas. GreenSkin put money into bio-composite materials even though there weren't many suppliers and the materials hadn't been tested in the field yet. The company's philosophy of iterative prototyping and being okay with failure helped it improve its product and gain trust in public-sector undertakings.

"We failed with two prototypes, but we learnt which materials reacted to heat and humidity." — Founder, GreenSkin

"We worked without a blueprint. Each façade we installed taught us something new about materials, people, and policy." CEO of Aludecor Lamination

"There was no precedent for using agricultural waste in façade panels—we had to write our own test standards." CEO of Aludecor Lamination

In contrast, Aludecor Lamination put a lot of money on IoT sensors and algorithm-based shading systems in the early stages, even though these technologies were mostly new to the people they were trying to sell to at the time. The CEO said that the risk was worth it since it would lead to long-term improvements in performance distinction. This concept aligns with other concepts that link entrepreneurial leadership to innovation, which is enabled by taking risks and exploring unconventional technological approaches (Gupta et al., 2004).

4.3 Collaborative Stakeholder Management

Both companies began using stakeholder engagement as a strategic leadership tool. GreenSkin was able to successfully retrofit a city library by using design processes that included engineers, architects, and city officials. Their style of leadership focused on building trust and involving people in decision-making, especially in places where people didn't know much about it. Aludecor Lamination worked with international developers, technology partners, and LEED consultants to create façade systems that are perfect for smart city projects. The CEO's ability to get all the different stakeholders on the same page about sustainability goals was a key factor in their success.

"We didn't just push our green facade; we worked with consultants, architects, and city officials to come up with a solution." — Founder, GreenSkin.

"Rather than pitch to clients, we sat with them—understood their constraints, and co-designed around them." CEO of Aludecor Lamination

"The municipality gave us land to test our panels—but only after we brought in engineers, professors, and even local activists to back the idea." — Founder, GreenSkin

This theme is about transformational and relational leadership, which means that success depends on getting many people to agree on a common vision of sustainability (Avolio & Bass, 2004).

4.4 Sustainability-Driven Value Creation

One of the most obvious differences between entrepreneurial leaders in this field is that they include sustainability in their basic business strategies. GreenSkin's consumer value proposition was based on being affordable and sensitive to the environment. This allowed them to work with public-sector clients that had strict budgets but still had to follow strict environmental rules. Aludecor Lamination included sustainability in their brand and sales strategy by showing off energy statistics, adaptive behaviour, and savings on the product's cost throughout its lifetime. This helped clients change their thinking about what was important, from upfront costs to long-term value. This was a big change for people who work in typical construction contracts.

"We always said, 'Pay a little more now and save on heating and cooling for 20 years.'" — CEO of Aludecor Lamination.

"We weren't just selling a product—we were telling a story about how even small buildings could be climate champions." — Founder, GreenSkin

"We chose not to compete on cost alone. Instead, we showed clients how our façades could repay themselves in three summers." — Founder, GreenSkin

This philosophy fits with the triple bottom line paradigm, which says that business executives should try to create economic, environmental, and social benefit at the same time (Elkington, 1997).

4.5 Technological Integration for Façade Performance

The last theme is about how technology can help leaders do their jobs better. Aludecor Lamination used BIM, real-time sensor data, and performance simulations to show clients how well their façades worked, which helped build trust and set them apart from other companies. GreenSkin used digital storytelling and energy modelling software to explain benefits and get approvals, even though they were less high-tech. In both cases, business leaders saw technology as more than just a tool; they saw it as a way to improve performance.

"Data helped us tell the story. We showed the building owners how our façades change every hour, and that won them over." — CEO of Aludecor Lamination.

"We made our façade smart enough to talk to the building's HVAC. That's the kind of leadership buildings now demand."— CEO, Aludecor Lamination

"We don't just install façades. We deliver energy reduction, carbon savings, and comfort—packaged in glass and steel." — Founder, GreenSkin

"Our competitive edge is that we sell performance, not just a product. That mindset came from how we lead internally." — Founder, GreenSkin

This case study shows how entrepreneurial leadership and digital transformation skills can work together, especially in developing economies where awareness and adoption are still growing.

4.6 Cross-Case Synthesis and Implications

Even though the two cases have different scales, markets, and levels of technology, they both show how entrepreneurial leadership can help make facades last longer. Vision, risk-taking, stakeholder inclusion, and tech fluency are key pillars. These results back up and build on existing theoretical models of sustainable entrepreneurial leadership, especially in applied engineering settings. The results show that, in practice, leadership in façade innovation cannot be separated from larger goals for sustainability. In this changing environment, business owners who make these values central to their plans, models, and customer interactions are more likely to succeed.

5. MANAGERIAL IMPLICATIONS

The study gives professionals in the façade and construction industry several useful tips: Leaders of businesses must think about how design innovation affects the environment, how it educates clients, and how it saves money in the long run. Startups and small and medium-sized businesses (SMEs) in the façade space should focus on co-creation strategies that involve a wide range of stakeholders, such as city regulators, engineers, sustainability consultants, and building users. Such collaboration will help make sure that new solutions are relevant and accepted. Also, using digital tools like BIM, IoT, and energy modelling should not just be seen as technical improvements, but also as strategic tools for telling stories, making decisions based on data, and standing out in the market. Policymakers and ecosystem enablers (like incubators, urban innovation hubs, and green building councils) should help these kinds of businesses by giving them more freedom with rules, pilot platforms, and networks for sharing information. Leadership development programs in engineering and architecture education also need to change so that students can learn how to be sustainable leaders and think like entrepreneurs.

6. THEORETICAL IMPLICATIONS

This study offers several theoretical contributions at the intersection of entrepreneurial leadership, sustainability, and façade engineering.

6.1 Expanding the Theory of Entrepreneurial Leadership

The results add to the body of research on entrepreneurial leadership by showing how leadership behaviours like visionary framing, taking calculated risks, and working together across sectors are used in very technical and sustainability-driven settings. Most existing models, like Gupta et al., 2004, and Renko et al., 2015, concentrate on entrepreneurial behaviour within general corporate environments. However, this study shows that these models are also useful in the constructed environment, where innovation is limited by rules, materials, and performance issues. The theme analysis shows that entrepreneurial leadership is a complex idea that comprises recognising opportunities, gathering resources, and creating value that lasts.

6.2 Combining Frameworks for Sustainability and Innovation

This chapter connects entrepreneurial leadership to façade performance and sustainability, filling in gaps between leadership philosophy and studies of sustainable innovation (Dean & McMullen, 2007). The results show that sustainability in façade systems is not just a technical or architectural issue; it is also heavily influenced by leadership initiatives that see environmental responsibility as a way to gain an edge over competitors. This study adds to the growing corpus of research on sustainable business, which says that leadership may provide both environmental and economic benefit.

6.3 Contribution to Façade Engineering Scholarship

Much façade engineering research prioritises material science, thermal performance, or digital modelling (Poirazis, 2006), frequently overlooking the human and leadership aspects of innovation. This study presents a conceptual paradigm (refer to Figure 2) that positions entrepreneurial leadership as a pivotal facilitator connecting enhancements in façade performance to overarching sustainability results. This introduces a strategic and organisational perspective to the literature on façade innovation.

6.4 Methodological Implications

Thematic analysis, in conjunction with case study data, illustrates how qualitative research methodologies may elucidate the intricate relationship among leadership behaviour, technological uptake, and sustainable practices. This methodological technique enhances theory-building by offering profound, contextual insights instead of only performance-based measures.

7. CONCLUSIONS

This chapter looked at how entrepreneurial leadership can help façade engineering become more sustainable and innovative. The study found five important aspects of leadership through thematic analysis of qualitative data and in-depth case studies. These were framing visionary opportunities, taking calculated risks, involving stakeholders, creating value that lasts, and integrating technology. These traits help entrepreneurial leaders push the limits of traditional façade design, turning façades from static parts of buildings into dynamic systems that improve building performance and the environment. These entrepreneurs are not only meeting market demands, but they are also actively shaping the future of sustainable construction by incorporating sustainability into their leadership styles and business models. The results add to the growing conversation about the links between entrepreneurial leadership, sustainable design, and new ideas in construction. They also show how vital it is to have a multidimensional leadership style that balances being good for the economy with being good for the environment and being able to use technology.

8. FUTURE RESEARCH DIRECTION

Future research could look at more areas by doing comparative studies across regions. For example, researchers could look at how entrepreneurial leadership practices in façade innovation differ between developed and developing economies. Longitudinal studies could also look at how leadership strategies change at different stages of a project, from coming up with ideas to looking at performance after installation. Subsequent study may integrate quantitative methodologies to assess the direct influence of entrepreneurial leadership on façade energy performance and material sustainability results. Longitudinal studies and performance-based measures would provide significant confirmation of the conceptual ideas articulated in this work. Quantitative studies that look at leadership metrics and façade performance indicators, such as thermal efficiency and energy savings, may also help confirm and expand on the ideas presented here. Further research is required to understand the impact of gender, organisational culture, and collaboration with individuals from diverse fields on entrepreneurial leadership in sustainable construction environments. By looking into these factors, all could learn more about how different types of leaders affect the paths of innovation in façade engineering and other fields.

ACKNOWLEDGMENT

The authors would like to express their sincere appreciation to the participating industry professionals and organisational representatives who contributed valuable insights through interviews and case-based inputs. Their practical perspectives significantly enriched the empirical depth of this study. Special acknowledgment is also extended to the institutions and research communities supporting advancements in façade engineering, sustainability, and entrepreneurial leadership.

REFERENCES

Al-Refaei, A. A. A., Abdulsamad, A., Ali, D. A., Ibrahim, A., Ateeq, A., & Al Balushi, F. K. (2024). A conceptual framework for the impact of entrepreneurial leadership on innovation work behavior and sustainable innovation performance. *Entrepreneurship Innovation and Education for Performance Improvement*, 577-598.

Attia, S., Lioure, R., & Declaude, Q. (2020). Future trends and main concepts of adaptive facade systems. *Energy Science & Engineering*, *8*(9), 3255–3272. DOI: 10.1002/ese3.725

Avolio, B. J., & Bass, B. M. (2004). *Multifactor Leadership Questionnaire: Manual and sampler set* (3rd ed.). Mind Garden.

Baburajan, M. (2025). Sustainable Sourcing: A comparative study of sourcing practices between Swedish and Qatari SMEs operating in the Façade industry.

Bagheri, A., & Harrison, C. (2020). Entrepreneurial leadership measurement: A multi-dimensional construct. *Journal of Small Business and Enterprise Development*, *27*(4), 659–679. DOI: 10.1108/JSBED-01-2019-0027

Bluyssen, P. M. (2010). Towards new methods and ways to create healthy and comfortable buildings. *Building and Environment*, *45*(4), 808–818. DOI: 10.1016/j.buildenv.2009.08.020

Boeske, J. (2023). Leadership towards sustainability: A review of sustainable, sustainability, and environmental leadership. *Sustainability (Basel)*, *15*(16), 12626. DOI: 10.3390/su151612626

Braun, V., & Clarke, V. (2006). Using thematic analysis in psychology. *Qualitative Research in Psychology*, *3*(2), 77–101. DOI: 10.1191/1478088706qp063oa

Darko, A., & Chan, A. P. C. (2017). Review of barriers to green building adoption. *Sustainable Development (Bradford)*, *25*(3), 167–179. DOI: 10.1002/sd.1651

Dean, T. J., & McMullen, J. S. (2007). Toward a theory of sustainable entrepreneurship: Reducing environmental degradation through entrepreneurial action. *Journal of Business Venturing*, *22*(1), 50–76. DOI: 10.1016/j.jbusvent.2005.09.003

Elkington, J. (1997). *Cannibals with forks: The triple bottom line of 21st century business*. Capstone Publishing.

Faton, C., & Hermans, J. (2025). Designing under Constraints: How Entrepreneurs Make the Future in Belgium's Construction Industry. Journal of Business Venturing Design, 100033.

Fernando, D., Navaratnam, S., Rajeev, P., & Sanjayan, J. (2023). Study of technological advancement and challenges of façade system for sustainable building: Current design practice. *Sustainability (Basel)*, *15*(19), 14319. DOI: 10.3390/su151914319

Gasparri, E., Brambilla, A., Lobaccaro, G., Goia, F., Andaloro, A., & Sangiorgio, A. (2022). Façade innovation: between 'product'and 'process. In *Rethinking Building Skins* (pp. 1–13). Woodhead Publishing. DOI: 10.1016/B978-0-12-822477-9.00025-5

Goh, C. S., Goh, H. H., Omar, M. F., & Toh, T. C. (2013). Leadership styles and project success in the construction industry. *International Journal of Project Management*, *31*(7), 998–1012.

Gupta, V., MacMillan, I. C., & Surie, G. (2004). Entrepreneurial leadership: Developing and measuring a cross-cultural construct. *Journal of Business Venturing*, *19*(2), 241–260. DOI: 10.1016/S0883-9026(03)00040-5

Hamann, R., Smith, J., Tashman, P., & Marshall, R. S. (2017). Why do SMEs go green? An analysis of wine firms in South Africa. *Business & Society*, *56*(1), 23–56. DOI: 10.1177/0007650315575106

Hargreaves, T. (2011). Practice-ing behaviour change: Applying social practice theory to pro-environmental behaviour change. *Journal of Consumer Culture*, *11*(1), 79–99. DOI: 10.1177/1469540510390500

He, L., Standen, P., & Coetzer, A. (2017). The perceived personal characteristics of entrepreneurial leaders. *Small Enterprise Research*, *24*(2), 97–119. DOI: 10.1080/13215906.2017.1334225

Hooi, H. C., Ahmad, N. H., Amran, A., & Rahman, S. A. (2016). The functional role of entrepreneurial orientation and entrepreneurial bricolage in ensuring sustainable entrepreneurship. *Management Research Review*, *39*(12), 1616–1638. DOI: 10.1108/MRR-06-2015-0144

Htet, A., Liana, S. R., Aung, T., & Bhaumik, A. (2023). Economic and environmental impact analysis of green building materials in facade engineering. *International Journal of Membrane Science and Technology, 10*(2), 2805–2814. DOI: 10.15379/ijmst.v10i2.2973

Ibrahim, A., Al-refaei, A. A. A., Alshuhumi, S., Al-Hidabi, D., & Ateeq, A. (2024). The effect of entrepreneurial leadership on employee's creativity and sustainable innovation performance in education sector: a literature review. *Business Development via AI and Digitalization: Volume 2*, 143-154.

Lahjiri, M., & Benhida, Z. (2025). Bridging the Sustainability Gap: A Behavioral Science Approach for Entrepreneurs and Managers. In Intersecting Corporate Governance, Financial Performance, and Long-Term Value Creation (pp. 293-320). IGI Global.

Lee, E., Selkowitz, S., Bazjanac, V., Inkarojrit, V., & Kohler, C. (2002). High-performance commercial building façades.

Lu, Y., & Weng, L. (2018). Smart construction: A new era of digital transformation. *Engineering, Construction, and Architectural Management, 25*(4), 495–513.

Nor-Aishah, H., Ahmad, N. H., & Thurasamy, R. (2020). Entrepreneurial leadership and sustainable performance of manufacturing SMEs in Malaysia: The contingent role of entrepreneurial bricolage. *Sustainability (Basel), 12*(8), 3100. DOI: 10.3390/su12083100

Poddar, S., & Guha, S. (2021). Daylight Illumination and Building Architecture-Effect at Workplace. [AMT]. *Advancement in Management and Technology, 2*(1), 11–16. DOI: 10.46977/apjmt.2021v02i01.002

Poirazis, H. (2006). Double skin façades for office buildings. *Energy and Building, 38*(3), 284–292.

Popoola, O., & Aluko, O. O. (2025). Entrepreneurial Enhancement through Commercial Architecture in Ondo State, Nigeria. *Journal of the Nigerian Institute of Town Planners, 30*(3), 43–63.

Prieto, A., Knaack, U., Auer, T., & Klein, T. (2019). COOLFACADE: State-of-the-art review and evaluation of solar cooling technologies on their potential for façade integration. *Renewable & Sustainable Energy Reviews, 101*, 395–414. DOI: 10.1016/j.rser.2018.11.015

Rae, D. (2017). Developing entrepreneurial leadership: The challenge for sustainable organisations. *International Journal of Work Innovation, 2*(1), 76–100. DOI: 10.1504/IJWI.2017.080725

Raiden, A., & King, A. (2021). Social value, organisational learning, and the sustainable development goals in the built environment. *Resources, Conservation and Recycling*, *172*, 105663. DOI: 10.1016/j.resconrec.2021.105663

Renko, M., El Tarabishy, A., Carsrud, A. L., & Brännback, M. (2015). Understanding and measuring entrepreneurial leadership style. *Journal of Small Business Management*, *53*(1), 54–74. DOI: 10.1111/jsbm.12086

Sainger, G. (2018). Leadership in digital age: A study on the role of leader in this era of digital transformation. *International Journal on Leadership*, *6*(1), 1.

Sangiorgio, A. (2022). Facade innovation: an industry perspective. In *Rethinking Building Skins* (pp. 15–58). Woodhead Publishing. DOI: 10.1016/B978-0-12-822477-9.00011-5

Shahruddin, S., & Husain, S. H. (2024). Navigating paradoxes of identity and leadership in the age of digital transformation of construction industry: Architects' experiences and perceptions. *Construction Management and Economics*, *42*(7), 591–609. DOI: 10.1080/01446193.2023.2293901

Valladares-Rendón, L. G., Schmid, G., & Lo, S. L. (2017). Review on energy savings by solar control techniques and optimal building orientation for the strategic placement of façade shading systems. *Energy and Building*, *140*, 458–479. DOI: 10.1016/j.enbuild.2016.12.073

Chapter 9
Façade Failures, Diagnostics, and Retrofitting Strategies:
Targeted Approaches for Sustainable and Extended Service Life

Arkar Htet
http://orcid.org/0000-0003-1301-3604
Lincoln University College, Malaysia

Theingi Aung
http://orcid.org/0009-0003-5063-575X
Lincoln University College, Malayisa

Nay Win
Mullion Facade Engineering School, Myanmar

Moe Min Min
Mullion Facade Engineering School, Myanmar

ABSTRACT

Façade deterioration affects energy performance, durability, comfort, and long-term building value. This chapter examines four recurring failure mechanisms-thermal bridging, sealant degradation, structural fatigue, and water ingress and explains how they develop through the interaction of exposure, detailing, material aging, and maintenance conditions. It reviews three practical diagnostic approaches: infrared thermography, drone-assisted visual inspection, and selective material sampling,

DOI: 10.4018/979-8-3373-6023-2.ch009

and proposes a staged workflow linking anomaly detection with technical confirmation. Building on this base, the chapter evaluates targeted retrofit strategies, including sealant rehabilitation, partial cladding replacement, thermal upgrades, and localized structural repair. Climate-responsive application is also considered to distinguish priorities in humid and temperate contexts. The chapter argues that sustainable façade renewal should be evidence-based, proportional, and lifecycle-oriented, extending service life while reducing waste, disruption, and unnecessary replacement.

1. INTRODUCTION

1.1 Role of Façades in Building Performance

Façades do more than enclose buildings. In contemporary construction, they must respond simultaneously to environmental, structural, operational, and architectural demands. Attia et al. (2018) note that façade systems have evolved from simple outer skins into performance-oriented assemblies, while Xu et al. (2021) show that their effectiveness depends on the interaction of glazing, insulation, cladding, joints, supports, and drainage rather than on any single component.

From a building-performance perspective, façades influence heat transfer, air leakage, moisture control, daylight, acoustics, and solar modulation. These effects shape indoor comfort, energy use, and maintenance demand. Wang et al. (2018) show that façade design choices can materially influence operational energy performance, especially in buildings with glazed or shaded envelope systems. A façade is therefore both a barrier and a long-term performance asset.

Façades also serve a protective and structural role. They must resist wind pressure, accommodate movement, and maintain functional integrity under environmental stress. Oesterle et al. (2014) describe façade engineering as the integration of architectural intent, material behavior, structural response, and environmental performance. This perspective is important because façade failure rarely results from a single defect; more often, it reflects the interaction of detailing, exposure, construction quality, and maintenance history.

Because façades affect durability, lifecycle cost, and sustainability, they should not be treated as superficial architectural features. They require a long-term management perspective rather than a purely reactive one. This view also aligns with emerging lifecycle and service-oriented approaches to building-envelope management (Aguerre et al., 2022).

1.2 Façade Degradation as a Lifecycle Problem

Façade degradation is unavoidable over time. Weather exposure, temperature fluctuation, moisture, ultraviolet radiation, pollution, and repeated structural movement gradually reduce the performance of materials, joints, and interfaces. In many cases, deterioration develops slowly through minor signs such as cracking, staining, sealant loss, local thermal anomalies, or intermittent leakage before it becomes a major defect. Hosseinalipour et al. (2024) suggest that early-stage underperformance, if ignored, can develop into more serious envelope failures with broader implications for comfort, durability, and maintenance cost.

Environmental exposure is one of the main drivers of this process. Repeated wetting and drying can weaken materials, encourage biological growth, and accelerate corrosion. Ultraviolet radiation can contribute to material aging, especially in polymeric envelope materials, while thermal cycling produces repeated expansion and contraction that can stress joints and supports. Riahinezhad et al. (2021) show that the long-term durability of polymer-based building-envelope materials depends heavily on exposure conditions, while Barrelas et al. (2021) similarly demonstrate that the service life of façade claddings is closely tied to environmental conditions and maintenance history.

This makes façade degradation a lifecycle issue rather than a simple repair issue. A purely reactive approach deals with visible damage after performance has already declined, whereas a lifecycle approach emphasizes early diagnosis, preventive intervention, and service-life extension. The broader influence of climatological stress on multiple building-system deficiencies, as noted by Mydin et al. (2024), reinforces the need to treat façade deterioration as part of long-term asset management rather than isolated defect correction.

1.3 Research and Practice Gap

Research on façade systems has expanded substantially, particularly in energy performance, adaptive envelopes, material innovation, and environmental response. Yet a gap remains between academic knowledge and maintenance decision-making in practice. Much of the literature is design-oriented or simulation-based, while real projects are shaped by constraints such as access, cost, uncertainty, safety, and operational continuity. As a result, research findings are not always translated into practical intervention frameworks.

Another limitation is fragmentation. Thermal bridges, sealant failure, structural fatigue, moisture ingress, diagnostic methods, and retrofit strategies are often discussed separately even though they are closely linked in real façade problems. A leaking interface may be related to joint failure, drainage defects, or movement. A thermal anomaly may indicate missing insulation, moisture, or flawed detailing. Treating these issues in isolation weakens the usefulness of the literature for diagnosis and repair planning.

This fragmentation is particularly evident in retrofit research. Many studies examine upgrading strategies, but fewer connect failure mechanisms, diagnostic evidence, and selective intervention in one coherent framework. Recent work, including Htet et al. (2025) and Luo et al. (2026), points to the need for more integrated approaches linking inspection findings with service-life and repair decisions. This chapter responds to that need by combining failure analysis, diagnostics, and retrofit strategy within a lifecycle perspective.

1.4 Chapter Scope and Structure

This chapter focuses on four recurring façade failure mechanisms that have strong implications for performance and service life: thermal bridging, sealant degradation, structural fatigue in connections and panels, and water ingress through joints and interfaces. These failure types were selected because they are common across many building types and because they frequently trigger wider durability or performance problems if not addressed in time (Straube & Burnett, 2005; Oesterle et al., 2014).

To support intervention planning, the chapter examines three practical diagnostic categories: infrared thermography, drone-assisted visual inspection, and selective material sampling and testing. These methods are not presented as isolated tools. Instead, they are treated as parts of a staged diagnostic workflow. The retrofit discussion then focuses on targeted interventions, especially sealant rehabilitation, partial cladding replacement, thermal upgrades, and localized structural repair. Together, these sections aim to show how façade renewal can be made more selective, evidence-based, and sustainable. In that broader sustainability context, circular-economy thinking in the built environment also remains relevant when discussing long-term resource value and service-life extension (Pomponi & Moncaster, 2017).

The chapter is structured as follows. Section 2 examines common façade failure mechanisms. Section 3 discusses diagnostic tools and integrated assessment. Section 4 evaluates retrofit strategies for service-life extension. Section 5 considers how these strategies vary across climatic and façade contexts. Section 6 presents best-practice recommendations, and Section 7 concludes the chapter.The overall logic of the chapter, linking façade failures, diagnostics, and retrofit strategies, is summarized in Figure 1.

Figure 1. Conceptual framework linking façade failure mechanisms, diagnostic methods, and retrofit strategies for service-life extension

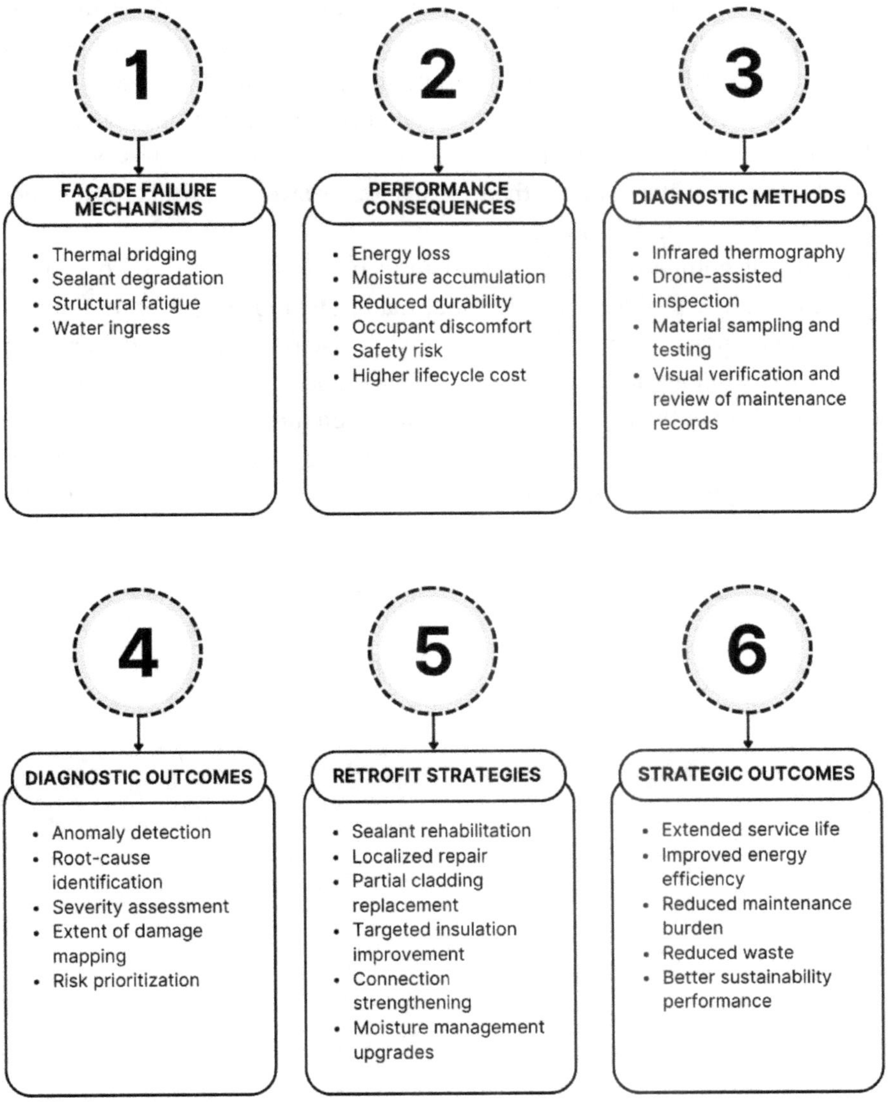

(Developed by the author based on the chapter synthesis)

2. COMMON FAÇADE FAILURE MECHANISMS

Façade systems are exposed to a demanding combination of environmental, structural, and operational stresses throughout their service life. Unlike many interior components, they must perform under fluctuating temperature, moisture, wind, ultraviolet radiation, pollution, and building movement. Souza et al. (2024) note that façade degradation is strongly influenced by degree of exposure, while Sandak et al. (2019) remind us that building skins should be understood as long-term performance systems rather than passive outer layers. For this reason, façade failure is often cumulative and multi-causal.

Recognizing common failure mechanisms is important because early symptoms can point to deeper problems in detailing, materials, movement accommodation, or maintenance practice. The four mechanisms discussed in this section thermal bridging, sealant degradation, structural fatigue, and water ingress were selected because they frequently appear in practice and often interact with one another. For clarity, Table 1 summarizes the main façade failure mechanisms, their common causes, typical symptoms, performance impacts, and suitable diagnostic methods.

Table 1. Summary of common façade failure mechanisms, causes, symptoms, impacts, and suitable diagnostic methods

Failure mechanism	Common causes	Typical symptoms	Main impacts	Suitable diagnostic methods
Thermal bridging	Discontinuous insulation, conductive subframes, poor detailing at junctions, installation defects	Localized cold or hot spots, condensation marks, mold growth, unexplained heat loss	Higher energy consumption, condensation risk, reduced thermal comfort, localized material deterioration	Infrared thermography, visual inspection, selective opening-up
Sealant degradation	UV exposure, thermal cycling, movement fatigue, pollution, material incompatibility, poor joint preparation	Cracking, hardening, debonding, joint gaps, discoloration	Water ingress, air leakage, reduced envelope continuity, reduced durability	Close visual inspection, drone imaging for upper zones, material sampling, joint probing
Structural fatigue in connections and panels	Repeated wind loading, thermal movement, building sway, corrosion, inadequate movement allowance, installation defects	Loose panels, rattling, local deformation, joint widening, corrosion marks, panel misalignment	Safety risk, loss of stiffness, detachment risk, associated water ingress	Drone inspection, direct access inspection, material/ fastener sampling, corrosion assessment
Water ingress through joints and interfaces	Failed sealants, poor flashing, blocked drainage, capillary paths, pressure differences, aging details	Damp patches, staining, mold, efflorescence, wet insulation, corrosion	Durability loss, indoor air quality issues, insulation performance reduction, hidden deterioration	Infrared thermography, moisture-oriented inspection, joint inspection, targeted opening-up

Source: Developed by the author based on synthesized literature on façade deterioration, diagnostics, and rehabilitation.

2.1 Thermal Bridging

Thermal bridging refers to localized areas in the façade where heat transfer is greater than intended because of conductive penetrations, material discontinuities, or defects in the insulation layer. These zones reduce thermal efficiency and can also create condensation and comfort problems. Ujma and Pomada (2019) show that thermal bridges are especially common around façade fixings, metallic subframes, brackets, anchors, and complex junctions.

Poor detailing is one of the main causes. Gaps in insulation, weak thermal breaks, and poorly resolved interfaces between façade and structure can create unintended heat-flow paths. Construction quality further affects the problem. Fylan and Glew (2021) note that deviations in installation can produce significant performance gaps even where the design itself is conceptually sound.

The implications extend beyond energy loss. Bienvenido-Huertas (2020) argues that thermal bridges increase heating and cooling demand and also raise environmental impact at building scale. Low internal surface temperatures can create condensation, which in turn contributes to mold growth, staining, and local material deterioration. Petresevics and Nagy (2022) further show that fastening systems in ventilated façades can significantly affect localized thermal transmittance.

Thermal bridging also affects occupant comfort. Buday et al. (2014) demonstrated that inadequate thermal breaks in projecting building elements can lower internal surface temperature and increase local discomfort. In façade systems with extensive glazing or metallic supports, these localized effects may be particularly noticeable. The thermal response of façade surfaces is also influenced by optical properties such as reflectance and solar interaction, as shown by Azarnejhad and Mahdavi (2019), although this becomes more relevant at broader façade-design level than at defect-diagnosis level.

Detection is often difficult through visual observation alone. While condensation marks or staining may be visible, many thermal anomalies remain concealed within the assembly. For this reason, infrared thermography is frequently used to identify suspect zones. Mirzabeigi et al. (2025) show that vision-based and thermal anomaly detection methods can reveal hidden problem areas that would otherwise remain unnoticed. Alhawari and Mukhopadhyaya (2018) similarly note that thermal bridges should be considered both a performance problem and a durability concern.

2.2 Sealant Degradation

Sealants are critical to façade performance because they help maintain weather tightness, air tightness, and movement accommodation at joints and interfaces (Htet et al., 2025a). Elguezabal and Garay (2015) describe them as key components in envelope upgrading because their condition directly affects leakage resistance and continuity. Despite this importance, sealants are among the most vulnerable façade elements because they are continuously exposed to ultraviolet radiation, movement cycles, pollution, and moisture.

Prolonged UV exposure is a major cause of deterioration. Over time, sealants may harden, discolor, crack, or lose adhesion. Asphaug et al. (2021), in their work on façade moisture exposure, show that climatic aging conditions can strongly affect façade performance and moisture accumulation. Mechanical fatigue is another major factor. Repeated thermal movement, wind action, and settlement gradually stress sealant joints, increasing the likelihood of cohesive or adhesive failure. Jiao et al. (2021) also demonstrate that mechanical properties in sealant materials are highly dependent on formulation and behavior under stress.

Installation quality and compatibility are equally important. Poor joint geometry, contaminated substrates, unsuitable primers, or incompatible adjacent materials can reduce service life substantially even when the sealant product itself is nominally appropriate. Ding and Liu (2006) emphasized that durability must be assessed in relation to aging conditions rather than product identity alone.

The consequences of sealant degradation are broad. Failed joints permit air leakage and water entry, which may lead to wet insulation, corrosion, staining, reduced thermal performance, and lower indoor environmental quality. Although the specific study by Sivaji et al. (2022) is not a core façade durability reference, it still supports the broader point that façade material performance should be evaluated comparatively rather than assumed. In practice, however, sealant failure is better understood as a trigger for wider envelope underperformance rather than as an isolated joint defect.

2.3 Structural Fatigue in Connections and Panels

Structural fatigue in façades develops through repeated mechanical stress acting on anchors, brackets, fasteners, subframes, and panels. It is typically progressive rather than sudden, beginning with local stress concentrations, looseness, or micro-damage and gradually developing into deformation, instability, or connection weakness. Bhering et al. (2023) show that interface performance under aging conditions can deteriorate in ways that are not immediately visible but become critical over time.

Façade systems are particularly vulnerable because they are repeatedly exposed to wind loading, movement, and temperature variation. Bedon and Amadio (2017) note that glazed and lightweight façade systems often rely on slender support arrangements that can be sensitive to repeated stress. Popovic and Vla (2020) similarly point out that façade assessment and repair must consider the cumulative effects of loading and deterioration rather than only visible damage.

Thermal cycling adds to this problem by imposing repeated expansion and contraction at support points and interfaces. Widomski et al. (2025) show that climatic stress can degrade the performance of structural materials, especially when environmental exposure is prolonged and severe. When movement is restrained or detailing is poor, repeated cycles may gradually weaken fixings and support arrangements.

Corrosion intensifies fatigue risk. Moisture, condensation, and aggressive exposure can reduce the effective cross-section of metallic elements and therefore weaken their resistance to repeated loading. Straube and Burnett (2005) emphasize that moisture-related deterioration in enclosure systems can turn manageable serviceability defects into serious performance and safety issues if left undetected.

Installation and maintenance also matter. Sezegen and Edis (2020) argue that façade product innovation must be considered alongside practical behavior in use, which indirectly reinforces the need to evaluate actual installation and long-term performance rather than nominal design only. In practice, misalignment, inadequate movement allowance, and poor fastening can introduce residual stresses from the start, making later deterioration more likely.

2.4 Water Ingress through Joints and Interfaces

Water ingress is one of the most common and damaging façade failure mechanisms. It occurs when protective layers, drainage pathways, or interface details fail to prevent or manage moisture penetration. Straube and Burnett (2005) stress that successful enclosure design does not rely on a single waterproof surface; it depends on a system of layers, cavities, flashings, sealants, and barriers that together manage water entry and drainage.

Several mechanisms contribute to this problem. Water may move through capillary paths, enter through poorly sealed joints, or be driven inward by pressure differences during wind-driven rain. Linden and Bossche (2025) demonstrate that ventilated façade systems with open joints can be highly sensitive to rainwater intrusion when detailing and pressure response are not well controlled. Similar hygrothermal concerns under present and future climate conditions have also been discussed by Bas et al. (2023), whose work reinforces the importance of climate-responsive envelope assessment.

Drainage failure is another frequent cause. When weep holes are blocked, cavities are poorly designed, or moisture barriers are discontinuous, infiltrated water may become trapped within the assembly. Baptista et al. (2024) emphasize that façade system selection should consider lifecycle effects, which is relevant here because trapped moisture can reduce durability, weaken materials, and increase future intervention needs. Construction quality remains decisive. Douglas and Ransom (2013) show that failures often emerge from detailing defects and workmanship issues that remain hidden until environmental exposure reveals them. Aging further increases vulnerability, as joints lose elasticity and materials become less resilient.

The consequences are substantial. Wet insulation, corrosion, mold, staining, and reduced thermal performance are common outcomes. Asphaug et al. (2021) show that moisture accumulation in façades can escalate quickly under severe exposure conditions. For that reason, water ingress should be treated as a major durability and management issue rather than only a localized leakage problem.

3. DIAGNOSTIC TOOLS AND ASSESSMENT METHODS

Reliable diagnosis is the basis of effective façade intervention. Many façade problems begin within concealed layers or at interfaces that cannot be assessed accurately through visual observation alone. A sound assessment process must therefore gather evidence in stages and combine methods according to the type of defect involved. Dias et al. (2021) argue that technically informed rehabilitation depends on identifying probable causes rather than reacting only to visible symptoms. Pereira et al. (2020) make a similar point by showing that organized inspection systems improve consistency when defects, diagnosis methods, and repair choices are linked systematically. This is also consistent with broader work on building inspection systems that stresses structured evaluation of technical performance in existing buildings (Bortolini & Forcada, 2018).

This chapter focuses on three practical diagnostic categories: infrared thermography, drone-assisted visual inspection, and selective material sampling and testing. These methods were selected because they are widely applicable in façade practice and because they provide complementary evidence. Used together, they can help bridge the gap between visible deterioration and defensible retrofit decisions. The growing emphasis on façade inspection also reflects safety concerns, especially in tall buildings where delayed maintenance can increase the risk of falling-object incidents and related hazards (Chew & Gan, 2022).

3.1 Infrared Thermography for Thermal and Moisture Anomalies

Infrared thermography is one of the most useful non-destructive tools in façade diagnostics because it can identify surface temperature variations associated with hidden defects. Ficapal and Mutis (2019) describe it as a method for detecting thermal irregularities linked to heat loss and envelope underperformance. In façade applications, this makes it valuable for identifying thermal bridges, insulation discontinuities, moisture-affected zones, and air leakage paths.

Its main strength lies in its ability to detect anomalies that may not yet be visible. Nardi et al. (2018) explain that infrared thermography is particularly useful for evaluating heat losses through building envelopes and for guiding intervention priorities. Tejedor et al. (2017) similarly show that in-situ infrared thermography can support the assessment of façade thermal behaviour under real conditions. In practical façade assessment, this means that missing insulation, conductive supports, or damp areas can often be located before secondary damage becomes obvious.

Thermography is also useful in moisture diagnosis. Mendes et al. (2022) show that moisture alters surface thermal behavior and can therefore be identified through thermographic patterns when interpretation is supported by contextual knowledge of materials and exposure. This is especially relevant where water ingress remains concealed until staining or mold develops. Moisture diagnosis in building components has also been discussed more broadly by Barreira and Delgado (2017), whose work supports the use of non-destructive techniques in identifying hidden damp-related problems.

Another advantage is efficiency. Large façade areas can be screened relatively quickly, which helps identify priorities for closer inspection. Kim et al. (2023) note that infrared thermography has become increasingly important in building diagnostics because it supports rapid, non-invasive assessment of insulation defects, moisture intrusion, and related anomalies.

However, the method has limitations. Thermal readings are highly sensitive to weather conditions, emissivity, solar radiation, reflections, and timing. Ficapal and Mutis (2019) and Nardi et al. (2018) both emphasize that thermography requires careful interpretation and cannot by itself confirm the exact physical cause of every anomaly. For that reason, it should be treated as a screening and diagnostic support tool rather than a stand-alone verdict.

3.2 Drone-Assisted Façade Inspection

Drone-assisted inspection has become increasingly valuable in façade assessment, particularly for tall or difficult-to-access buildings. Falorca and Lanzinha (2020) show that drones can provide a practical alternative to conventional access systems during preliminary or periodic inspections. Their main advantage is improved access to elevations, parapets, corners, and other hard-to-reach zones.

With high-resolution imaging, drones can record cracks, open joints, staining, panel displacement, corrosion marks, and other visible symptoms. Dias et al. (2021) note that recent inspection technologies have improved both the reach and quality of façade documentation, especially when paired with structured assessment methods.

Drone inspection also supports long-term documentation. Repeated image capture can help track deterioration over time and improve defect mapping. Mendes et al. (2022) demonstrate how image-based recording and façade mapping can support systematic condition assessment in a non-destructive way.

Safety is another advantage. Drones reduce the need for immediate manual access during initial surveys, which can be especially useful where façade conditions are uncertain. Ficapal and Mutis (2019) further show that drone platforms can be paired with thermal imaging to strengthen large-area screening.

Yet limitations remain. Drone inspection is primarily surface-based unless paired with thermal or other advanced sensing. It cannot by itself confirm concealed moisture, hidden material decay, or internal connection problems. Falorca and Lanzinha

(2020) caution that drones should therefore be integrated into a broader inspection workflow rather than treated as complete substitutes for technical examination.

3.3 Material Sampling and Testing

Although non-destructive methods are increasingly advanced, some façade conditions still require selective material sampling and testing. This is particularly true where there is uncertainty about sealant aging, corrosion, compatibility, hidden moisture, or localized structural degradation. In such cases, sampling provides direct evidence that can confirm or refine conclusions drawn from imaging or visual inspection.

Sampling may involve removing sealant specimens, opening selected joints, extracting small pieces of cladding or substrate, or examining fasteners and support elements. Dias et al. (2021) suggest that advanced and conventional methods are often most effective when used together, especially where the mechanism of deterioration remains uncertain after non-destructive screening.

This method is particularly useful in confirming whether deterioration is localized or systemic. That distinction matters because it directly affects retrofit scope. Where the problem is confined, targeted intervention may be enough. Where degradation is widespread or material incompatibility is involved, broader replacement may be justified. Sampling also has clear limitations. It is partly invasive, can require temporary removal of façade elements, and should therefore be used strategically rather than routinely. The best practice is to select sampling locations based on prior evidence from visual review, thermography, or known defect history.

3.4 Integrated Diagnostic Workflow

In practice, façade problems are rarely diagnosed well through one method alone. More reliable assessment comes from combining tools in sequence, with each method contributing a different form of evidence. Dias et al. (2021) emphasize that effective façade diagnosis must connect observed anomalies with appropriate diagnostic techniques and later repair decisions. Pereira et al. (2020) likewise show that inspection systems become more useful when defect identification, verification, and intervention logic are structured together. The value of integrating non-destructive techniques with broader diagnostic interpretation has also been demonstrated in building-diagnosis studies such as Diz-Mellado et al. (2021).

A practical workflow begins with preliminary review. Drawings, maintenance records, repair history, complaints, and exposure conditions help identify likely defect zones before site investigation starts. Douglas and Ransom (2013) and Straube and Burnett (2005) both show that building failure analysis is stronger when historical and contextual information is considered early. The next stage is broad visual screening, supported where needed by drone-assisted imaging. This identifies visible cracks, staining, open joints, displacement, or surface deterioration and helps prioritize where deeper investigation is needed. Where access is difficult, drone inspection can significantly improve coverage.

The third stage is anomaly detection through thermography and related non-destructive screening. At this point, hidden thermal or moisture irregularities can be mapped and compared against visible symptoms. Ficapal and Mutis (2019) and Nardi et al. (2018) both support the value of this step in identifying likely problem zones before invasive testing.

The fourth stage is targeted confirmation. This may include close-up inspection, probing, selective opening-up, or material sampling. At this point, the aim is to move from suspicion to technical confidence by identifying the actual mechanism involved. Dias et al. (2021) argue that this is the stage at which a technically informed rehabilitation strategy becomes possible.

The final stage is diagnostic synthesis. Evidence from review, visual inspection, thermal screening, and targeted confirmation is brought together to define likely causes, severity, extent, and urgency. A good diagnostic process should therefore end not only in defect description, but in repair logic. This is what allows retrofit to remain selective, proportionate, and evidence-based. The staged relationship between review, screening, confirmation, and intervention planning is illustrated in Figure 2.

Figure 2. Integrated façade diagnostic workflow for evidence-based maintenance and retrofit planning

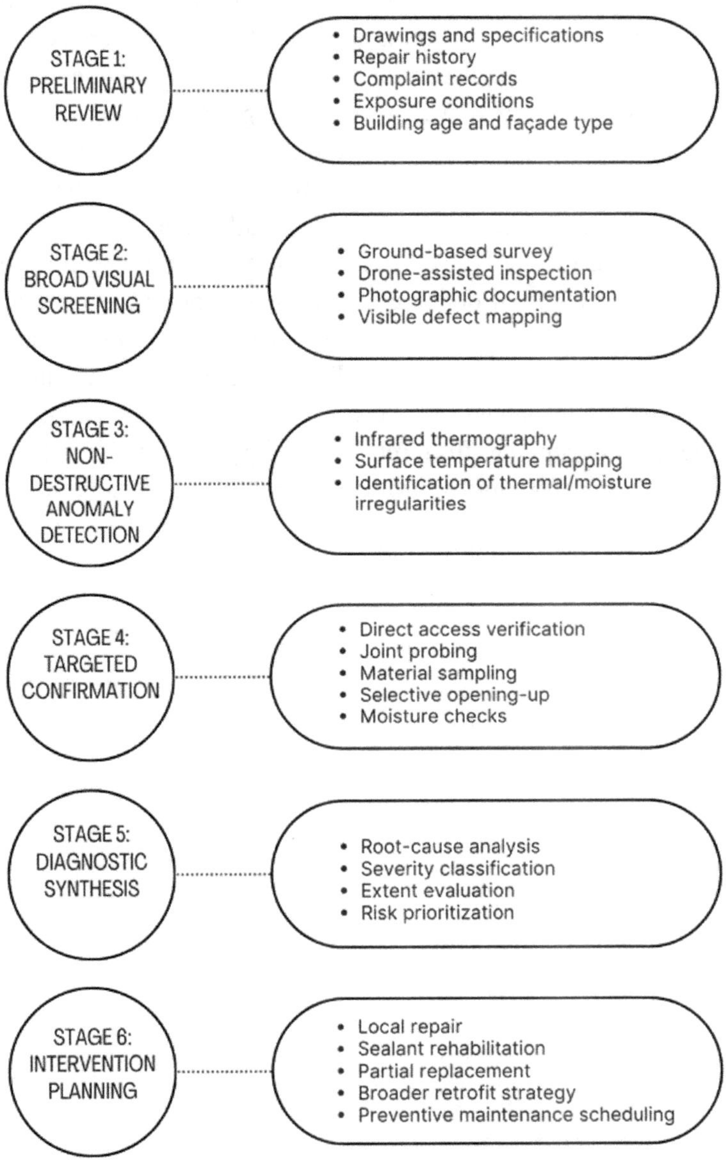

(Developed by the author based on integrated façade diagnostic principles)

3.5 Strengths and Limitations of the Main Diagnostic Methods

Each diagnostic method discussed above has clear value, but none is sufficient in isolation. Infrared thermography is highly effective for identifying hidden thermal and moisture anomalies, but interpretation depends heavily on survey conditions and technical expertise. Drone inspection improves access and documentation, but remains mainly surface-oriented. Material sampling provides strong confirmation, but is localized and partly invasive. Dias et al. (2021) and Kim et al. (2023) both indicate that modern building diagnostics are most reliable when they rely on complementary methods rather than single-technique judgment.

In practical terms, thermography is strongest as a screening tool, drones are strongest in access and visual documentation, and sampling is strongest in confirmation. Their real value lies in combination. This is important because retrofit quality depends directly on diagnostic quality. A misread thermal anomaly or an unverified visible defect can lead to unnecessary replacement or incomplete repair. Diagnosis, then, is not separate from retrofit strategy. It is the basis for it. Table 2 compares their purposes, strengths, limitations, and best-use scenarios.

Table 2. Comparative overview of key façade diagnostic methods

Diagnostic method	Main purpose	Key strengths	Main limitations	Invasiveness	Best use cases
Infrared thermography	Detect thermal anomalies, insulation defects, moisture-related irregularities	Non-destructive, rapid, broad coverage, useful for hidden anomalies	Sensitive to weather, emissivity, reflections, and interpretation quality	Non-invasive	Thermal bridging, moisture suspicion, envelope performance screening
Drone-assisted inspection	Visual assessment of difficult-to-reach façade areas	Safer access, rapid documentation, useful for high-rise façades, repeatable image records	Primarily surface-based, may miss concealed defects, subject to weather/ regulations	Non-invasive	Crack mapping, joint review, staining, panel displacement, initial condition survey
Material sampling and testing	Confirm material degradation, aging, compatibility, corrosion, and concealed failure	High diagnostic confidence, direct evidence, supports precise intervention decisions	Localized, partly destructive, more time-consuming, not suitable for whole-façade screening	Semi-invasive	Sealant failure confirmation, corrosion assessment, moisture-damaged build-up verification
Direct access close-up inspection	Detailed on-site observation of critical locations	Strong visual confirmation, supports tactile inspection and precise condition judgment	Access cost, safety considerations, slower for large buildings	Low to moderate	Confirmation of defects identified during screening, joint and fixing inspection

Source: Developed by the author based on the diagnostic methods discussed in this chapter.

3.6 Transition to Retrofit Strategy Selection

Once façade deterioration has been identified and interpreted, the next step is to define the level of intervention required. This is where diagnosis becomes operational. Thermal bridging may support local insulation improvement or detail correction; sealant failure may justify joint rehabilitation; widespread moisture or support deterioration may point toward broader renewal. The following section examines how retrofit strategies can be selected in relation to defect type, service-life extension, and sustainability.

4. RETROFITTING STRATEGIES FOR FAÇADE RENEWAL AND SERVICE LIFE EXTENSION

Façade retrofitting is no longer treated only as a response to visible deterioration. In current practice, it is increasingly understood as a strategic intervention that can improve durability, restore performance, reduce maintenance demand, and extend service life. Because façades strongly influence heat transfer, moisture control, comfort, and long-term operational quality, retrofit decisions must balance technical need with environmental, economic, and practical considerations (Kamel & Memari, 2022; Dadzie et al., 2019). This is especially important in existing buildings, where full replacement may not always be justified.

A key principle in façade rehabilitation is proportionality. Some defects can be addressed through local repair or joint renewal, while others indicate broader underperformance requiring more extensive upgrading. Intervention should therefore follow verified condition rather than default assumptions about replacement. As shown in Table 3, retrofit strategies differ in scope, trigger condition, disruption level, and sustainability contribution, making strategy selection a matter of judgment rather than routine prescription.

Table 3. Comparison of façade retrofit strategies by scope, suitability, disruption, and sustainability contribution

Retrofit strategy	Typical trigger condition	Scope of work	Main strengths	Main limitations	Disruption level	Sustainability contribution
Sealant rehabilitation and joint renewal	Cracked, hardened, debonded, or leaking joints with otherwise serviceable façade components	Localized	Low cost, fast, low material use, restores weather tightness and air tightness	Limited effect if deeper moisture or substrate damage exists	Low	High when deterioration is truly joint-based
Partial cladding replacement	Localized panel damage, recurrent leakage in specific areas, corrosion around fixing zones, limited detachment risk	Moderate/localized	Preserves serviceable façade areas, allows targeted upgrading, reduces waste compared with full replacement	Requires careful compatibility management and accurate damage mapping	Medium	Moderate to high if retained façade areas remain durable
Thermal performance upgrade	Significant heat loss, thermal bridging, poor insulation, condensation-related envelope problems	Moderate to broad	Improves energy performance, comfort, and often durability	Risk of unintended moisture consequences if poorly designed	Medium to high	High if climate-responsive and properly detailed

continued on following page

Table 3. Continued

Retrofit strategy	Typical trigger condition	Scope of work	Main strengths	Main limitations	Disruption level	Sustainability contribution
Localized structural strengthening	Fatigued or corroded anchors, bracket weakness, connection-related deformation or safety concern	Localized to moderate	Addresses safety-critical support issues without full façade replacement	Needs close engineering review and may not suit widespread support deterioration	Medium	Moderate to high when it preserves the main façade system
Full façade replacement / total recladding	Systemic failure, widespread deterioration, very limited residual service life, major code/performance gap	Broad	Allows comprehensive redesign and high performance improvement	High cost, high waste, major disruption, higher embodied impact	High	Variable; justified mainly when systemic failure makes selective intervention inadequate

Source: Developed by the author based on synthesized literature on façade retrofitting and service-life extension.

4.1 Principles of Targeted Façade Retrofitting

Targeted retrofitting begins with the idea that interventions should respond to confirmed failure mechanisms rather than visible symptoms alone. This distinction matters because façade distress is often cumulative. A stained joint may suggest sealant failure, but it may also indicate drainage defects or recurring moisture exposure behind the cladding. Similarly, localized thermal discomfort may result from thermal bridges, air leakage, glazing weakness, or insulation defects. For this reason, retrofit choice should be linked directly to diagnosis rather than separated from it (Dias et al., 2021; Pereira et al., 2020).

Another important principle is service-life extension. In many existing buildings, large portions of the façade still retain useful performance even when specific joints, panels, or interfaces have begun to deteriorate. Replacing the entire façade in such cases may introduce unnecessary cost, waste, and embodied carbon. Procaccini and Monticelli (2024) argue that retrofit strategies should increasingly support resilience through selective intervention, while Dadzie et al. (2019) likewise emphasize the value of upgrading viable assets rather than replacing them prematurely. This way of thinking is also compatible with circular-economy perspectives in the built environment, which emphasize retaining value in existing building systems rather than relying only on replacement-based logic (Pomponi & Moncaster, 2017).

Practical execution must also be considered. Existing buildings are often occupied during retrofit works, so access, safety, dust, noise, weather exposure, and installation speed all influence the choice of intervention. Chung-Camargo et al. (2024) note that retrofit planning must reflect local climate and implementation realities, not only technical ambition. In façade projects, the best strategy is often the one that improves performance while remaining manageable in cost, access, and disruption.

4.2 Sealant Rehabilitation and Joint Renewal

Sealant rehabilitation is one of the most common and most efficient retrofit strategies where deterioration is concentrated at joints, perimeter interfaces, or movement gaps. Because sealants are central to air and water control, their failure can lead to leakage, reduced thermal performance, corrosion, and secondary material deterioration. Where the surrounding façade remains sound, targeted joint renewal can restore performance without major replacement work (Beech & Beasley, 1994; Ding & Liu, 2006).

Its main strength lies in proportionality. Instead of replacing serviceable cladding or glazing, the intervention focuses on the most vulnerable lines of defense. When diagnosis confirms that deterioration is concentrated in cracked, hardened, debonded, or chemically aged sealants, careful removal, substrate preparation, compatibility review, and reapplication can significantly improve envelope continuity (Elguezabal & Garay, 2015; Ding & Liu, 2006).

Even so, sealant renewal should not be treated as a cosmetic operation. Its long-term success depends on joint geometry, movement allowance, substrate condition, and installation quality. If detailing remains inadequate, resealing may only delay repeated failure. This is one reason why joint rehabilitation must be based on diagnosis rather than routine maintenance assumptions.

From a sustainability perspective, this strategy is favorable because it is localized, uses relatively little material, and causes limited disruption. However, it is only appropriate where deterioration has not already progressed into deeper substrate or moisture-related failure. Table 3 highlights this point by comparing sealant renewal with broader interventions that may be needed when damage is more extensive.

4.3 Partial Cladding Replacement

Partial cladding replacement is appropriate when deterioration extends beyond isolated joint failure but remains localized enough that full replacement would be excessive. Typical cases include cracked panels, local detachment risk, corrosion near fixings, or recurring leakage confined to limited areas. In such situations, selective replacement can restore performance while preserving serviceable parts of the façade (Kamel & Memari, 2022; Fiore et al., 2021).

This strategy reflects condition-based intervention. Façade deterioration is often uneven across elevations and details because exposure is not uniform. Some zones may be heavily affected by wind-driven rain or solar stress, while others remain relatively sound. Chung-Camargo et al. (2024) show that climatic exposure strongly affects retrofit suitability, which supports selective rather than uniform replacement.

Its main challenge is compatibility. New and existing façade sections must work together without creating new interface problems in appearance, movement, fixing, or moisture behavior. Partial replacement therefore requires careful damage mapping and continuity in support, waterproofing, and joint detailing. Where done well, it can also introduce local upgrades such as better thermal breaks or more durable materials.

4.4 Thermal Performance Upgrades

Where diagnosis identifies heat loss, thermal bridging, or poor insulation as major sources of underperformance, thermal upgrading becomes a central retrofit strategy. This may involve improved insulation, better thermal breaks, glazing replacement, air-tightness enhancement, or ventilated façade systems. Because the envelope is a major determinant of heating and cooling demand, thermal retrofits can significantly improve operational performance when they are well matched to building type and climate (Kamel & Memari, 2022; Fiore et al., 2021).

Thermal upgrading should not be reduced to the simple idea of adding more insulation. The effectiveness of the intervention depends on the specific failure mechanism. In some buildings, the problem lies in conductive support systems or slab-edge details; in others, it lies in glazing performance or degraded air barriers. Nardi et al. (2018) and Ficapal and Mutis (2019) both show that thermographic evidence can help identify where heat-loss pathways are actually occurring, which is critical for choosing the right intervention. The effectiveness of thermal insulation also depends on where it is positioned within the envelope build-up, as demonstrated by Roque and Santos (2017).

Different retrofit systems also perform differently under different climates. Fiore et al. (2021) compare opaque-envelope interventions and show that performance gains must be considered alongside environmental effects. Chung-Camargo et al. (2024) make a related point in tropical conditions, where retrofit choices must balance insulation, shading, ventilation, and moisture behavior rather than maximizing only one performance variable.

Thermal retrofits can also contribute to durability by reducing condensation and stabilizing envelope conditions. Yet they must be designed carefully, because poorly considered upgrades may create trapped moisture, internal condensation planes, or incompatibility within the assembly. This is one reason thermal upgrading should remain diagnosis-led rather than product-led.

4.5 Localized Structural Strengthening and Connection Repair

When diagnosis identifies fatigue, corrosion, looseness, or deformation in anchors, brackets, rails, or fasteners, localized structural strengthening may be more appropriate than surface repair alone. In many cases, visible distress such as joint widening or panel movement reflects deeper weakness in support components, and treating only the external symptom is unlikely to be effective (Douglas & Ransom, 2013; Bedon & Amadio, 2017).

Structural repair at this level may involve replacing corroded fasteners, strengthening brackets, correcting misalignment, or redesigning movement allowances. Because these problems are often concealed, close-up inspection and targeted confirmation are essential. Image-based screening can identify suspect zones, but it usually cannot define the condition of internal supports with enough confidence on its own (Dias et al., 2021; Falorca & Lanzinha, 2020).

The strength of this approach is that it can reduce safety risk while preserving the façade system where the cladding itself remains serviceable. However, it must be carefully engineered. Poorly conceived local restraints may unintentionally increase stress elsewhere in the system. The strategy therefore works best where deterioration is limited and the main façade still retains adequate residual durability.

4.6 Criteria for Selecting the Most Appropriate Retrofit Strategy

Selecting a retrofit strategy involves balancing several factors rather than optimizing one alone. Technical suitability is the first requirement, but it is not enough by itself. Decision-makers must also consider extent of deterioration, residual service life, disruption, cost, access difficulty, environmental impact, and future maintenance implications (Dadzie et al., 2019; Fiore et al., 2021).

Extent matters because localized failures often justify selective interventions, whereas widespread or systemic deterioration may require broader renewal. Project objectives matter as well. A façade upgrade intended only to stop leakage will differ from one intended to improve durability, thermal performance, visual quality, and market value at the same time. As Table 3 indicates, these strategies should be compared against both technical need and practical consequence rather than chosen on cost alone.

Sustainability should be treated as a core criterion. In existing buildings, the most sustainable solution is often the one that achieves the necessary performance with the least unnecessary waste and disruption (Htet et al.,2025b). Procaccini and Monticelli (2024) argue that retrofit should increasingly align with resilience and lifecycle thinking, while Kamel and Memari (2022) reinforce the need to evaluate interventions through long-term building performance rather than short-term repair logic.

4.7 Toward a Sustainable Façade Retrofit Approach

A sustainable retrofit approach does not assume that the newest or largest intervention is automatically the best. It begins by preserving serviceable components, addressing verified defects, and improving performance in proportion to need. This is particularly important because replacement-heavy strategies can generate substantial waste and embodied carbon even where they improve operational performance (Dadzie et al., 2019; Procaccini & Monticelli, 2024). This also aligns with emerging service-oriented views of façade systems, where long-term performance delivery and resource retention are becoming increasingly important (Aguerre et al., 2022).

For this chapter, the main point is clear: retrofit should be evidence-based, selective, and lifecycle-oriented. Diagnosis defines what is wrong; retrofit determines how far intervention should go. When these stages are closely connected, service life can be extended more responsibly and sustainably.

5. ILLUSTRATIVE APPLICATIONS ACROSS CLIMATIC AND FAÇADE CONTEXTS

Façade diagnosis and retrofitting are not climate-neutral. The same defect can develop differently, progress at different speeds, and require different responses depending on exposure, material system, and building type. Moisture-related deterioration may dominate in humid climates, while thermal inefficiency and condensation may be more prominent in cooler or mixed climates. As recent literature shows, retrofit strategies should therefore be interpreted through climatic context rather than transferred mechanically from one setting to another (Chung-Camargo et al., 2024; Lee & Ng, 2024). This climate sensitivity is also reflected in hygrothermal studies showing that wall and façade systems may behave differently under current and future climatic conditions (Bas et al., 2023).

This section does not present detailed project case studies. Instead, it outlines climate-based application scenarios that help connect the chapter's diagnostic and retrofit logic to practical conditions. The aim is to show how the same framework can lead to different priorities depending on environment and façade type. The main contrasts are summarized in Table 4, which compares common issues, diagnostic priorities, and retrofit implications across climatic contexts.

5.1 Tropical and Humid Climate Contexts

In tropical and humid climates, façade deterioration is often closely linked to moisture, solar exposure, biological growth, and repeated wetting drying cycles. High humidity, intense rainfall, and prolonged surface moisture can accelerate staining, sealant degradation, corrosion, and water penetration through joints and interfaces. Athauda et al. (2023) note that changes in humidity and precipitation can intensify deterioration where moisture protection is already weak. Lee and Ng (2024) similarly emphasize the importance of precipitation and humidity in façade underperformance.

In such settings, diagnostic patterns often begin with visible staining, mold, sealant cracking, algae growth, or localized corrosion near panel edges and fixings. These signs usually point toward joint failure, drainage weakness, or interface problems rather than purely thermal underperformance. Because moisture may remain hidden behind cladding, diagnosis typically combines visual review with thermographic screening and selective confirmation (Mendes et al., 2022; Dias et al., 2021).

Under these conditions, the first priority is often moisture control rather than full replacement. Sealant rehabilitation, drainage correction, localized panel replacement, and detail improvement can be effective where the main system remains serviceable. As Table 4 indicates, tropical retrofit priorities often focus on water management and exposure control before broader thermal intervention is considered.

Thermal upgrading in humid climates also requires caution. The goal is not only to reduce heat gain, but also to avoid moisture entrapment and unintended hygrothermal side effects. Chung-Camargo et al. (2024) show that shading, ventilation, and selective envelope improvement may be more appropriate than insulation-heavy solutions in moisture-sensitive assemblies.

5.2 Temperate and Mixed Climate Contexts

In temperate and mixed climates, retrofit priorities often shift toward heat loss, condensation risk, aging insulation performance, and the interaction between energy upgrading and moisture safety. Seasonal temperature variation introduces stronger cycles of heating demand, cooling demand, and thermal movement. In these settings, thermal bridging, glazing weakness, cold-surface condensation, and localized leakage may appear together rather than separately (Nardi et al., 2018; Wójcik & Bomberg, 2016).

A common diagnostic pattern begins with complaints about discomfort, recurring condensation near façade edges, mold growth at corners, or increased heating demand. Infrared thermography is particularly valuable here because it can reveal insulation defects, conductive junctions, and moisture-related temperature anomalies that are not otherwise visible. Nardi et al. (2018) and Ficapal and Mutis (2019) both support the use of thermography as a bridge between occupant symptoms and technical diagnosis.

Retrofit responses in these climates often include glazing replacement, improved thermal breaks, selective overcladding, ventilated façade upgrades, or insulation enhancement. But such interventions must be designed carefully because altering the thermal regime of the façade can also alter moisture behavior. Wójcik and Bomberg (2016) show that rehabilitation decisions may introduce new moisture risks if building-physics consequences are not considered.

Predominantly glazed façades bring an additional layer of complexity. In those systems, energy performance, solar control, frame behavior, and sealant condition are closely interconnected. Čikić Tovarović et al. (2017) show that glass façade renovation can address both energy and functional performance, but only when the diagnosis distinguishes clearly between glazing, joint, and support-related weaknesses.

5.3 Comparison of Climate-Responsive Retrofit Priorities

The comparison between humid and temperate contexts shows that retrofit priorities depend on dominant exposure risks. In humid climates, moisture ingress, joint durability, drainage, and biological deterioration often lead the agenda. In temperate and mixed climates, condensation, heat loss, thermal bridging, and energy-related underperformance become more prominent, even though moisture remains important. These differences are summarized in Table 4, which provides a concise comparison of likely issues, diagnostic focus, and retrofit implications.

This distinction matters because the same intervention can behave differently across climates. Increasing insulation may be effective in one context but problematic in another if moisture risk is not addressed. Similarly, a façade that performs adequately in a dry temperate environment may deteriorate more quickly in a tropical one because of persistent humidity and UV exposure. Chung-Camargo et al. (2024) and Lee and Ng (2024) both reinforce the importance of climate-sensitive retrofit planning.

Table 4. Comparison of climate-responsive façade deterioration patterns, diagnostic priorities, and retrofit responses

Climate context	Common façade issues	Main diagnostic priorities	Typical retrofit priorities	Key cautions
Tropical / humid	Water ingress, sealant degradation, mold/algae, corrosion, surface weathering, solar exposure damage	Joint assessment, moisture mapping, drainage review, thermography for damp zones, image-based deterioration mapping	Sealant rehabilitation, drainage correction, localized panel replacement, shading improvement, ventilated solutions where suitable	Avoid trapping moisture; do not overemphasize insulation without hygrothermal review
Temperate / mixed	Thermal bridging, condensation, heat loss, localized leakage, glazing inefficiency, thermal movement stress	Thermography, condensation-risk areas, glazing/joint review, targeted opening-up, support/ detail verification	Thermal upgrades, glazing replacement, selective recladding, thermal break improvement, localized repair of moisture-affected details	Avoid creating new condensation planes or hidden moisture risks during insulation upgrade
Predominantly glazed façades	Air leakage, sealant failure, thermal discomfort, solar gain, frame thermal bridging, aging curtain wall components	Thermography, joint and perimeter seal review, close-up inspection, component-specific assessment	Sealant renewal, selective curtain wall replacement, improved solar control, upgraded glazing	Ensure compatibility between old and new components
Aging mixed-material façades	Differential movement, interface failure, localized cracking, patchy deterioration, uneven performance by elevation	Interface inspection, movement-related cracking review, mapping of zone-by-zone condition	Partial cladding replacement, local strengthening, joint redesign, selective energy upgrade	Do not assume uniform deterioration across the whole façade

Source: Developed by the author based on synthesized literature and climate-responsive façade retrofit logic.

5.4 Implications for Practice

For practitioners, the main lesson is that façade retrofit should begin with problem framing rather than solution selection. Moisture symptoms should not be dismissed as cosmetic, and condensation should not be attributed only to occupant behavior without investigating the envelope. Reliable diagnosis remains the basis of effective rehabilitation, while climatic context helps explain why similar symptoms may require different responses in different settings (Dias et al., 2021).

A second lesson is that service-life extension is often best achieved through staged and selective intervention rather than immediate whole-façade replacement.

When deterioration is localized and diagnosis is sound, joint renewal, local repair, or partial replacement may offer a better balance between performance, disruption, and sustainability than complete renewal. This is especially important in existing-building portfolios where intervention priorities must be justified over time.

6. BEST PRACTICES AND STRATEGIC RECOMMENDATIONS

Effective façade renewal depends on linking diagnosis, retrofit choice, and future maintenance into one process. Repair quality often declines when inspections are superficial, when interventions are chosen before root causes are confirmed, or when maintenance planning is separated from rehabilitation. Dias et al. (2021) emphasize that reliable diagnosis is essential for technically informed rehabilitation, while Kamel and Memari (2022) note that retrofit strategies are most effective when they follow actual envelope condition rather than generic assumptions.

A first best practice is staged diagnosis. Preliminary review, visual assessment, thermographic screening, and targeted confirmation should be used in sequence so that intervention responds to verified mechanisms rather than appearances. This reduces the risk of unnecessary replacement and improves confidence in the repair scope (Dias et al., 2021; Pereira et al., 2020).

A second recommendation is proportional intervention. Where deterioration is localized, sealant renewal, local strengthening, or partial replacement is often more sustainable than full recladding. Dadzie et al. (2019) argue that sustainable upgrading depends on balancing technical, environmental, and practical factors, while Procaccini and Monticelli (2024) support retrofit approaches that preserve viable envelope components wherever possible.

A third recommendation is climate-aware design. Moisture, solar exposure, condensation risk, and movement should be assessed in relation to environmental context rather than through one standard retrofit model. Chung-Camargo et al. (2024) and Lee and Ng (2024) both show that climatic conditions influence which façade problems dominate and which retrofit measures are likely to work well.

A fourth recommendation is documentation. Inspection findings, verified defects, repaired locations, materials used, and unresolved monitoring points should all be recorded clearly. This supports future maintenance and reduces uncertainty in later assessments. In practical terms, documentation is part of service-life extension, not an administrative afterthought. More broadly, structured assessment and maintenance thinking also reflect the logic of formal building inspection systems and performance tracking in existing buildings (Bortolini & Forcada, 2018).

Overall, the most reliable façade strategy is one that is evidence-based, selective, and maintainable. Diagnosis should define the problem, retrofit should match the scale of the problem, and maintenance should preserve the value of the intervention over time.

7. CONCLUSION

Façade failures are rarely isolated technical events. More often, they emerge through the combined effects of exposure, aging, detailing weaknesses, workmanship, and maintenance history. This chapter has shown that thermal bridging, sealant degradation, structural fatigue, and water ingress are especially important because they influence durability, performance, comfort, and safety at the same time.

The discussion also showed that effective intervention depends on structured diagnosis. Infrared thermography, drone-assisted inspection, close-up review, and selective testing each offer different kinds of evidence, but their value is strongest when they are used together in a staged workflow. A façade should therefore be assessed through systematic investigation rather than reactive observation alone.

A further conclusion is that retrofit should be proportional to verified condition. Localized deterioration may justify joint renewal, partial replacement, or local strengthening, while broader system weakness may require more extensive upgrading. The best intervention is not necessarily the largest one, but the one that responds most accurately to the actual defect pattern and service-life objective.

In the end, sustainable façade renewal depends on connecting diagnosis, climate-aware retrofit design, and future maintenance into one continuous strategy. When those stages are aligned, façade rehabilitation can do more than correct defects. It can extend service life, improve performance, reduce waste, and preserve the long-term value of existing building envelopes.

REFERENCES

Aguerre, J., Klein, T., Konstantinou, T., & Veerman, M. (2022). Facades-as-a-service: The role of technology in the circular servitisation of the building envelope. *Applied Sciences (Basel, Switzerland)*, *12*(3), 1267. DOI: 10.3390/app12031267

Alhawari, A., & Mukhopadhyaya, P. (2018). Thermal bridges in building envelopes An overview of impacts and solutions. *International Review of Applied Sciences and Engineering*, *9*(1), 31–40. DOI: 10.1556/1848.2018.9.1.5

Asphaug, S. K., Time, B., & Kvande, T. (2021). Moisture accumulation in building façades exposed to accelerated artificial climatic ageing: A complementary analysis to NT Build 495. *Buildings (Basel, Switzerland)*, *11*(12), 568. DOI: 10.3390/buildings11120568

Athauda, R. S., Asmone, A. S., & Conejos, S. (2023). Climate change impacts on facade building materials: A qualitative study. *Sustainability (Basel)*, *15*(10), 7893. DOI: 10.3390/su15107893

Attia, S., Bilir, S., Safy, T., Struck, C., Loonen, R., & Goia, F. (2018). Current trends and future challenges in the performance assessment of adaptive façade systems. *Energy and Building*, *179*, 165–182. DOI: 10.1016/j.enbuild.2018.09.017

Azarnejhad, A., & Mahdavi, A. (2019). On the impact of building façades' color on thermal building performance and outdoor thermal comfort. *Applied Mechanics and Materials*, *887*, 189–195. . DOI: 10.4028/www.scientific.net/AMM.887.189

Baptista, J., Kokare, S., Godina, R., & Aelenei, D. (2024). A comparative life cycle assessment of ETICS and ventilated facade systems with timber cladding. *Energy and Building*, *304*, 4375760. DOI: 10.1016/j.enbuild.2023.113842

Barreira, E., & Delgado, J. (2017). Diagnosis of rising damp and other kinds of moisture in building components using non-destructive techniques. *Diffusion Foundations*, *10*, 120–131. . DOI: 10.4028/www.scientific.net/DF.10.120

Barrelas, J., Dias, I. S., Silva, A., de Brito, J., Flores-Colen, I., & Tadeu, A. (2021). Impact of environmental exposure on the service life of façade claddings A statistical analysis. *Buildings (Basel, Switzerland)*, *11*(12), 615. DOI: 10.3390/buildings11120615

Bas, C., Türkeri, A. N., & Lacasse, M. (2023). Hygrothermal performance of external wall systems under current climatic conditions and future climate projections of Turkey. In *DBMC 2023 Proceedings*. DOI: 10.23967/c.dbmc.2023.082

Bedon, C., & Amadio, C. (2017). Passive control systems for the blast enhancement of glazing curtain walls under explosive loads. *The Open Civil Engineering Journal, 11*(1), 396–419. DOI: 10.2174/1874149501711010396

Beech, J., & Beasley, J. L. (1994). Further studies of cure and durability of building sealants. In *Science and Technology of Building Seals, Sealants, Glazing, and Waterproofing* (Vol. 3, pp. 33-50). ASTM International. https://doi.org/DOI: 10.1520/STP13998S

Bhering, A. F., Franco, R. N., Santos, M. A., Sathler, L. M., Pedroti, L. G., Varum, H., & Paes, J. L. R. (2023). Experimental analysis of watertightness performance of interfaces between masonry and steel structures subjected to accelerated aging. *Buildings (Basel, Switzerland), 13*(9), 2123. DOI: 10.3390/buildings13092123

Bienvenido-Huertas, D. (2020). Analysis of the relationship of the improvement of façades and thermal bridges of Spanish building stock with the mitigation of its energy and environmental impact. *Energies, 13*(17), 4499. DOI: 10.3390/en13174499

Bortolini, R., & Forcada, N. (2018). Building inspection system for evaluating the technical performance of existing buildings. *Journal of Performance of Constructed Facilities, 32*(5), 04018073. Advance online publication. DOI: 10.1061/(ASCE)CF.1943-5509.0001220

Buday, P., Ingeli, R., & Čekon, M. (2014). Influence of thermal break element applied in balcony slab on internal surface temperature. *Advanced Materials Research, 1057*, 79–86. . DOI: 10.4028/www.scientific.net/AMR.1057.79

Chew, M. Y. L., & Gan, V. J. (2022). Long-standing themes and future prospects for the inspection and maintenance of façade falling objects from tall buildings. *Sensors (Basel), 22*(16), 6070. DOI: 10.3390/s22166070 PMID: 36015831

Chung-Camargo, K., González, J., Chen Austin, M., Carpino, C., Mora, D., & Arcuri, N. (2024). Advances in retrofitting strategies for energy efficiency in tropical climates: A systematic review and analysis. *Buildings (Basel, Switzerland), 14*(6), 1633. DOI: 10.3390/buildings14061633

Čikić Tovarović, J., Ivanović-Šekularac, J., & Šekularac, N. (2017). Renovation of existing glass facade in order to implement energy efficiency and media facade. *Energy and Building, 152*, 653–666. DOI: 10.1016/j.enbuild.2017.07.049

Dadzie, J., Runeson, G., & Ding, G. (2019). Assessing determinants of sustainable upgrade of existing buildings. *Journal of Engineering. Defence Technology, 18*(1), 270–292. DOI: 10.1108/JEDT-09-2018-0148

Dias, I. S., Flores-Colen, I., & Silva, A. (2021). Critical analysis about emerging technologies for building's façade inspection. *Buildings (Basel, Switzerland)*, *11*(2), 53. DOI: 10.3390/buildings11020053

Ding, S., & Liu, D. (2006). Accelerated aging on sealants and a quantificational way to evaluate the durability. *Journal of Testing and Evaluation*, *34*(2), 79–82. DOI: 10.1520/JTE14067

Diz-Mellado, E., Mascort-Albea, E., Romero-Hernández, R., Galán-Marín, C., Rivera-Gómez, C., Ruiz-Jaramillo, J., & Jaramillo-Morilla, A. (2021). Non-destructive testing and finite element method integrated procedure for heritage diagnosis: The Seville Cathedral case study. *Journal of Building Engineering*, *37*, 102134. DOI: 10.1016/j.jobe.2020.102134

Douglas, J., & Ransom, B. (2013). *Understanding building failures* (4th ed.). Routledge. DOI: 10.4324/9780203125175

Elguezabal, P., & Garay, R. (2015). Experiences when employing different alternatives for envelope upgrading. *Journal of Facade Design and Engineering*, *3*(1), 81–89. DOI: 10.3233/FDE-150032

Falorca, J. F., & Lanzinha, J. (2020). Facade inspections with drones theoretical analysis and exploratory tests. *International Journal of Building Pathology and Adaptation*, *39*(2), 235–258. DOI: 10.1108/IJBPA-07-2019-0063

Ficapal, A., & Mutis, I. (2019). Framework for the detection, diagnosis, and evaluation of thermal bridges using infrared thermography and unmanned aerial vehicles. *Buildings (Basel, Switzerland)*, *9*(8), 179. DOI: 10.3390/buildings9080179

Fiore, P., Donnarumma, G., Falce, C., & Sicignano, G. (2021). A tool for the evaluation of energy-environmental retrofit interventions on opaque walls using ETICS. *Sustainability (Basel)*, *13*(7), 4063. DOI: 10.3390/su13074063

Fylan, F., & Glew, D. (2021). Barriers to domestic retrofit quality: Are failures in retrofit standards a failure of retrofit standards? *Indoor and Built Environment*, *31*(3), 710–718. DOI: 10.1177/1420326X211027197

Hosseinalipour, S. M., Al-Taee, A. A. H., & Asiaei, S. (2024). Improving energy consumption for dynamic intake and exhaust façades: A new setup design: Part B. *Building Services Engineering Research and Technology*, *45*(5), 613–639. DOI: 10.1177/01436244241264273

Htet, A., Gupta, S. K., & Oruganti, S. K. (2025). Modular adaptive facades: Designing a scalable framework for retrofitting urban buildings. *SGS Engineering & Sciences, 1*(2), NA.

Htet, A., Liana, S. R., Aung, T., Bhaumik, A., & Giri, O. P. (2025). The Role of Silicone in Enhancing the Weatherability of Facade Materials. *Journal of Innovations in Business and Industry*, *3*(4), 201–214. DOI: 10.61552/JIBI.2025.04.00

Jiao, Y., Chu, L., Zhu, M., Wei, Y., Wang, W., Wang, H., & Wang, H. (2021). UV-curable hyperbranched polyester sealant with tunable mechanical properties. *Journal of Biomaterials Applications*, *36*(4), 731–739. DOI: 10.1177/08853282211000527 PMID: 33940972

Kamel, E., & Memari, A. M. (2022). Residential building envelope energy retrofit methods, simulation tools, and example projects: A review of the literature. *Buildings (Basel, Switzerland)*, *12*(7), 954. DOI: 10.3390/buildings12070954

Kim, H., Lamichhane, N., Kim, C.-S., & Shrestha, R. (2023). Innovations in building diagnostics and condition monitoring: A comprehensive review of infrared thermography applications. *Buildings (Basel, Switzerland)*, *13*(11), 2829. DOI: 10.3390/buildings13112829

Lee, Y., & Ng, E. (2024). Critical review of current understanding of passive façade design in residential buildings. *Environmental Research: Infrastructure and Sustainability*, *4*(3), 032001. DOI: 10.1088/2634-4505/ad78fc

Linden, S., & Bossche, N. V. (2025). Experimental and numerical assessment of rainwater intrusion in ventilated facades with open joints. *Journal of Building Physics*, *49*(2), 173–206. DOI: 10.1177/17442591251324122

Luo, D., Zhang, T., Zheng, W., & Nie, Q. (2026). Intelligent detection and energy-driven repair of building envelope defects for improved thermal and energy performance. *Energies*, *19*(2), 351. DOI: 10.3390/en19020351

Mendes, M. V. A. da S., Pavón, E., & Melo, M. B. (2022). Damages mapping of façade using aerophotogrammetry and thermography inspection: Zoroastro Artiaga Museum - Art Deco in Brazil. *Matéria (Rio de Janeiro)*, *27*(3), e20220031. Advance online publication. DOI: 10.1590/1517-7076-rmat-2022-0031

Mirzabeigi, S., Razkenari, R., & Crovella, P. (2025). Building envelope thermal anomaly detection using an integrated vision-based technique and semantic segmentation. *Buildings (Basel, Switzerland)*, *15*(15), 2672. DOI: 10.3390/buildings15152672

Mydin, M. A. O., Sarpin, N., Zainol, R. M., Odeh, R. A., & Nawi, M. N. M. (2024). The impact of climatological factors on the multifaceted and multisystemic deficiencies of building anatomy. *Journal of Advanced Research in Applied Sciences and Engineering Technology*, *50*(1), 308–329. DOI: 10.37934/araset.50.1.308329

Nardi, I., Lucchi, E., de Rubeis, T., & Ambrosini, D. (2018). Quantification of heat energy losses through the building envelope: A state-of-the-art analysis with critical and comprehensive review on infrared thermography. *Building and Environment*, *146*, 190–205. DOI: 10.1016/j.buildenv.2018.09.050

Oesterle, E., Lieb, R.-D., Lutz, M., & Heusler, W. (2014). *Double-skin façades: Integrated planning*. Prestel.

Pereira, C., de Brito, J., & Silvestre, J. D. (2020). Harmonised classification of the causes of defects in a global inspection system: Proposed methodology and analysis of fieldwork data. *Sustainability (Basel)*, *12*(14), 5564. DOI: 10.3390/su12145564

Petresevics, F., & Nagy, B. (2022). FEM-based evaluation of the point thermal transmittance of various types of ventilated façade cladding fastening systems. *Buildings (Basel, Switzerland)*, *12*(8), 1153. DOI: 10.3390/buildings12081153

Pomponi, F., & Moncaster, A. (2017). Circular economy for the built environment: A research framework. *Journal of Cleaner Production*, *143*, 710–718. DOI: 10.1016/j.jclepro.2016.12.055

Popovic, P. L., & Vla, D. (2020). Assessment and repair of exterior facades. *Scientific Journal of Civil Engineering*, *9*(1), 43–51. DOI: 10.55302/SJCE2091043p

Procaccini, G., & Monticelli, C. (2024). The application and potentialities of textile facade retrofit strategies for energy-efficient and resilient buildings. *E3S Web of Conferences, 546*, 03005. https://doi.org/DOI: 10.1051/e3sconf/202454603005

Riahinezhad, M., Hallman, M., & Masson, J. (2021). Critical review of polymeric building envelope materials: Degradation, durability and service life prediction. *Buildings (Basel, Switzerland)*, *11*(7), 299. DOI: 10.3390/buildings11070299

Roque, E., & Santos, P. (2017). The effectiveness of thermal insulation in lightweight steel-framed walls with respect to its position. *Buildings (Basel, Switzerland)*, *7*(1), 13. DOI: 10.3390/buildings7010013

Sandak, A., Sandak, J., Brzezicki, M., & Kutnar, A. (2019). *Bio-based building skin*. Springer. DOI: 10.1007/978-981-13-3747-5

Sezegen, A., & Edis, E. (2020). Product innovation types: A discussion considering building facade products. *Engineering, Construction, and Architectural Management*, *27*(9), 2379–2408. DOI: 10.1108/ECAM-10-2018-0454

Sivaji, C., Chidambaram, P., Ramachandran, M., & Murugan, A. (2022). Performance analysis of facade materials using VIKOR method. *REST Journal on Advances in Mechanical Engineering*, *1*(2), 41–49. DOI: 10.46632/jame/1/2/6

Souza, A. L. R., de, , & Romeiro, T. R. (2024). Influence of the exposure degree on the degradation of facades of buildings in Brasília, Brazil. *Buildings (Basel, Switzerland)*, *14*(1), 133. DOI: 10.3390/buildings14010133

Straube, J. F., & Burnett, E. F. P. (2005). *Building science for building enclosure design*. Building Science Press.

Tejedor, B., Casals, M., Gangolells, M., & Roca, X. (2017). Quantitative internal infrared thermography for determining in-situ thermal behaviour of façades. *Energy and Building*, *151*, 187–197. DOI: 10.1016/j.enbuild.2017.06.040

Ujma, A., & Pomada, M. (2019). Analysis of the temperature distribution in the place of fixing the ventilated facade. *E3S Web of Conferences, 97*, 01041. https://doi.org/DOI: 10.1051/e3sconf/20199701041

Wang, T., Ward, G. J., & Lee, E. S. (2018). Efficient modeling of optically-complex, non-coplanar exterior shading: Validation of matrix algebraic methods. *Energy and Building*, *174*, 464–483. DOI: 10.1016/j.enbuild.2018.06.022

Widomski, P., Maksymowicz, P., Trzaska, O., Mayer, P., Kaczyński, P., Berbesz-Wyrodek, A., & Kogut, M. S. (2025). The impact of destructive climatic factors on the mechanical and performance properties of structural materials. *Materials (Basel)*, *18*(13), 2970. DOI: 10.3390/ma18132970 PMID: 40649459

Wójcik, R., & Bomberg, M. (2016). On interior rehabilitation of buildings with historic facades. *Journal of Building Physics*, *40*(2), 144–161. DOI: 10.1177/1744259116662280

Xu, C., Xi, Q., & Ma, Q. (2021). Technical characteristics and energy performance of double skin faces: A review. *E3S Web of Conferences, 248*, 01029. https://doi.org/DOI: 10.1051/e3sconf/202124801029

Chapter 10

Developing a Façade Risk Evaluation Framework (F-REF) Based on Failure Mode and Effects Analysis (FMEA)

Theingi Aung
http://orcid.org/0009-0003-5063-575X
Lincoln University College, Malayisa

Arkar Htet
http://orcid.org/0000-0003-1301-3604
Lincoln University College, Malaysia

Sui Reng Liana
Lincoln University College, Malayisa

Yin Hlaing Min
http://orcid.org/0009-0004-2266-6935
Azteca University, Mexico

ABSTRACT

This chapter develops a Façade Risk Evaluation Framework (F-REF) grounded in Failure Mode and Effects Analysis (FMEA) for the systematic identification, comparison, and prioritization of façade risks across design, construction, and operation. The framework extends conventional FMEA through façade-specific severity, occurrence, and detection rubrics for unitized curtain walls and ventilated

DOI: 10.4018/979-8-3373-6023-2.ch010

rainscreen systems; a façade-weighted priority index used alongside the traditional Risk Priority Number; dependency-based treatment of coupled failures; and full-matrix application across three façade scenarios with sensitivity testing under alternative weighting schemes. The chapter also positions F-REF against Fault Tree Analysis, Event Tree Analysis, Bow-Tie analysis, and qualitative checklist methods. Although further expert calibration is needed, the framework provides a more rigorous and actionable basis for façade risk screening, inspection planning, and lifecycle decision support.

1. INTRODUCTION

1.1 Importance of Risk Management in Façade Engineering

Façade systems have evolved from exterior building skins into complex engineering assemblies that influence structural behavior, environmental separation, energy performance, durability, and architectural expression. In contemporary buildings, especially high-rise and technically demanding projects, the façade acts as the primary interface between interior and exterior conditions and therefore plays a decisive role in weather protection, thermal regulation, occupant comfort, and operational performance (Kassem & Mitchell, 2015; Fernando et al., 2023). As integrated façade engineering has developed, designers have increasingly been required to balance structural, environmental, and architectural objectives within the same envelope system (Takeuchi et al., 2006; Kaneki et al., 2008; Inoue et al., 2003; Laufs & Verboon, 2013).

This increasing performance burden has been accompanied by greater technical complexity. Modern façades frequently incorporate unitized curtain walls, high-performance glazing, adaptive components, ventilated cavities, and advanced thermal interfaces. Attia et al. (2018) noted that contemporary façade systems are no longer judged only by enclosure function, but also by their contribution to comfort, energy efficiency, and responsive building performance. These developments create substantial opportunities, but they also increase the number of interfaces, tolerances, and failure pathways that must be controlled during design, fabrication, installation, and operation (Fernando et al., 2023).

Façade failures can have consequences that extend far beyond appearance. Water ingress, glazing breakage, detachment of cladding units, thermal bridging, sealant degradation, and concealed corrosion may affect safety, comfort, durability, energy use, and lifecycle cost. Straube and Burnett (2005) explained that enclosure failures are often difficult and expensive to diagnose because the visible symptoms may emerge only after hidden deterioration has advanced. Similarly, Santamaria-Fernandez et al.

(2020) showed that façade pathologies can affect not only durability but also user comfort, reinforcing the need to treat façade risk as a performance issue rather than a purely technical defect issue. Similar observations have been reported in studies of curtain wall sustainability and façade technology, where performance weaknesses often extend beyond aesthetics to long-term operational and environmental implications (Wu & Flemmer, 2020).

For this reason, risk management in façade engineering should be viewed as a strategic design and lifecycle function. A structured risk perspective helps project teams identify vulnerable details, prioritize failure-sensitive interfaces, and reduce the likelihood of expensive remedial work after handover. In this sense, façade risk management supports not only defect prevention, but also long-term resilience, maintainability, and project value (Kassem & Mitchell, 2015; Vandi et al., 2024).

1.2 Limitations of Current Risk Assessment Practices

Despite the importance of façade performance, risk assessment in façade projects often remains informal and fragmented. In many projects, risks are identified through checklists, precedent-based judgment, or experience-driven review without being translated into a structured prioritization process. Although expert judgment is indispensable, reliance on unstructured assessment can result in inconsistent decisions and weak transparency, particularly when multiple competing risks must be evaluated at the same time (Liu et al., 2013; Kassem & Mitchell, 2015; Silva et al., 2014). Research on standardized building condition assessment similarly shows that consistent inspection criteria are important for converting observed defects into usable maintenance and asset-management data (Straub, 2009).

Another limitation arises from the fragmented nature of façade project delivery. Architects, structural engineers, façade consultants, specialist contractors, suppliers, and installers may each focus on different aspects of performance. As Kassem and Mitchell (2015) observed, this separation often produces information gaps and delayed coordination. These coordination limitations are also increasingly discussed in BIM-related façade studies, particularly where design information and specialist knowledge are not integrated early enough (Htet et al., 2023a). In façade engineering, such fragmentation is especially problematic because many failures originate at interfaces between materials, between trades, and between design intent and site execution (Orlowski et al., 2018).

Current practice also lacks a widely accepted risk evaluation framework tailored specifically to façade systems. General construction risk tools exist, but they do not always capture façade-specific concerns such as moisture migration, thermal discontinuity, concealed support vulnerability, movement accommodation, fire-interface continuity, and maintenance accessibility. This gap is significant because façade performance is shaped by system interactions rather than by isolated components alone (Fernando et al., 2023; Santamaria-Fernandez et al., 2020; Inoue et al., 2003), while façade inspection research also shows the need for reliable diagnosis, hazard-sensitive assessment, and systematic defect information (Dias et al., 2021; Ruiz et al., 2019).

These limitations point to the need for a more structured and façade-sensitive method of analysis. Such a method should identify failure modes systematically, link them to likely causes and consequences, and support consistent prioritization across the façade life cycle. This need provides the rationale for adapting Failure Mode and Effects Analysis to façade engineering.

1.3 Potential of Reliability Engineering Tools

Reliability engineering offers useful tools for dealing with complex systems in which failures may emerge from interacting technical and process-related factors. Among these tools, Failure Mode and Effects Analysis (FMEA) is one of the most established approaches for identifying how systems may fail, evaluating the likely effects of those failures, and prioritizing risk before problems become critical (Stamatis, 2003; Bowles, 2004; Ashley & Armitage, 2010). Its preventive logic makes it particularly relevant for façade engineering, where many significant failures are difficult to detect early and costly to correct later.

In broader engineering practice, FMEA has been used across manufacturing, aerospace, process systems, and quality management because it transforms dispersed expert knowledge into a structured decision framework. Liu et al. (2013) showed that FMEA remains widely valued despite methodological limitations because it offers a practical way to compare risks systematically. Reviews of FMEA methodology likewise confirm its continuing relevance across sectors because of its flexibility as a preventive decision-support tool (Silva et al., 2014). This is important in façade engineering, where risks such as leakage, anchorage weakness, thermal defects, and concealed deterioration often compete for attention but are rarely assessed in a unified way.

The logic of FMEA is also compatible with the way façade failures occur in reality. Façade problems typically develop through interactions among detailing, materials, movement, environment, and workmanship rather than through a single isolated defect. A structured method that focuses on failure modes, causes, and consequences is therefore well suited to façade risk analysis. However, direct transfer of generic FMEA is insufficient; façade conditions require contextual adaptation of both the categories and the rating logic (Fernando et al., 2023; Straube & Burnett, 2005).

1.4 Research Gap and Chapter Aim

Although façade engineering literature has expanded in areas such as performance design, innovative systems, durability, and monitoring, there remains a lack of a dedicated risk evaluation framework specifically structured for façade applications. Existing studies address important issues such as weatherproofing, adaptive envelope performance, structural behavior, and maintenance but they do not yet converge into a practical and façade-centered method for failure-based prioritization (Attia et al., 2018; Orlowski et al., 2018; Laufs & Verboon, 2013; Kültür et al., 2019).

The gap is therefore twofold. First, façade engineering still lacks a clear adaptation of reliability-based assessment tools such as FMEA. Second, the industry needs a framework that is rigorous enough to support systematic analysis while remaining practical for multidisciplinary use in design, construction, and operation. Without such a framework, façade risk assessment is likely to remain reactive and inconsistent (Kassem & Mitchell, 2015; Straube & Burnett, 2005).

This chapter addresses that gap by proposing a Façade Risk Evaluation Framework (F-REF) based on the principles of Failure Mode and Effects Analysis. The chapter aims to:

1. clarify the nature of façade risks and failure modes,
2. explain the relevance of FMEA to façade engineering,
3. develop the conceptual structure of F-REF, and
4. discuss its use across design, construction, and maintenance contexts.

2. FUNDAMENTALS OF FAÇADE RISKS AND FAILURE MODES

2.1 Understanding Risk in Façade Engineering

In façade engineering, risk refers to the possibility that a façade system, component, or interface will fail to meet its intended performance requirements and thereby create adverse consequences for safety, serviceability, durability, comfort, energy efficiency, or cost. Unlike single-function building elements, façades must perform as integrated systems that respond simultaneously to structural, environmental, thermal, and operational demands. This makes façade risk inherently multidimensional and highly dependent on system integration rather than on isolated material behavior (Kassem & Mitchell, 2015; Fernando et al., 2023).

Integrated façade engineering has reinforced this complexity by combining architectural design, structural performance, and environmental response within the same envelope system. Takeuchi et al. (2006) and Kaneki et al. (2008) showed that façade design increasingly operates as a coordinated engineering discipline rather than a finishing activity. As façades become more performance-driven, the number of technical interfaces, tolerance-sensitive details, and lifecycle dependencies also increases, which raises the likelihood of failure if risks are not systematically managed.

A defining feature of façade risk is that many failures originate at interfaces. Problems often emerge at joints, sealants, anchors, glazing supports, slab edges, drainage paths, thermal breaks, and transitions between façade and structure. Orlowski et al. (2018) demonstrated that the weatherproofing performance of façade joints is highly sensitive to detailing, workmanship, and movement, while Straube and Burnett (2005) emphasized that enclosure failures often arise where control layers become discontinuous or poorly coordinated.

Another important characteristic is delayed visibility. Many façade defects are latent during design or installation and only become evident under long-term exposure to wind, moisture, thermal cycling, structural movement, or maintenance neglect. Santamaria-Fernandez et al. (2020) showed that glass-envelope pathologies can affect both durability and user comfort, illustrating how façade defects often evolve from hidden technical weaknesses into broader performance problems. Service-life research on façade claddings also confirms that environmental exposure, orientation, and degradation history influence the timing and severity of façade deterioration (Barrelas et al., 2021; Silva et al., 2015). For this reason, façade risk should be treated as a lifecycle concern rather than a short-term construction issue. In severe cases, latent façade deterioration may also escalate into direct safety concerns that require urgent intervention (Niedostatkiewicz, 2022; Ruiz et al., 2019).

2.2 Major Categories of Façade Risk

Façade risks can be grouped into several broad categories that help structure failure analysis and later support the development of the proposed F-REF.

Table 1. Major categories of façade risk

Risk Category	Description	Typical Examples	Main Impact
Structural and Mechanical	Inability to safely resist loads or movement	Anchor distress, connector failure, panel instability	Safety, stability
Weathertightness	Failure to control water or air penetration	Leakage, drainage failure, joint distress	Durability, internal damage
Thermal and Energy	Failure to regulate heat transfer efficiently	Thermal bridging, condensation, energy loss	Comfort, efficiency
Material Durability	Degradation over time under exposure	Corrosion, sealant aging, coating failure	Service life, maintenance

continued on following page

Table 1. Continued

Risk Category	Description	Typical Examples	Main Impact
Fire and Life Safety	Failure to resist or limit fire spread	Cavity barrier discontinuity, combustible façade spread	Safety, compliance
Operational and Maintenance	Inadequate monitoring, inspection, or replacement capacity	Hidden deterioration, access difficulty	Lifecycle performance

Structural and mechanical risks arise when façade systems cannot adequately accommodate wind pressure, differential movement, thermal expansion, seismic demand, or service loads. This includes anchor distress, bracket deformation, glazing instability, or detachment risk. Ji et al. (2024) showed that movement-sensitive connection behavior can govern façade failure progression, particularly in stone curtain wall systems where connector disengagement may precede wider instability. Similar concerns have also been noted in studies of glazed curtain wall vulnerability under seismic action, where façade movement and connection behavior are directly linked to safety performance (Djuric-Mijovic et al., 2018). Hazard-based façade condition assessment further supports giving high priority to defects that may expose pedestrians or occupants to injury from falling façade elements (Ruiz et al., 2019). This confirms that support details and movement compatibility are central to façade reliability (Takeuchi et al., 2006).

Weathertightness risk remains one of the most frequent and consequential categories of façade failure. Leakage typically develops through the combined effects of poor detailing, inadequate drainage, workmanship defects, movement, or sealant incompatibility. Orlowski et al. (2018) noted that sealant-only solutions are often insufficient when inspection access is limited and movement is not fully controlled. Because moisture ingress can lead to corrosion, hidden wetting, insulation damage, and occupant discomfort, water control remains a core risk category in façade assessment (Straube & Burnett, 2005).

Thermal and energy-performance risks are often less visible but strategically important. Inadequate façade design can increase heat gain, reduce thermal comfort, create condensation risk, and raise building operating energy. Latifah et al. (2022) showed that envelope configuration, glazing choice, and façade geometry directly influence operational thermal performance. Similarly, Pracucci et al. (2020) demonstrated that high-performance façade design increasingly requires balancing thermal efficiency, durability, and replacement strategy, especially in advanced glazed systems (Li et al., 2015; Wu & Flemmer, 2020).

Material durability risk concerns long-term degradation caused by moisture, UV exposure, pollution, cyclic temperature change, fatigue, or incompatibility between materials. Sealants may lose adhesion, metals may corrode, protective coatings may fail, and glazing-associated elements may deteriorate under repeated exposure. Santamaria-Fernandez et al. (2020) highlighted that these pathologies can affect both durability and occupant experience, which makes durability risk highly relevant to lifecycle-based façade management. Material selection also has broader implications for environmental and lifecycle efficiency in façade engineering (Htet et al., 2023b).

Fire-related risk, although less frequent in everyday service conditions, represents a high-consequence category. Façade systems can contribute to rapid vertical fire spread when combustible components, inadequate firestopping, or discontinuous cavity barriers are present. Srivastava et al. (2020) showed that full-scale façade fire behavior may differ significantly from simplified certification assumptions, emphasizing the need to treat concealed interface protection as a serious façade risk issue.

Operational and maintenance risk relates to the difficulty of detecting, accessing, and managing deterioration once the façade is in service. Attia et al. (2018) argued that the long-term performance of advanced façades is strongly influenced by maintenance strategy and operational control. Condition-assessment and façade-inspection studies similarly emphasize that systematic diagnosis is needed to support rational maintenance and rehabilitation decisions (Straub, 2009; Dias et al., 2021). Vandi et al. (2024) and Wei et al. (2023) likewise demonstrated that façade monitoring and structural-health-oriented assessment are becoming increasingly important for proactive maintenance and defect detection.

2.3 Typical Façade Failure Modes

To move from broad risk categories to usable analysis, it is necessary to identify specific failure modes. A failure mode describes the way a façade system loses its intended performance. Typical façade failure modes include water leakage through joints, loss of seal adhesion, blocked drainage routes, panel detachment, anchor corrosion, glass breakage, excessive frame deflection, thermal bridging, discontinuity of insulation, and failure of fire barriers. These failure modes differ in visibility and consequence, but all are relevant because they compromise core façade functions (Straube & Burnett, 2005; Orlowski et al., 2018).

Table 2. Typical façade failure modes and associated causes

Failure Mode	Likely Cause	Typical Effect
Water leakage	Joint failure, blocked drainage, workmanship error	Wetting, staining, corrosion
Sealant degradation	UV exposure, movement, incompatibility	Loss of watertightness
Glass breakage	Thermal stress, edge damage, restraint	Safety hazard, replacement need
Anchor corrosion	Moisture ingress, poor protection	Structural weakening
Panel detachment	Connector failure, inadequate fixing	Falling hazard
Thermal bridging	Discontinuous insulation, poor detailing	Heat loss, condensation
Fire barrier discontinuity	Installation defect, coordination gap	Fire spread risk

Some failure modes are immediately visible, such as cracked glass, open joints, staining, or displaced panels. Others remain concealed until secondary effects appear. Hidden corrosion, wet insulation, discontinuous fire barriers, or thermal defects within slab-edge interfaces may remain undetected for long periods. This distinction is important because low detectability often increases the seriousness of the risk even when visible damage is initially limited (Liu et al., 2013). Façade inspection research also shows that reliable diagnosis depends on combining visual assessment with systematic procedures and, where suitable, emerging inspection technologies (Dias et al., 2021). Case-based pathology studies likewise show that both visible and concealed façade defects can significantly affect user comfort, durability, and safety (Santamaria-Fernandez et al., 2020; Niedostatkiewicz, 2022).

Moisture-related failure modes are especially significant because they often initiate chains of secondary deterioration. Straube and Burnett (2005) emphasized that bulk water penetration can rapidly damage enclosure systems when drying and drainage are inadequate. Likewise, Santamaria-Fernandez et al. (2020) showed that envelope pathologies can compromise both comfort and durability, illustrating that the consequence of failure is often broader than the original defect itself.

Structural failure modes may be less frequent, but they are often more severe. Ji et al. (2024) demonstrated that connector disengagement and related support distress can trigger instability in façade systems exposed to movement. This reinforces the idea that façade failure should be interpreted as a system problem rather than a simple material defect.

2.4 Interdependence of Failure Mechanisms

Façade failures rarely occur in isolation. Instead, one weakness often intensifies another. Water ingress may accelerate corrosion; corrosion may weaken fixings;

weakened fixings may increase movement; and increased movement may cause seal rupture or panel instability. In the same way, poor thermal detailing may create condensation, which then contributes to material degradation and hidden moisture accumulation. These relationships show that façade failure is often cascading rather than independent (Straube & Burnett, 2005; Fernando et al., 2023).

This interdependence is one reason why fragmented risk assessment is insufficient. If one stakeholder focuses only on structure, another only on aesthetics, and another only on sealing, important combined risks may remain unrecognized. Attia et al. (2018) noted that performance assessment of advanced façade systems must address interactions among multiple performance demands rather than treating them separately. The same principle applies to façade risk, especially where multiple performance criteria must be balanced simultaneously within one envelope system (Pracucci et al., 2024).

Figure 1. Conceptual interaction of façade failure mechanisms

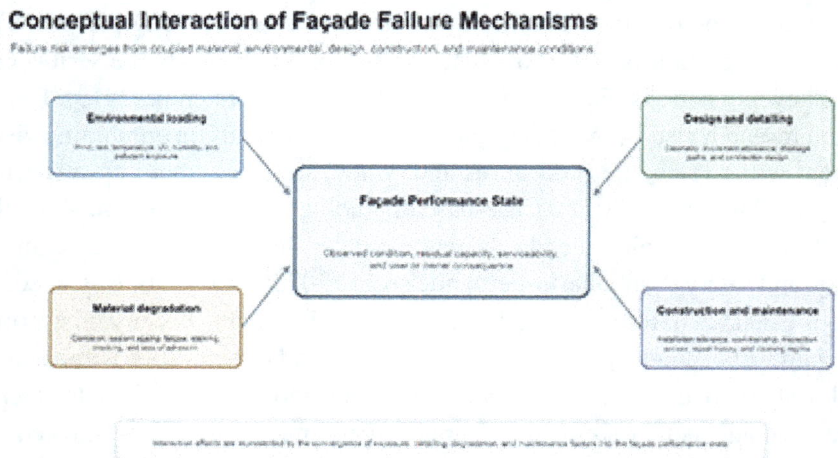

Source: Developed by the author based on façade failure mechanisms discussed in this chapter.

3. FUNDAMENTALS OF FAILURE MODE AND EFFECTS ANALYSIS AND ITS ADAPTATION TO FAÇADE ENGINEERING

3.1 Overview of FMEA

Failure Mode and Effects Analysis (FMEA) is a structured preventive method used to identify how a system may fail, assess the effects of those failures, and prioritize corrective action before significant damage occurs. Rather than reacting after a defect becomes visible, FMEA encourages early analysis of likely failure pathways, their causes, and their consequences. Stamatis (2003) described FMEA as a practical reliability tool because it translates technical knowledge into a formal process of risk prioritization.

Originally developed in high-reliability sectors, FMEA later expanded into manufacturing, quality management, healthcare, and engineering decision-making. Its continued use is largely due to its simplicity and adaptability. Broader reviews of FMEA methodology have similarly shown that its long-term adoption is strongly linked to this adaptability (Silva et al., 2014). As Liu et al. (2013) noted, FMEA remains one of the most widely used methods for risk evaluation despite the existence of more advanced alternatives. This relevance is important for façade engineering, where multiple performance risks must often be compared in a structured way.

3.2 Core Logic of FMEA

The logic of FMEA is built around a small set of linked questions:

1. What can fail?
2. Why might it fail?
3. What happens if it fails?
4. How serious is that effect?
5. How likely is it to occur, and how easily can it be detected?

These questions are translated into a standard structure of failure mode, cause, effect, existing control, and risk rating. The strength of the method is that it does not treat defects as isolated events; instead, it connects technical causes with practical consequences. In façade engineering, this is particularly useful because problems such as leakage, anchor distress, or thermal bridging often arise through a chain of related factors rather than a single isolated defect (Fernando et al., 2023).

Figure 2. Basic analytical logic of FMEA adapted for façade risk evaluation

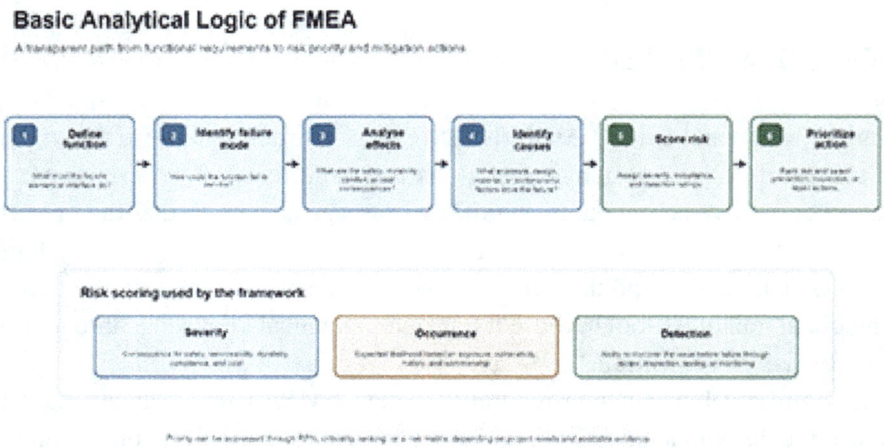

Source: Developed by the author based on FMEA principles and façade engineering literature.

3.3 Severity, Occurrence, and Detection

Traditional FMEA prioritizes risk through three parameters: severity, occurrence, and detection. Severity refers to the seriousness of the consequence if failure occurs. Occurrence refers to the likelihood of the failure. Detection refers to the likelihood that the failure will be identified before it causes major consequences (Stamatis, 2003).

In façade engineering, these parameters need contextual interpretation. A façade failure with low frequency may still deserve high priority if the consequence is severe, such as panel detachment or fire-interface failure. Likewise, a defect with moderate severity may still become important if it is difficult to detect, as in the case of concealed corrosion or hidden moisture damage. This is one reason façade risk cannot be assessed adequately through informal judgment alone (Liu et al., 2013).

Table 3. FMEA parameters in façade engineering

Parameter	General Meaning	Façade Interpretation
Severity	Seriousness of effect	Safety, durability, comfort, compliance, cost impact
Occurrence	Likelihood of failure	Probability under design, exposure, workmanship, and operation
Detection	Likelihood of early discovery	Ease of identifying the defect before serious damage develops

3.4 Risk Priority Number and Its Limitations

In traditional FMEA, the three parameters are combined into a Risk Priority Number (RPN):

RPN = Severity × Occurrence × Detection

This calculation provides a familiar screening device for comparing failure modes and deciding which deserve earlier intervention. Its practical advantage is clarity: project teams can rank multiple risks within one framework even when the technical nature of those risks differs substantially.

However, the traditional RPN is not without problems. Different combinations of severity, occurrence, and detection can produce the same score even though their real risk implications differ. Anes et al. (2018), Dong (2007), von Ahsen (2008), and Liu et al. (2013) noted that the conventional RPN approach can obscure differences in risk meaning, especially where ordinal scores are multiplied, ties are frequent, or weighting assumptions remain implicit. For this reason, RPN should be treated as a prioritization aid rather than a precise measure of absolute risk (Bowles, 2004).

For this chapter, the importance of FMEA lies less in the exact mathematics of a single product score and more in the structured logic it offers. That logic remains highly relevant to façade systems, but it can be strengthened by making the weighting assumptions explicit.

A practical companion measure for façade applications is therefore a façade-weighted priority index (FWPI):

$$FWPI_j = w_s S_j + w_o O_j + w_d D_j,$$

with

$$w_s + w_o + w_d = 1$$

where FWPI is the façade-weighted priority index for failure mode j; S , C , and D are the severity, occurrence, and detection ratings for failure mode j; and w , w_O, and w are the weighting coefficients assigned to severity, occurrence, and detection, respectively.

For routine façade screening, a balanced starting scheme may be defined as w = 0.45, w_O = 0.30, and w = 0.25, reflecting the practical importance of consequence while still accounting for recurrence and inspectability. For life-safety critical details, a safety-led scheme (0.60, 0.20, 0.20) may be more appropriate; for concealed interfaces with difficult access, a maintenance-led scheme (0.40, 0.20, 0.40) can be used to test whether detectability should dominate the ranking.

Where coupled failures are expected, occurrence may also be adjusted by a dependency term:

$$O_j^* = \min(10, O_j + \sum_i d_{ij} a_i)$$

where O* is the dependency-adjusted occurrence rating for failure mode j; O is the baseline occurrence rating; d represents the influence of upstream failure mode i on failure mode j; a indicates whether the upstream condition is present; and the min function limits the adjusted occurrence rating to a maximum value of 10. This additional scaffold does not replace conventional RPN; rather, it complements it by reducing tied rankings and by making façade-specific assumptions transparent.

3.5 Relevance of FMEA to Façade Engineering

FMEA is relevant to façade engineering because façade failures are often:

- multi-causal,
- interface-driven,
- difficult to detect early,
- and capable of producing cascading consequences.

These characteristics are consistent with the type of systems for which FMEA is most useful. Kassem and Mitchell (2015) showed that façade projects are affected by fragmented decision-making and coordination challenges, while Fernando et al. (2023) highlighted the growing technical complexity of façade systems. A structured failure-based method can therefore improve clarity, prioritization, and communication across the façade life cycle.

The method is also compatible with the way façade defects are managed in practice. Designers, contractors, and operators already think in terms of what might go wrong, what the likely consequence is, and what should be checked first. FMEA formalizes that thinking into a more consistent process. In this respect, its practical strength lies in structuring professional judgment rather than replacing it (Ashley & Armitage, 2010).

3.6 Need for Adaptation to Façade Conditions

Despite its usefulness, generic FMEA cannot simply be transferred into façade engineering without modification. Traditional FMEA was developed largely for industrial and process settings, whereas façade systems are exposed to weather, movement, maintenance limitations, regulatory interfaces, and multidisciplinary delivery conditions. As a result, façade-specific adaptation is necessary.

For façade applications, the method should:

- focus more strongly on interfaces and performance functions,
- interpret severity in terms of safety, durability, comfort, and compliance,
- interpret occurrence in relation to exposure, detailing, workmanship, and service conditions,
- and give particular importance to detectability because many façade failures are concealed.

It is also important to situate F-REF against alternative building-risk methods. Fault Tree Analysis is valuable when the analyst needs a top-down causal map from a defined top event, whereas Event Tree Analysis is stronger for tracing consequence pathways once an initiating event has already been identified (Yazdi et al., 2023; Shan et al., 2017). Bow-Tie analysis combines these logics and is especially effective for communicating preventive and protective barriers, but it is usually most useful when the critical event is already clear (Shan et al., 2017; Şeker, 2019).

By contrast, F-REF is most useful earlier in façade decision-making, when multiple failure modes must be screened, compared, and communicated across design, construction, and maintenance. In that sense, it should be treated as a coordinating framework that can later be complemented by FTA, ETA, Bow-Tie analysis, mock-up testing, or forensic investigation when a specific high-consequence pathway requires deeper study.

This need for contextualization provides the basis for the proposed Façade Risk Evaluation Framework (F-REF), which is developed in the next section.

Table 4. From generic FMEA to façade-oriented assessment

Aspect	Generic FMEA	Façade-Oriented Adaptation
Failure focus	Product/process deviation	Interface and performance failure
Severity basis	General defect impact	Safety, leakage, durability, comfort, compliance
Occurrence basis	Process frequency	Exposure, detailing, workmanship, lifecycle conditions
Detection basis	Inspection/test discovery	Visibility of hidden façade defects
Main purpose	Process improvement	Design review, quality control, maintenance prioritization

4. DEVELOPMENT OF THE FAÇADE RISK EVALUATION FRAMEWORK (F-REF)

4.1 Conceptual Basis of F-REF

The Façade Risk Evaluation Framework (F-REF) is proposed as a façade-specific adaptation of Failure Mode and Effects Analysis for the structured identification, classification, and prioritization of façade-related risks. Its purpose is to provide a practical method for organizing façade risks that are often addressed informally in current practice. Earlier work on curtain wall and cladding evaluation also supports the need for structured technical criteria in façade-related decision-making (Chew et al., 2004). Rather than treating defects as isolated events, F-REF links façade functions, failure modes, causes, effects, and priority rankings in a single assessment structure.

The framework is based on three principles. First, façade failures should be assessed in relation to functions and interfaces, not only materials or individual components. Second, risk prioritization should reflect both technical consequence and detectability, since many façade defects remain hidden until secondary damage appears. Third, the method must remain usable in practice, especially in multidisciplinary façade projects where full statistical data are rarely available (Stamatis, 2003; Liu et al., 2013; Silva et al., 2014).

Figure 3. Overall structure of the Façade Risk Evaluation Framework (F-REF)

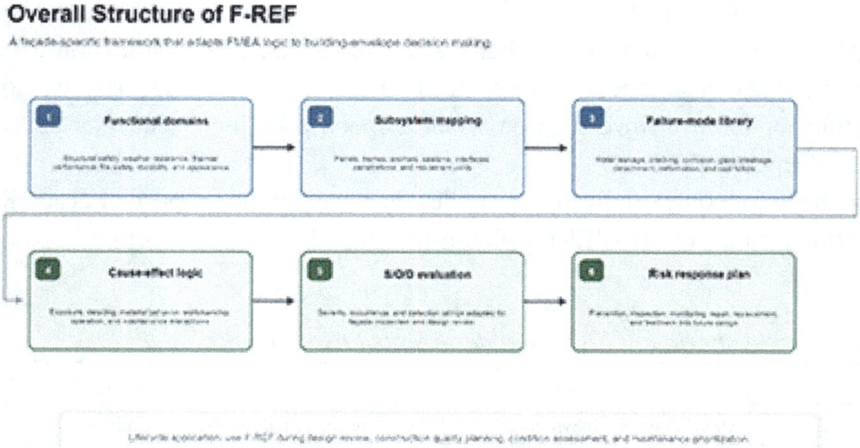

Source: Developed by the author as the proposed F-REF structure.

4.2 Main Structure of the Framework

F-REF is organized into six main analytical layers:

1. **Function identification**: Define what the façade must do, such as provide structural stability, water control, thermal regulation, fire-interface continuity, durability, or maintainability.
2. **Subsystem and interface mapping**: Break the façade into practical units for analysis, such as glazing, framing, anchors, brackets, joints, drainage zones, thermal breaks, and slab-edge interfaces.
3. **Failure mode definition**: Identify how each subsystem or interface may fail, such as leakage, corrosion, detachment, thermal bridging, or barrier discontinuity.
4. **Cause–effect interpretation**: Link each failure mode to likely root causes and likely consequences.
5. **Risk rating and prioritization**: Evaluate severity, occurrence, and detection using façade-oriented criteria.
6. **Response planning**: Define the action needed, such as redesign, enhanced inspection, testing, or maintenance intervention.

Table 5. Main components of F-REF

Layer	Purpose	Output
Function identification	Define required façade performance	Functional criteria
Subsystem/interface mapping	Identify analyzable façade zones	Interface map
Failure mode definition	Describe how failure may occur	Failure inventory
Cause–effect interpretation	Connect defects to causes and consequences	Failure chain
Risk rating	Prioritize failure modes	Ranked risks
Response planning	Define mitigation measures	Action plan

4.3 Functional Basis of the Framework

A key feature of F-REF is that it starts with façade functions rather than directly with defects. This helps keep the assessment performance-oriented. In façade engineering, a defect only becomes meaningful when related to an intended function. Water leakage is a failure because the façade must control moisture; panel detachment is a failure because the façade must remain stable; thermal bridging is a failure because the façade must regulate heat transfer.

For practical use, façade functions may be grouped into six core domains:

- structural stability and movement accommodation,
- water and air control,
- thermal and energy performance,
- fire and life-safety support,
- durability under exposure,
- inspection and maintenance capability.

This functional basis also improves communication among project participants, because different stakeholders may use different technical language but can usually agree on what the façade is expected to achieve (Kassem & Mitchell, 2015; Fernando et al., 2023). Recent façade system studies likewise support the need to evaluate envelope performance across several simultaneous criteria rather than isolated technical checks (Pracucci et al., 2024).

4.4 Failure Mode Classification in F-REF

To keep the framework consistent, failure modes should be classified systematically. F-REF proposes a three-part classification:

(a) **by performance domain**: Structural, moisture, thermal, fire, durability, or operational performance.
(b) **by lifecycle stage**: Design, fabrication, installation, operation, or maintenance.
(c) **by visibility type**: Visible, partially observable, or concealed.

This approach improves clarity. For example, anchor corrosion belongs mainly to the structural domain, often emerges during operation, and may remain concealed for long periods. A leakage defect may belong to the moisture domain, originate from design or installation, and be only partially observable. Such classification supports both technical diagnosis and prioritization.

Table 6. Proposed failure mode classification in F-REF

Dimension	Categories	Example
Performance domain	Structural, moisture, thermal, fire, durability, operational	Thermal bridge
Lifecycle stage	Design, fabrication, installation, operation, maintenance	Installation-induced drainage defect
Visibility type	Visible, partial, concealed	Concealed anchor corrosion

4.5 Adapting Severity, Occurrence, and Detection

F-REF retains the traditional FMEA parameters but interprets them for façade applications and for distinct façade typologies. Severity reflects the seriousness of the consequence in façade terms, including falling hazard, loss of weather protection, hidden moisture accumulation, energy-performance loss, user disruption, compliance exposure, and repair burden.

Occurrence reflects the likelihood of failure under the combined influence of detailing, exposure, workmanship, movement demand, maintenance conditions, and repetition across the façade field.

Detection reflects how easily the failure can be identified before serious consequences develop. This parameter is especially important in façade systems because many defects remain concealed behind finishes, insulation layers, or cavity zones, and because inspection reliability depends on the quality of diagnostic method, access, documentation, and available detection technology (Dias et al., 2021; Vandi et al., 2024).

To make the framework more actionable, Table 7 translates S/O/D scoring into façade-specific rubrics for two widely used façade typologies: unitized curtain walls and ventilated rainscreen systems. The intention is not to fix universal scores, but to show how calibration becomes more reliable when the rating language is tied to actual façade interfaces, inspection constraints, and failure consequences.

A practical way to calibrate these rubrics is a two-round Delphi or expert-panel exercise in which façade consultants, specialist contractors, building-envelope inspectors, and maintenance engineers score shared failure exemplars and then reconcile divergent ratings. Delphi-based inspection studies in the building sector show that structured expert rounds are useful when formal datasets remain incomplete and when consistent inspection criteria must be established across heterogeneous building conditions (Jiménez-Pulido et al., 2020; Sulakatko & Vogdt, 2018).

Table 7. Suggested S/O/D interpretation for F-REF

Façade typology	Parameter	Low (1-3)	Medium (4-7)	High (8-10)
Unitized curtain wall	Severity	Local staining or gasket wear without concealed wetting	Recurring leakage, occupant discomfort, localized insulation wetting	Glass fallout risk, anchor instability, major moisture damage, or fire-interface noncompliance
Unitized curtain wall	Occurrence	Isolated detail in sheltered zone with verified QA and mock-up confirmation	Movement-sensitive joint or drainage path with moderate repetition/ exposure	Repeated edge, corner, or stack-joint issue across elevations or known workmanship variability
Unitized curtain wall	Detection	Visible during routine walkdown, chamber test, or pressure test	Needs selective opening, IR scan, or targeted moisture survey	Concealed anchor, spandrel, or perimeter condition inaccessible after closure
Ventilated rainscreen	Severity	Minor panel distortion or isolated finish blemish	Hidden moisture behind cladding, local insulation wetting, moderate repair campaign	Panel detachment hazard, severe subframe corrosion, or rapid cavity fire spread potential
Ventilated rainscreen	Occurrence	Sheltered elevation with robust fixing pattern and documented installation control	Moderate exposure with localized fixing variability or drainage vulnerability	Highly exposed elevation, repeated fixing corrosion, weak compartmentation, or systemic installation inconsistency
Ventilated rainscreen	Detection	Loose panel or open joint visible in routine inspection	Requires borescope, sample panel removal, or targeted pull-check	Hidden subframe, barrier, or cavity condition only visible after intrusive access

4.6 Capturing Coupled and Cascading Risks

The framework explicitly recognizes that façade failure modes may not be independent. Moisture ingress may accelerate corrosion; corrosion may reduce bracket stiffness; reduced stiffness may amplify movement; and movement may trigger additional seal failure or glazing distress. In rainscreen and cavity-based systems, discontinuities in fire, moisture, and thermal control layers may likewise propagate risk across adjacent assemblies rather than remaining confined to a single detail.

Accordingly, F-REF can include a simple dependency matrix in which each assessed failure mode is checked against upstream and downstream links. A low dependency score indicates negligible interaction; a medium score indicates a credible escalation pathway; and a high score indicates that the mode is likely to intensify occurrence or consequence elsewhere unless arrested. The adjusted occurrence term O*_j proposed in Section 3.4 offers one transparent way to reflect this coupling without abandoning the familiar S/O/D structure.

In practical terms, the dependency review is most valuable for hidden supports, joint networks, spandrel zones, slab-edge closures, cavity barriers, and interfaces where multiple trades contribute to one performance line. These are the conditions in which a façade risk register should move beyond isolated defects and record how one deficiency may activate another.

4.7 Workflow of F-REF

For practical use, F-REF may be applied through the following sequence, including explicit review of coupled risks and weighting assumptions:

1. Define façade scope and context
2. Identify key façade functions
3. Map subsystems and interfaces
4. List likely failure modes
5. Identify causes and effects

   Map dependency links and upstream triggers

6. Review existing controls
7. Assign severity, occurrence, and detection ratings
8. Calculate RPN and, where useful, FWPI

   Test sensitivity under alternative weighting schemes

9. Recommend actions
10. Reassess after mitigation

Figure 4. Proposed workflow of the façade risk evaluation framework

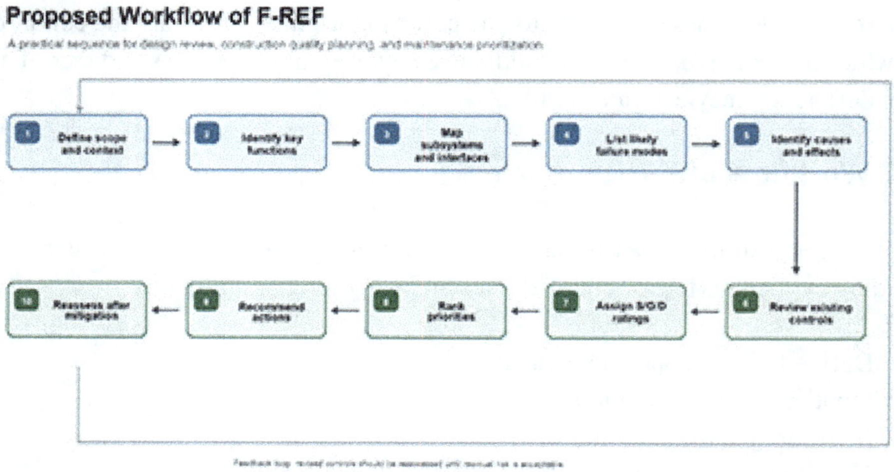

Source: Developed by the author as the proposed F-REF workflow.

This workflow remains intentionally simple, but the added dependency and sensitivity steps help project teams show why specific façade risks stay at the top of the action list under different assumptions rather than relying on a single static score.

4.8 Prototype Assessment Matrix

A simplified assessment matrix helps show how the framework may be used in practice.

Table 8. Prototype F-REF assessment matrix

Function	Interface	Failure mode	Cause	Effect	S	O	D	Dependency note	Action
Water control	Joint	Leakage	Seal failure / blocked drainage	Wetting, staining	7	6	5	May elevate corrosion risk at anchors and subframes	Review detail and inspect drainage
Structural stability	Anchor bracket	Corrosion / distress	Moisture ingress	Support weakening	9	4	8	Often triggered by repeated leakage or trapped condensation	Improve inspection and protection
Thermal control	Slab edge	Thermal bridging	Poor continuity	Condensation, energy loss	6	7	7	Can amplify moisture accumulation in adjacent finishes	Redetail insulation continuity
Fire safety	Cavity barrier	Discontinuity	Installation defect	Fire spread risk	10	3	8	May coexist with moisture and air-barrier discontinuities at closures	Add hold-point inspection

This table is illustrative rather than prescriptive, but it now demonstrates how F-REF can register not only a primary failure mode but also the coupled pathways through which one façade deficiency may elevate another.

4.9 Advantages and Boundaries of the Framework

The main strength of F-REF is that it provides a structured façade risk language across design, construction, and operation. It can support earlier intervention, improve coordination, and make risk prioritization more transparent. The framework is especially useful for interface-sensitive and hidden defects because it combines façade-specific S/O/D rubrics, a companion weighted prioritization index, and explicit dependency mapping for coupled failures.

At the same time, F-REF remains a conceptual framework. It does not replace detailed engineering analysis, code compliance checks, mock-up testing, or forensic investigation. Like other FMEA-derived methods, it still depends partly on expert judgment, and its practical robustness would benefit from Delphi-style calibration, project-based testing, and wider case benchmarking (Anes et al., 2018; Dong, 2007; von Ahsen, 2008; Bowles, 2004; Jiménez-Pulido et al., 2020).

Even with these limitations, the framework offers a useful basis for more consistent façade risk evaluation and provides the foundation for its illustrative application in the next section.

5. ILLUSTRATIVE APPLICATION OF F-REF ACROSS THE FAÇADE LIFECYCLE

5.1 Purpose of the Illustrative Application

Because this chapter proposes a conceptual framework rather than a full empirical case study, it is useful to demonstrate how F-REF can be applied to representative façade problems. The purpose of this section is not to claim numerical validation, but to show how the framework may support risk thinking in realistic façade conditions. The examples translate the framework from abstract structure into practical use, which is especially important in façade engineering, where innovation and system diversity are often discussed more extensively than structured risk application (Laufs & Verboon, 2013).

Three representative scenarios are used here:

1. joint-related weathertightness failure,
2. anchorage and movement-related structural risk,
3. concealed fire-interface or inspection-sensitive risk.

These examples reflect recurring façade concerns identified in the literature and are sufficient to illustrate how F-REF may support decision-making at different lifecycle stages (Fernando et al., 2023; Orlowski et al., 2018).

5.2 Scenario 1: Joint Weathertightness Failure

The first scenario concerns repeated water ingress at façade joints after handover. In F-REF terms, the relevant façade function is water control and drainage continuity. The likely failure mode is loss of watertightness at the joint interface.

Possible causes include sealant incompatibility, poor workmanship, insufficient joint width, blocked drainage, or movement beyond design expectation. Orlowski et al. (2018) showed that façade sealing performance is highly dependent on detailing, tolerances, and movement. The likely effects include staining, wet insulation, corrosion of concealed elements, interior damage, and user complaints (Santamaria-Fernandez et al., 2020).

In this case, the matrix shows how ranking emerges within one scenario rather than from a single defect label alone. The most urgent item is not necessarily the most visible one: blocked drainage outranks a localized sealant defect because recurrence and concealment make the downstream consequences more persistent.

Table 9. Illustrative F-REF Application: Joint Weathertightness Failure

Function	Interface	Failure mode	Cause	Effect				RPN	Rank	Action
Water control	Horizontal stack joint	Blocked drainage path	Debris / sealant squeeze-out	Hidden wetting, staining, corrosion	7	7	7	343	1	Open joint, verify drainage route, adjust inspection frequency
Water control	Corner joint	Movement-induced opening	Insufficient joint allowance	Recurring leakage, wet insulation	8	5	8	320	2	Recheck movement design and corner detailing
Water control	Vertical mullion joint	Sealant discontinuity	Compatibility/workmanship error	Localized leakage and complaints	8	6	6	288	3	Review sealant system and mock-up record

5.3 Scenario 2: Anchorage and Movement-Related Risk

The second scenario concerns a façade system in which anchors, brackets, or connectors experience distress because façade movement is not adequately accommodated. The relevant function is structural stability and movement accommodation.

The likely failure mode is connector distress or reduced support stability. Possible causes include underestimated movement, inaccurate installation tolerance, corrosion, fatigue, or poor support geometry. Ji et al. (2024) demonstrated that connector behavior can play a decisive role in façade instability under dynamic loading. The possible effects include misalignment, cracking, local loss of support, or falling hazard.

Under F-REF, this type of failure would usually receive high priority because even if occurrence is moderate, severity and detectability may justify early action. Earlier integrated façade research also supports evaluating structural response together with broader envelope performance implications (Takeuchi et al., 2006). Appropriate measures would include detail review, hidden support inspection, movement verification, and where necessary, monitoring of sensitive zones.

Table 10. Illustrative F-REF application: Anchorage / movement failure

Function	Interface	Failure mode	Cause	Effect	S	O	D	RPN	Rank	Action
Structural stability	Bracket-to-slab connection	Movement restraint / overstress	Tolerance mismatch and differential movement	Distortion, cracking, support distress	8	6	7	336	1	Verify movement path and re-detail brackets
Structural stability	Concealed bracket zone	Bracket corrosion	Moisture ingress and poor protection	Loss of section and hidden instability	9	4	8	288	2	Open-up inspection and corrosion protection review
Structural stability	Anchor channel / fixing	Fixing loosening	Fatigue, installation error	Local instability and misalignment	8	5	6	240	3	Targeted torque checks and monitoring

5.4 Scenario 3: Concealed Interface Protection Failure

The third scenario concerns incomplete or defective cavity barriers, fire-stopping interfaces, or other concealed protective layers. The relevant function is fire and life-safety support through interface continuity.

The likely failure mode is loss of intended protective continuity at concealed interfaces. Causes may include poor installation, weak coordination, material substitution, or lack of inspection checkpoints. Srivastava et al. (2020) showed that façade systems can contribute to rapid fire spread when critical concealed elements are inadequate.

This kind of risk is important because severity is high while detectability is often low. The expanded matrix below shows that concealed fire-interface failures remain high priority even before visible damage is present; in this sense, F-REF prioritizes verification and documented hold points rather than post-closure assumptions.

Table 11. Illustrative F-REF application: Concealed interface protection failure

Function	Interface	Failure mode	Cause	Effect	S	O	D	RPN	Rank	Action
Fire and life safety	Slab-edge cavity barrier	Barrier discontinuity	Installation gap / omitted segment	Rapid cavity fire spread	10	4	9	360	1	Immediate hold-point verification and opening-up
Fire and life safety	Penetration / anchor interface	Fire-stop gap	Poor coordination / substitution	Loss of compartmentation	9	5	8	360	2	Verify interface details and material traceability
Moisture and fire support	Spandrel closure	Unsealed closure line	Incomplete closure / workmanship defect	Hidden moisture path and reduced barrier continuity	8	5	7	280	3	Targeted intrusive inspection and closure remediation

Table 12. Sensitivity of scenario rankings under alternative weighting schemes

Weighting scheme	Scenario 1 ranking	Scenario 2 ranking	Scenario 3 ranking
Balanced weights (0.45/0.30/0.25)	Blocked drainage > Movement opening > Sealant discontinuity	Movement restraint > Bracket corrosion > Fixing loosening	Barrier discontinuity > Fire-stop gap > Unsealed closure
Safety-led (0.60/0.20/0.20)	Movement opening > Sealant discontinuity > Blocked drainage	Bracket corrosion > Movement restraint > Fixing loosening	Barrier discontinuity > Fire-stop gap > Unsealed closure
Maintenance-led (0.40/0.20/0.40)	Movement opening > Blocked drainage > Sealant discontinuity	Bracket corrosion > Movement restraint > Fixing loosening	Barrier discontinuity > Fire-stop gap > Unsealed closure

5.5 Lifecycle Use of F-REF

A major advantage of F-REF is that it can be applied across the façade lifecycle rather than only after failure occurs. Weighting assumptions and dependency notes can also be carried forward from one phase to the next so that lessons from design review remain visible during installation and later inspection.

During the design phase, the framework can help identify vulnerable details, movement-sensitive interfaces, and high-risk performance conflicts before they become embedded in procurement or fabrication. This is important because many façade risks are effectively designed in at an early stage (Kassem & Mitchell, 2015). Digital coordination tools may strengthen this stage further by improving early integration of façade information and specialist analysis (Htet et al., 2023a).

During the construction phase, F-REF can guide inspection priorities. Rather than treating all façade details equally, project teams can focus on hidden supports, drainage-critical interfaces, cavity barriers, and movement-sensitive joints. This improves the usefulness of QA/QC by linking inspection effort to risk significance and by documenting where coupled failures may propagate if concealed details are closed too early.

During the operation and maintenance phase, F-REF can support targeted inspection and proactive upkeep. Standardized condition assessment, service-life modeling, and emerging inspection technologies all reinforce the need to connect observed defects with maintenance decisions rather than treating inspections as isolated records (Straub, 2009; Silva et al., 2015; Dias et al., 2021). Vandi et al. (2024) and Wei et al. (2023) emphasized the growing importance of façade monitoring and defect detection for long-term performance management. In this context, the framework can help owners and facility teams decide which zones require more frequent inspection, which failures should trigger intrusive verification, and which dependencies justify earlier intervention.

Table 13. Potential lifecycle applications of F-REF

Project Phase	Main Use of F-REF	Typical Output
Design	Identify vulnerable details and interfaces	Design-risk register
Fabrication / Installation	Prioritize hidden and critical details	Inspection checkpoints
Handover	Record known risk zones	Façade risk log
Operation / Maintenance	Guide inspection and maintenance priority	Lifecycle maintenance plan

5.6 Practical Implications

The examples above show that F-REF can support clearer communication among designers, façade specialists, contractors, and operators. Instead of treating façade issues as isolated complaints, the framework encourages teams to think in terms of function, failure mode, cause, effect, and action.

Its main practical value lies in helping project teams:

- identify high-risk façade conditions earlier,
- prioritize limited inspection and maintenance resources,
- and improve transparency in façade-related decision-making.

In this sense, F-REF is valuable not because it produces a number alone, but because it organizes façade risk into a more disciplined and actionable form.

6. DISCUSSION, IMPLICATIONS, LIMITATIONS, AND CONCLUSION

6.1 Discussion and Implications

The proposed Façade Risk Evaluation Framework (F-REF) addresses a practical gap in façade engineering by offering a structured method for identifying and prioritizing façade-related risks across the building life cycle. In current practice, façade risks are often recognized but assessed in fragmented ways, typically through discipline-specific review, informal experience, or reactive problem-solving. F-REF contributes by organizing those risks around functions, failure modes, causes, effects, and priority logic.

From a theoretical perspective, the framework extends reliability-oriented thinking into façade engineering, where failures are frequently interface-driven, exposure-sensitive, and difficult to detect early. Liu et al. (2013) noted that FMEA remains valuable because it creates a clear structure for comparing risks, even in complex systems. In façade engineering, this is especially useful because problems such as leakage, anchor distress, thermal bridging, and concealed interface failure often compete for attention but do not receive consistent prioritization.

Practically, the framework can support different stakeholder groups. Designers may use it to review vulnerable details and performance conflicts during early stages. Contractors and façade specialists may use it to improve inspection priorities during fabrication and installation. Owners and facility teams may use it to support more risk-sensitive maintenance planning. Attia et al. (2018) emphasized that long-term façade performance depends not only on design intent but also on operation, maintenance, and monitoring. In this regard, F-REF supports a more lifecycle-oriented view of façade reliability. This is also consistent with broader sustainability-oriented discussions of façade systems, where long-term performance, material choice, and operational efficiency must be considered together (Wu & Flemmer, 2020; Htet et al., 2023b).

This positioning makes clear that F-REF is not proposed as a substitute for all other methods. Rather, it offers the most value when teams need a common façade-specific screening platform before they commit resources to deeper fault-tree, event-tree, bow-tie, testing, or forensic work.

Table 14. Positioning F-REF against complementary building-risk methods

Method	Primary logic	Strength in façade engineering	Limitation	Best use
F-REF	Failure-mode screening across S/O/D criteria	Transparent prioritization across design, construction, and maintenance	Depends on rating discipline and expert calibration	Early screening, coordination, inspection planning
FTA	Top-down causation from a defined top event	Strong for tracing why a critical façade failure occurred	Less efficient for broad multi-mode screening	Deep causal analysis of one high-consequence event
ETA	Forward consequence branching from an initiating event	Useful for mapping escalation after breach or ignition	Needs a clearly defined initiating event	Scenario escalation and contingency planning
Bow-Tie	Combined cause-and-consequence barrier logic	Effective for communicating preventive and protective barriers	Can become cumbersome when many façade modes are still undefined	Life-safety communication and barrier management
Checklist / qualitative review	Experience-led screening	Fast and familiar during early review	Weak prioritization transparency and traceability	Quick preliminary screening before formal assessment

6.2 Limitations

The framework also has important limitations. First, F-REF is still a conceptual model rather than a fully validated industry tool. Its usefulness in practice depends on expert interpretation, clarity of rating criteria, and disciplined application across project stages.

Second, like other FMEA-based approaches, it retains a degree of subjectivity in scoring severity, occurrence, and detection. This is difficult to avoid in façade engineering, where large comparative datasets are often unavailable and project conditions vary significantly. Anes et al. (2018), Dong (2007), and von Ahsen (2008) highlighted that this is a common limitation in risk-prioritization methods derived from traditional FMEA. The new façade-specific rubrics and weighting schemes improve transparency, but they do not remove the need for calibration.

Third, the framework does not replace detailed engineering analysis, code compliance review, mock-up testing, or forensic investigation. Instead, it should be treated as a decision-support layer that helps determine where attention, testing, and control should be concentrated and which coupled risks justify deeper follow-up.

6.3 Future Directions

Future research should focus on validating and refining the framework through practical application. A first priority is expert-based calibration of severity, occurrence, and detection criteria for different façade typologies through a structured Delphi panel. Building-inspection Delphi studies suggest that even a modest expert panel can improve rating consistency when the scoring language is tied to concrete defect examples and repeated feedback rounds (Jiménez-Pulido et al., 2020; Sulakatko & Vogdt, 2018). A second priority is case-based testing, where the framework is applied to real façade projects or known failure cases.

Further development should also test the modified scoring logic presented in this chapter. Sensitivity studies can examine how priority order shifts when severity, occurrence, or detection are weighted differently for life-safety details, durability-critical interfaces, or maintenance-driven inspection planning. Longitudinal inspection and degradation datasets could also be used to improve occurrence assumptions for different façade materials and exposure conditions (Barrelas et al., 2021; Silva et al., 2015). In parallel, digital integration with monitoring, sensing, defect detection, and BIM-enabled coordination could help move F-REF from a static assessment table toward a more adaptive lifecycle tool (Vandi et al., 2024; Wei et al., 2023; Htet et al., 2023a; Pracucci et al., 2024).

6.4 Conclusion

This chapter developed a Façade Risk Evaluation Framework (F-REF) based on the principles of Failure Mode and Effects Analysis to address the lack of a structured façade-specific risk prioritization method. The chapter showed that façade systems involve interrelated structural, environmental, durability, fire, and operational risks that are often managed in fragmented ways despite their strong interactions.

By adapting FMEA to façade functions, interfaces, and lifecycle conditions, the chapter proposes a framework that can identify likely failure modes, connect them to causes and effects, and support clearer risk prioritization. This contribution is strengthened through façade-specific S/O/D rubrics, a weighted prioritization option, dependency-aware occurrence adjustment, sensitivity testing, and a clearer explanation of how F-REF should sit alongside complementary risk methods.

Its main contribution lies in providing a clearer and more disciplined basis for risk-informed façade engineering without sacrificing practical usability. With further calibration and validation, F-REF may offer a useful step toward more consistent and lifecycle-oriented façade risk management.

REFERENCES

Anes, V., Henriques, E., Freitas, M., & Reis, L. (2018). A new risk prioritization model for failure mode and effects analysis. *Quality and Reliability Engineering International*, *34*(4), 516–528. DOI: 10.1002/qre.2269

Ashley, L., & Armitage, G. (2010). Failure mode and effects analysis. *Journal of Patient Safety*, *6*(4), 210–215. DOI: 10.1097/PTS.0b013e3181fc98d7 PMID: 21500607

Attia, S., Bilir, S., Safy, T., Struck, C., Loonen, R., & Goia, F. (2018). Current trends and future challenges in the performance assessment of adaptive façade systems. *Energy and Building*, *179*, 165–182. DOI: 10.1016/j.enbuild.2018.09.017

Barrelas, J., Dias, I. S., Silva, A., de Brito, J., Flores-Colen, I., & Tadeu, A. (2021). Impact of environmental exposure on the service life of façade claddings A statistical analysis. *Buildings (Basel, Switzerland)*, *11*(12), 615. DOI: 10.3390/buildings11120615

Bowles, J. B. (2004). An assessment of RPN prioritization in a failure modes effects and criticality analysis. *Proceedings of the Annual Reliability and Maintainability Symposium*, 380–386. DOI: 10.17764/jiet.47.1.y576m26127157313

Chew, M. Y. L., Tan, S., & Kang, K. H. (2004). A technical evaluation index for curtain wall and cladding facades. *Structural Survey*, *22*(4), 210–227. DOI: 10.1108/02630800410563750

Dias, I. S., Flores-Colen, I., & Silva, A. (2021). Critical analysis about emerging technologies for building's façade inspection. *Buildings (Basel, Switzerland)*, *11*(2), 53. DOI: 10.3390/buildings11020053

Djuric-Mijovic, D., Savić, J., Milanović, D., & Cilic, A. (2018). Seismic damage mitigation of the glazed building façade. Facta Universitatis, Series. *Architecture and Civil Engineering*, *16*(3), 343–353. DOI: 10.2298/FUACE171115013D

Dong, C. (2007). Failure mode and effects analysis based on fuzzy utility cost estimation. *International Journal of Quality & Reliability Management*, *24*(9), 958–971. DOI: 10.1108/02656710710826216

Fernando, D., Navaratnam, S., Rajeev, P., & Sanjayan, J. (2023). Study of technological advancement and challenges of façade system for sustainable building: Current design practice. *Sustainability (Basel)*, *15*(19), 14319. DOI: 10.3390/su151914319

Htet, A., Liana, S. R., Aung, T., & Bhaumik, A. (2023a). The intersection of facade engineering and building information modeling: Opportunities and challenges. *Journal of Technology Innovations and Energy*, *2*(3), 94–110. DOI: 10.56556/jtie.v2i3.643

Htet, A., Liana, S. R., Aung, T., & Bhaumik, A. (2023b). Economic and environmental impact analysis of green building materials in facade engineering. *International Journal of Membrane Science and Technology*, *10*(2), 2805–2814. DOI: 10.15379/ijmst.v10i2.2973

Inoue, T., Matsumura, S., & Seike, T. (2003). Analysis of façade engineering in Japan. *Journal of Asian Architecture and Building Engineering*, *2*(2), b77–b82. DOI: 10.3130/jaabe.2.b77

Ji, X., Luo, R., Zhuang, Y., Gao, X., Lim, W., & Qu, Z. (2024). Seismic behavior of a stone curtain wall system with undercut bolt anchorage under various loading protocols. *Earthquake Engineering & Structural Dynamics*, *54*(2), 706–723. DOI 10.1002/eqe.4277

Jiménez-Pulido, C., Jiménez-Rivero, A., & García-Navarro, J. (2020). Sustainable management of the building stock: A Delphi study as a decision-support tool for improved inspections. *Sustainable Cities and Society*, *61*, 102184. DOI: 10.1016/j.scs.2020.102184

Kaneki, Y., Takeuchi, T., Miyazaki, K., & Iwata, M. (2008). Studies on integrated facade engineering Structural performance of integrated facades. *AIJ Journal of Technology and Design*, *14*(27), 137–142. DOI: 10.3130/aijt.14.137

Kassem, M., & Mitchell, D. M. (2015). Bridging the gap between selection decisions of facade systems at the early design phase: Issues, challenges and solutions. *Journal of Facade Design and Engineering*, *3*(2), 165–183. DOI: 10.3233/FDE-150037

Kültür, S., Türkeri, N., & Knaack, U. (2019). A holistic decision support tool for facade design. *Buildings (Basel, Switzerland)*, *9*(8), 186. DOI: 10.3390/buildings9080186

Latifah, N. L., Zulwaqar, K. H., Andini, K., & Nisa, A. S. (2022). Building envelope design with glass curtain wall to reduce OTTV, study case: WU Tower Building at Bandung, Indonesia. *International Journal of Built Environment and Scientific Research*, *6*(2), 97–110. DOI: 10.24853/ijbesr.6.2.97-110

Laufs, W., & Verboon, E. (2013). Innovative façade design and products. In Façade Tectonics Institute / ASCE proceedings (pp. 154-193). American Society of Civil Engineers. DOI: 10.1061/9780784412701.ch10

Li, R., Dai, Y., & Wang, R. Z. (2015). Experimental and theoretical analysis on thermal performance of solar thermal curtain wall in building envelope. *Energy and Building*, *87*, 324–334. DOI: 10.1016/j.enbuild.2014.11.029

Liu, H.-C., Liu, L., & Liu, N. (2013). Risk evaluation approaches in failure mode and effects analysis: A literature review. *Expert Systems with Applications*, *40*(2), 828–838. DOI: 10.1016/j.eswa.2012.08.010

Niedostatkiewicz, M. (2022). Emergency condition of the ceramic facade of a residential building. *Inżynieria Bezpieczeństwa Obiektów Antropogenicznych*, *2*(2), 32–38. DOI: 10.37105/iboa.140

Orlowski, K., Shanaka, K., & Mendis, P. (2018). Design and development of weatherproof seals for prefabricated construction: A methodological approach. *Buildings (Basel, Switzerland)*, *8*(9), 117. DOI: 10.3390/buildings8090117

Pracucci, A., Magnani, S., & Casadei, O. (2020). The integration of vacuum insulated glass in unitized façade for the development of innovative lightweight and highly insulating energy efficient building envelope The results of Eensulate façade system design. *Designs*, *4*(4), 40. DOI: 10.3390/designs4040040

Pracucci, A., Vandi, L., Morganti, L., Fernández, A. G., Nunez Diaz, M., Navarro Muedra, A., Győri, V., Kouyoumji, J.-L., & Astudillo Larraz, J. (2024). Design and simulation for technological integration of bio-based components in façade system modules. *Buildings (Basel, Switzerland)*, *14*(4), 1114. DOI: 10.3390/buildings14041114

Ruiz, F., Aguado, A., Serrat, C., & Casas, J. R. (2019). Condition assessment of building façades based on hazard to people. *Structure and Infrastructure Engineering*, *15*(10), 1346–1365. DOI: 10.1080/15732479.2019.1621907

Santamaria-Fernandez, S., Eizaguirre-Iribar, A., & Olano-Azkune, X. (2020). Pathologies of a glass building envelope that affect durability and comfort. DBMC 2020 Proceedings. DOI: 10.23967/dbmc.2020.105

Şeker, Ş. (2019). Analyzing occupational risks of pharmaceutical industry under uncertainty using a Bow-Tie analysis. *International Journal of Optimization and Control: Theories & Applications*, *9*(2), 113–124. DOI: 10.11121/ijocta.01.2019.00655

Shan, X., Liu, K., & Sun, P.-L. (2017). Risk analysis on leakage failure of natural gas pipelines by fuzzy Bayesian network with a Bow-Tie model. *Scientific Programming*, *3639524*, 1–11. Advance online publication. DOI: 10.1155/2017/3639524

Silva, A., Gaspar, P. L., de Brito, J., & Neves, L. C. (2015). Probabilistic analysis of degradation of façade claddings using Markov chain models. *Materials and Structures*, *49*(7), 2871–2892. DOI: 10.1617/s11527-015-0692-5

Silva, M. M., Ferreira, F. A. F., & Santos, S. P. (2014). Methodologies for failure mode and effect analysis: A literature review. *Quality and Reliability Engineering International*, *30*(7), 989–1005.

Srivastava, G., Nakrani, D., & Ghoroi, C. (2020). Performance of combustible facade systems with glass, ACP and firestops in full-scale, real fire experiments. *Fire Technology*, *56*(4), 1575–1598. DOI: 10.1007/s10694-019-00943-4

Stamatis, D. H. (2003). *Failure mode and effect analysis: FMEA from theory to execution* (2nd ed.). ASQ Quality Press.

Straub, A. (2009). Dutch standard for condition assessment of buildings. *Structural Survey*, *27*(1), 23–35. DOI: 10.1108/02630800910941665

Straube, J. F., & Burnett, E. F. P. (2005). *Building science for building enclosures*. Building Science Press.

Sulakatko, V., & Vogdt, F. U. (2018). Construction process technical impact factors on degradation of the external thermal insulation composite system. *Sustainability (Basel)*, *10*(11), 3900. DOI: 10.3390/su10113900

Takeuchi, T., Koyano, K., Yasuda, K., Yuasa, K., & Iwata, M. (2006). Studies on integrated facade engineering: Proposal for integrated facade and its evaluation. [Transactions of AIJ]. *Journal of Environmental Engineering*, *71*(601), 81–88. DOI: 10.3130/aije.71.81_1

Takeuchi, T., Yasuda, K., & Iwata, M. (2006). Studies on integrated building facade engineering with high-performance structural elements. IABSE Reports, 33-40. DOI: 10.2749/222137806796185526

Vandi, L., Calcagni, M. T., Belletti, F., Pandarese, G., Martarelli, M., Revel, G. M., Docter, V., & Pracucci, A. (2024). Structural health monitoring for prefabricated building envelope under stress tests. *Applied Sciences (Basel, Switzerland)*, *14*(8), 3260. DOI: 10.3390/app14083260

von Ahsen, A. (2008). Cost-oriented failure mode and effects analysis. *International Journal of Quality & Reliability Management*, *25*(5), 466–476. DOI: 10.1108/02656710810873871

Wei, G., Wan, F., Zhou, W., Xu, C., Ye, Z., Liu, W., Lei, G., & Xu, L. (2023). BFD-YOLO: A YOLOv7-based detection method for building façade defects. *Electronics (Basel), 12*(17), 3612. DOI: 10.3390/electronics12173612

Wu, Y., & Flemmer, C. L. (2020). Glass curtain wall technology and sustainability in commercial buildings in Auckland, New Zealand. *International Journal of Built Environment and Sustainability, 7*(2), 57–65. DOI: 10.11113/ijbes.v7.n2.495

Yazdi, M., Mohammadpour, J., Li, H., Huang, H.-Z., Zarei, E., Pirbalouti, R. G., & Adumene, S. (2023). Fault tree analysis improvements: A bibliometric analysis and literature review. *Quality and Reliability Engineering International, 39*(5), 1639–1659. DOI: 10.1002/qre.3271

Compilation of References

Abdelrady, A., Abdelhafez, M. H. H., & Ragab, A. (2021). Use of insulation based on nanomaterials to improve energy efficiency of residential buildings in a hot desert climate. *Sustainability (Basel)*, *13*(19), 10856. DOI: 10.3390/su13095266

Abediniangerabi, B., Makhmalbaf, A., & Shahandashti, M. (2021). Deep learning for estimating energy savings of early-stage facade design decisions. *Energy and AI*, *5*, 100077. DOI: 10.1016/j.egyai.2021.100077

Addington, M., & Schodek, D. (2005). *Smart Materials and Technologies in Architecture*. Architectural Press.

Adegoke, M., Alaka, H., & Ajayi, S. (2022). Application of multilayer extreme learning machine for efficient building energy prediction. *Energies*, *15*(24), 9512. DOI: 10.3390/en15249512

Adio, S. A., Atofarati, E. O., Muritala, A. O., Huan, Z., & Veeredhi, V. R. (2025). Nanofluids flow boiling and convective heat transfer in microchannels: A systematic review. *Journal of Thermal Analysis and Calorimetry*, *150*(12), 8879–8911. DOI: 10.1007/s10973-025-14265-x

Aelenei, D., Gonçalves, H., & Pereira, R. (2016). Adaptive facades for sustainable buildings. *Energy Procedia*, *91*, 269–275. DOI: 10.1016/j.egypro.2016.06.218

Aguerre, J., Klein, T., Konstantinou, T., & Veerman, M. (2022). Facades-as-a-service: The role of technology in the circular servitisation of the building envelope. *Applied Sciences (Basel, Switzerland)*, *12*(3), 1267. DOI: 10.3390/app12031267

Ahmed, H., & Podder, C. (2024). Mixed convection heat transfer and flow of Al_2O_3–water nanofluid in a square enclosure with heated obstacles and varied boundary conditions. *arXiv preprint arXiv:2401.05497*.

Ahmed, Z., & Bhargav, A. (2020). Thermal interfacial resistance and nanolayer effect on the thermal conductivity of Al_2O_3–CO_2 nanofluid: a molecular dynamics approach. arXiv preprint arXiv:2006.12805.

Ahmed, Z., & Bhargav, A. (2021). Effect of aggregation morphology on thermal conductivity and viscosity of Al_2O_3–CO_2 nanofluid: a molecular dynamics approach. *Nanoscience and Technology: An International Journal, 12*(1).

Ahmed, M. G., & Gawad, M. A. (2022). Architecture sustainability and energy efficiency. *International Journal of Energy Production and Management*, *7*(4), 320–330. https://elar.urfu.ru/handle/10995/126075

Ahmed, M., Meteab, M. M., Salih, Q. O., Mohammed, H. A., & Alawi, O. A. (2022). Experimental investigation of aqueous hybrid nanofluid in flat plate solar collectors. *Energies*, *15*(22), 8541. DOI: 10.3390/en15228541

Ahmed, S. F., Mofijur, M., Rafa, N., Chowdhury, A. T., Nuzhat, S., Rahman, S. A., Ong, H. C., & Mahlia, T. M. I. (2022). Green approaches in synthesising nanomaterials for environmental nanobioremediation: Technological advancements, applications, benefits and challenges. *Environmental Research*, *204*, 112320. DOI: 10.1016/j.envres.2021.111967 PMID: 34450159

Akimov, L., Bezborodov, A. A., & Badenko, V. (2023). Example-based dynamic facade design using the facade daylighting performance improvement (FDPI) indicator. *Building Simulation*, *16*(12), 2261–2283. DOI: 10.1007/s12273-023-1073-6

Alexandri, E., & Jones, P. (2008). Temperature decreases in urban canyons due to green walls and green roofs in diverse climates. *Building and Environment*, *43*(4), 480–493. DOI: 10.1016/j.buildenv.2006.10.055

Alhawari, A., & Mukhopadhyaya, P. (2018). Thermal bridges in building envelopes An overview of impacts and solutions. *International Review of Applied Sciences and Engineering*, *9*(1), 31–40. DOI: 10.1556/1848.2018.9.1.5

Alioglu, T. U. G. B. A., & Sirel, A. Y. S. E. (2018). The Use of Textile-Based Materials in Shell System Design in Architecture and an Evaluation in Terms of. *J. Contemp. Urban Aff*, *2*(3), 88–94. DOI: 10.25034/ijcua.2018.4722

Almeida, R., Rodrigues, H., Gomes, M. G., & Furtado, A. (2025). Financial incentives for energy and seismic retrofitting of reinforced concrete buildings: A review of an integrated approach. *Innovative Infrastructure Solutions*, *10*(8), 329. DOI: 10.1007/s41062-025-02123-6

Almohaimeed, N., & Pérez-Villalba, M. (2025). The UAE's Journey to Sports Sustainability: Commitment in Action.

Al-Refaei, A. A. A., Abdulsamad, A., Ali, D. A., Ibrahim, A., Ateeq, A., & Al Balushi, F. K. (2024). A conceptual framework for the impact of entrepreneurial leadership on innovation work behavior and sustainable innovation performance. *Entrepreneurship Innovation and Education for Performance Improvement*, 577-598.

Al-Yasiri, Q., & Szabó, M. (2021). Incorporation of phase change materials into building envelope for thermal comfort and energy saving: A comprehensive analysis. *Journal of Building Engineering, 44*, 102943. DOI: 10.1016/j.jobe.2020.102122

American Institute of Steel Construction. (2016). *ANSI/AISC 360-16: Specification for Structural Steel Buildings*. American Institute of Steel Construction.

American Institute of Steel Construction. (2017). *Steel Construction Manual* (15th ed.). American Institute of Steel Construction.

Anes, V., Henriques, E., Freitas, M., & Reis, L. (2018). A new risk prioritization model for failure mode and effects analysis. *Quality and Reliability Engineering International, 34*(4), 516–528. DOI: 10.1002/qre.2269

Antypa, D., Vlysidis, A., Gkika, A., et al. (2022). Life cycle assessment of advanced building materials towards NZEBs. E3S Web of Conferences, 349, 04001. DOI: 10.1051/e3sconf/202234904001

Architektur & Der Technik. (2017). Textile-based architecture.

Arnaud, L., & Gourlay, E. (2012). Experimental study of parameters influencing mechanical properties of hemp concrete. *Construction & Building Materials, 28*(1), 50–56. DOI: 10.1016/j.conbuildmat.2011.07.052

Arnesano, M., Pandarese, G., Martarelli, M., Naspi, F., Gurunatha, K. L., Sol, C., Portnoi, M., Ramirez, F. V., Parkin, I. P., Papakonstantinou, I., & Revel, G. M. (2021). Optimization of the thermochromic glazing design for curtain wall buildings based on experimental measurements and dynamic simulation. *Solar Energy, 230*, 155–168. DOI: 10.1016/j.solener.2021.01.013

Arowoiya, V. A., Moehler, R. C., & Fang, Y. (2024). Digital twin technology for thermal comfort and energy efficiency in buildings: A state-of-the-art and future directions. *Energy and Built Environment, 5*(1), 1–19. DOI: 10.1016/j.enbenv.2023.05.004

Asadi, E., da Silva, M. G., Antunes, C. H., & Dias, L. (2012). Multi-objective optimization for building retrofit strategies: A model and an application. *Energy and Building, 44*, 81–87. DOI: 10.1016/j.enbuild.2011.10.016

Asadi, H., Uhlemann, J., Stranghoener, N., & Ulbricht, M. (2021). Water influence on the uniaxial tensile behavior of polytetrafluoroethylene-coated glass fiber fabric. *Materials (Basel), 14*(4), 846. DOI: 10.3390/ma14040846 PMID: 33578878

Asadi, H., Uhlemann, J., & Stranghöner, N. (2018). Water infiltration impact on tensile strength and breaking strain of architectural fabrics. *Advances in Structural Engineering, 21*(16), 2605–2616. DOI: 10.1177/1369433218756005

Asdrubali, F., Ferracuti, B., Lombardi, L., Guattari, C., Evangelisti, L., & Grazieschi, G. (2015). A review of structural, thermo-physical, acoustical, and environmental properties of wooden materials for building applications. *Building and Environment, 114*, 307–332. DOI: 10.1016/j.buildenv.2016.12.033

Ashby, M. F. (2013). *Materials and the Environment: Eco-informed Material Choice* (2nd ed.). Butterworth-Heinemann.

Ashley, L., & Armitage, G. (2010). Failure mode and effects analysis. *Journal of Patient Safety, 6*(4), 210–215. DOI: 10.1097/PTS.0b013e3181fc98d7 PMID: 21500607

Asphaug, S. K., Time, B., & Kvande, T. (2021). Moisture accumulation in building façades exposed to accelerated artificial climatic ageing: A complementary analysis to NT Build 495. *Buildings (Basel, Switzerland), 11*(12), 568. DOI: 10.3390/buildings11120568

Atewi, Y. R., Hasan, M. F., & Güneyisi, E. (2019). Fracture and permeability properties of glass fiber reinforced self-compacting concrete with and without nanosilica. *Construction & Building Materials, 226*, 993–1005. DOI: 10.1016/j.conbuildmat.2019.08.029

Athauda, R. S., Asmone, A. S., & Conejos, S. (2023). Climate change impacts on facade building materials: A qualitative study. *Sustainability (Basel), 15*(10), 7893. DOI: 10.3390/su15107893

Attia, S., Bilir, S., Safy, T., Struck, C., Loonen, R., & Goia, F. (2018). Current trends and future challenges in the performance assessment of adaptive façade systems. *Energy and Building, 179*, 165–182. DOI: 10.1016/j.enbuild.2018.09.017

Attia, S., Eleftheriou, P., Xeni, F., Morlot, R., Ménézo, C., Kostopoulos, V., Betsi, M., Kalaitzoglou, I., Pagliano, L., Cellura, M., Almeida, M., Ferreira, M., Baracu, T., Badescu, V., Crutescu, R., & Hidalgo-Betanzos, J. M. (2017). Overview and future challenges of nearly zero energy buildings (nZEB) design in Southern Europe. *Energy and Building, 155*, 439–458. DOI: 10.1016/j.enbuild.2017.09.043

Attia, S., Lioure, R., & Declaude, Q. (2020). Future trends and main concepts of adaptive facade systems. *Energy Science & Engineering*, *8*(9), 3255–3272. DOI: 10.1002/ese3.725

Avolio, B. J., & Bass, B. M. (2004). *Multifactor Leadership Questionnaire: Manual and sampler set* (3rd ed.). Mind Garden.

Azarnejhad, A., & Mahdavi, A. (2019). On the impact of building façades' color on thermal building performance and outdoor thermal comfort. *Applied Mechanics and Materials*, *887*, 189–195. . DOI: 10.4028/www.scientific.net/AMM.887.189

Aziz, H. N. A., & Abdelall, M. I. (2021). Responsive façades design using nanomaterials for optimizing buildings' energy performance. *WIT Transactions on Ecology and the Environment*, *253*, 321–332. https://books.google.com. DOI: 10.2495/SC210331

Baburajan, M. (2025). Sustainable Sourcing: A comparative study of sourcing practices between Swedish and Qatari SMEs operating in the Façade industry.

Bagheri, M. (2025). Advancing architecture through shape memory alloy actuators. *Virginia Tech ETDs*. https://vtechworks.lib.vt.edu/handle/10919/119126

Bagheri, A., & Harrison, C. (2020). Entrepreneurial leadership measurement: A multi-dimensional construct. *Journal of Small Business and Enterprise Development*, *27*(4), 659–679. DOI: 10.1108/JSBED-01-2019-0027

Balani, R. S., Patel, D., Barthwal, S., & Arun, J. (2021). Fabrication of Air Conditioning System Using the Engine Exhaust Gas. In *Advances in Interdisciplinary Research in Engineering and Business Management* (pp. 369–378). Springer Nature Singapore. DOI: 10.1007/978-981-16-0037-1_29

Baptista, J., Kokare, S., Godina, R., & Aelenei, D. (2024). A comparative life cycle assessment of ETICS and ventilated facade systems with timber cladding. *Energy and Building*, *304*, 4375760. DOI: 10.1016/j.enbuild.2023.113842

Barai, R., Kumar, D., & Wankhade, A. (2023). Heat transfer performance of nanofluids in heat exchangers: A review. *Journal of Thermal Engineering*, *9*(1), 86–106. DOI: 10.18186/thermal.1243398

Barozzi, M., Lienhard, J., Zanelli, A., & Monticelli, C. (2016). The sustainability of adaptive envelopes: Developments of kinetic architecture. *Procedia Engineering*, *155*, 275–284. DOI: 10.1016/j.proeng.2016.08.029

Barreira, E., & Delgado, J. (2017). Diagnosis of rising damp and other kinds of moisture in building components using non-destructive techniques. *Diffusion Foundations*, *10*, 120–131. . DOI: 10.4028/www.scientific.net/DF.10.120

Barrelas, J., Dias, I. S., Silva, A., de Brito, J., Flores-Colen, I., & Tadeu, A. (2021). Impact of environmental exposure on the service life of façade claddings A statistical analysis. *Buildings (Basel, Switzerland), 11*(12), 615. DOI: 10.3390/buildings11120615

Bas, C., Türkeri, A. N., & Lacasse, M. (2023). Hygrothermal performance of external wall systems under current climatic conditions and future climate projections of Turkey. In *DBMC 2023 Proceedings*. DOI: 10.23967/c.dbmc.2023.082

Bassir, D., Lodge, H., Chang, H., & Celik, T. (2023). Application of artificial intelligence and machine learning for BIM [Review]. *International Journal for Simulation and Multidisciplinary Design Optimization, 14*, 5. DOI: 10.1051/smdo/2023005

Bedon, C., & Amadio, C. (2017). Passive control systems for the blast enhancement of glazing curtain walls under explosive loads. *The Open Civil Engineering Journal, 11*(1), 396–419. DOI: 10.2174/1874149501711010396

Bedon, C., & Rajčić, V. (2019). Textiles and fabrics for enhanced structural glass facades: Potentials and challenges. *Buildings (Basel, Switzerland), 9*(7), 156. DOI: 10.3390/buildings9070156

Beech, J., & Beasley, J. L. (1994). Further studies of cure and durability of building sealants. In *Science and Technology of Building Seals, Sealants, Glazing, and Waterproofing* (Vol. 3, pp. 33-50). ASTM International. https://doi.org/DOI: 10.1520/STP13998S

Bhering, A. F., Franco, R. N., Santos, M. A., Sathler, L. M., Pedroti, L. G., Varum, H., & Paes, J. L. R. (2023). Experimental analysis of watertightness performance of interfaces between masonry and steel structures subjected to accelerated aging. *Buildings (Basel, Switzerland), 13*(9), 2123. DOI: 10.3390/buildings13092123

Bianchi, S., Andriotis, C. P., Klein, T., & Overend, M. (2024). Multi-criteria design methods in façade engineering: State-of-the-art and future trends. *Building and Environment, 250*, 111184. DOI: 10.1016/j.buildenv.2024.111184

Bienvenido-Huertas, D. (2020). Analysis of the relationship of the improvement of façades and thermal bridges of Spanish building stock with the mitigation of its energy and environmental impact. *Energies, 13*(17), 4499. DOI: 10.3390/en13174499

Bluyssen, P. M. (2010). Towards new methods and ways to create healthy and comfortable buildings. *Building and Environment, 45*(4), 808–818. DOI: 10.1016/j.buildenv.2009.08.020

Bocanegra, J. A., Marchitto, A., & Misale, M. (2025). Nanofluids in solar collectors: A comprehensive review focused on its sedimentation. *Clean Technologies and Environmental Policy*, *27*(4), 1753–1784. DOI: 10.1007/s10098-024-02964-2

Boeri, S. (2015). *Bosco Verticale: A home for trees, humans, and birds*. Corraini Edizioni.

Boeske, J. (2023). Leadership towards sustainability: A review of sustainable, sustainability, and environmental leadership. *Sustainability (Basel)*, *15*(16), 12626. DOI: 10.3390/su151612626

Bolek, B., Tutal, O., & Ozbasaran, H. (2023). A systematic review on artificial intelligence applications in architecture. *Journal of Design for Resilience in Architecture and Planning*, *4*(1), 91–104. DOI: 10.47818/DRArch.2023.v4i1085

Bortolini, R., & Forcada, N. (2018). Building inspection system for evaluating the technical performance of existing buildings. *Journal of Performance of Constructed Facilities*, *32*(5), 04018073. Advance online publication. DOI: 10.1061/(ASCE)CF.1943-5509.0001220

Both, I., Burcă, M., Benzar, ., & Ungureanu, V. (2023). Numerical study on the behaviour of built-up cold-formed steel corrugated web beams end connections. *Civil Engineering Journal*, *9*(4), 770–786. DOI: 10.28991/CEJ-2023-09-04-01

Bottger, W. (2019, June). Development of (interactive) façade elements on base of waste materials of (waste) water companies.

Bowles, J. B. (2004). An assessment of RPN prioritization in a failure modes effects and criticality analysis. *Proceedings of the Annual Reliability and Maintainability Symposium*, 380–386. DOI: 10.17764/jiet.47.1.y576m26127157313

Bozorgan, N., & Shafahi, M. (2015). Performance evaluation of nanofluids in solar energy: A review of the recent literature. *Micro and Nano Systems Letters*, *3*(1), 5. DOI: 10.1186/s40486-015-0014-2

Bran, F. (2014). *Green infrastructure for sustainable urban development.* Urbanism. Architecture. *Constructions*, *5*(2), 29–38.

Braun, V., & Clarke, V. (2006). Using thematic analysis in psychology. *Qualitative Research in Psychology*, *3*(2), 77–101. DOI: 10.1191/1478088706qp063oa

Bribian, I. Z., Capilla, A. V., & Usón, A. A. (2010). Life cycle assessment of building materials: Comparative analysis of energy and environmental impacts and evaluation of the eco-efficiency improvement potential. *Building and Environment*, *46*(5), 1133–1140. DOI: 10.1016/j.buildenv.2010.12.002

Brzezicki, M. (2021). A typology of adaptive façades. *An empirical study based on the morphology of glazed facades. Cogent Arts & Humanities*, *8*(1), 1960699. DOI: 10.1080/23311983.2021.1960699

Buday, P., Ingeli, R., & Čekon, M. (2014). Influence of thermal break element applied in balcony slab on internal surface temperature. *Advanced Materials Research*, *1057*, 79–86. . DOI: 10.4028/www.scientific.net/AMR.1057.79

Bueno, A. M., de Paula Xavier, A. A., & Broday, E. E. (2021). Evaluating the connection between thermal comfort and productivity in buildings: A systematic literature review. *Buildings (Basel, Switzerland)*, *11*(10), 446. DOI: 10.3390/buildings11060244

Building Research Establishment. (2018). *BREEAM UK new construction 2018 technical manual*. BRE.

Búryová, D., & Sedlák, P. (2021). Life cycle assessment of coated and thermally modified wood façades. *Coatings*, *11*(12), 1487. DOI: 10.3390/coatings11121487

Cabeza, L. F., Rincón, L., Vilariño, V., Pérez, G., & Castell, A. (2014). Life cycle assessment (LCA) and life cycle energy analysis (LCEA) of buildings and the building sector: A review. *Renewable & Sustainable Energy Reviews*, *29*, 394–416. DOI: 10.1016/j.rser.2013.08.037

Cabeza, L., Castell, A., Barreneche, C., De Gracia, A., & Fernández, A. (2011). Materials used as PCM in thermal energy storage in buildings: A review. *Renewable & Sustainable Energy Reviews*, *15*(3), 1675–1695. DOI: 10.1016/j.rser.2010.11.018

CEN. (2006). *EN 1993-1-5 Eurocode 3: Design of Steel Structures, Part 1-5: Plated Structural Elements*. European Committee for Standardization.

Chamilothori, K., Kampitaki, A.-M., & Oungrinis, K.-A. (2013, November). Climate-responsive shading systems with integrated shape memory alloys (SMA). In Proceedings of the 8th Energy Forum on Advanced Building Skins (pp. [page range if available]). Bressanone, Italy.

Charbonneau, L., Polak, M. A., & Penlidis, A. (2014). Mechanical properties of ETFE foils: Testing and modelling. *Construction & Building Materials*, *60*, 63–72. DOI: 10.1016/j.conbuildmat.2014.02.048

Chaudhari, K. S., & Walke, P. V. (2014). Applications of nanofluid in solar energy—A review. *International Journal of Engine Research*, *3*(3).

Chen, Y., Groll, M., Mertz, R., Maydanik, Y. F., & Vershinin, S. V. (2005, January). Steady-state and transient performance of a miniature loop heat pipe. In *International Conference on Nanochannels, Microchannels, and Minichannels* (Vol. 41855, pp. 183–189). DOI: 10.1115/ICMM2005-75122

Cherif, C., Diestel, O., Engler, T., Hufnagl, E., & Weiland, S. (2016). Processing aspects and application examples. *In Textile Materials for Lightweight Constructions: Technologies-Methods-Materials-Properties (pp.* 599-661). Berlin, Heidelberg: Springer Berlin Heidelberg.

Chew, M. Y. L., & Gan, V. J. L. (2022). Long-standing themes and future prospects for the inspection and maintenance of facade falling objects from tall buildings. *Sensors (Basel)*, *22*(16), 6070. DOI: 10.3390/s22166070 PMID: 36015831

Chew, M. Y. L., Tan, S., & Kang, K. H. (2004). A technical evaluation index for curtain wall and cladding facades. *Structural Survey*, *22*(4), 210–227. DOI: 10.1108/02630800410563750

Chilton, J., & Lau, B. (2015). Lighting and the visual environment in architectural fabric structures. *In J.* I. de Llorens (Ed.), Fabric Structures in Architecture (pp. 203-219). Woodhead Publishing. DOI: 10.1016/B978-1-78242-233-4.00007-3

Chilton, J., & Velasco, R. (2005). Applications of textile composites in the construction industry. In *Design and manufacture of textile composites* (pp. 424–435). Woodhead Publishing. DOI: 10.1533/9781845690823.424

Choi, S. U. (1995, November). Enhancing thermal conductivity of fluids with nanoparticles. In *ASME International Mechanical Engineering Congress and Exposition* (Vol. 17421, pp. 99–105). American Society of Mechanical Engineers. DOI: 10.1115/IMECE1995-0926

Chruściel, J. J. (2022). Modifications of textile materials with functional silanes, liquid silicone softeners, and silicone rubbers-A review. *Polymers*, *14*(20), 4382. DOI: 10.3390/polym14204382 PMID: 36297958

Chung-Camargo, K., González, J., Chen Austin, M., Carpino, C., Mora, D., & Arcuri, N. (2024). Advances in retrofitting strategies for energy efficiency in tropical climates: A systematic review and analysis. *Buildings (Basel, Switzerland)*, *14*(6), 1633. DOI: 10.3390/buildings14061633

Churkina, G., Organschi, A., Reyer, C. P. O., Ruff, A., Vinke, K., Liu, Z., & Schellnhuber, H. J. (2020). Buildings as a global carbon sink. *Nature Sustainability*, *3*(4), 269–276. DOI: 10.1038/s41893-019-0462-4

Čikić Tovarović, J., Ivanović-Šekularac, J., & Šekularac, N. (2017). Renovation of existing glass facade in order to implement energy efficiency and media facade. *Energy and Building*, *152*, 653–666. DOI: 10.1016/j.enbuild.2017.07.049

Corradi, M., Mustafaraj, E., & Speranzini, E. (2023). Sustainability considerations in remediation, retrofit, and seismic upgrading of historic masonry structures. *Environmental Science and Pollution Research International*, *30*(10), 25274–25286. DOI: 10.1007/s11356-021-17490-7 PMID: 34797539

Dabaon, M., El Hadidy, A. M., & Manaa, R. (2019). Shear behaviour of continuous tapered steel plate girders with corrugated webs. *Journal of Engineering Research*, *3*(6), 60–65. DOI: 10.21608/erjeng.2019.125506

Dadzie, J., Runeson, G., & Ding, G. (2019). Assessing determinants of sustainable upgrade of existing buildings. *Journal of Engineering. Defence Technology*, *18*(1), 270–292. DOI: 10.1108/JEDT-09-2018-0148

Dahmani, S. (2024). Sustainable approach to nanocomposites and nanocoatings: Research and industrial applications of hydrophobic nanocoatings. *Sustainable Approach to Protective Nanocoatings, IGI Global.* https://www.igi-global.com/chapter/sustainable-approach-to-nanocomposites-and-nanocoatings/330064

Darko, A., & Chan, A. P. C. (2017). Review of barriers to green building adoption. *Sustainable Development (Bradford)*, *25*(3), 167–179. DOI: 10.1002/sd.1651

Das, S. K., Putra, N. S. D., Thiesen, P., & Roetzel, W. (2003). Temperature dependence of thermal conductivity enhancement for nanofluids. *Journal of Heat Transfer*, *125*(4), 567–574. DOI: 10.1115/1.1571080

De Filippo, M., Asadiabadi, S., Kuang, J. S., Mishra, D. K., & Sun, H. (2023). AI-powered inspections of facades in reinforced concrete buildings. *HKIE Transactions*, *30*(1), 1–14. DOI: 10.33430/V30N1THIE-2020-0023

Dean, T. J., & McMullen, J. S. (2007). Toward a theory of sustainable entrepreneurship: Reducing environmental degradation through entrepreneurial action. *Journal of Business Venturing*, *22*(1), 50–76. DOI: 10.1016/j.jbusvent.2005.09.003

Dehghan, M., Akbari, G., Montazerin, N., & Maroufi, A. (2024). Thermoeconomic prioritization of heat transfer fluids in solar-driven tri-generation systems using MADM methods. *Energy Equipment & Systems, 12*(2).

Demirbas, A. (2004). Combustion characteristics of different biomass fuels. *Progress in Energy and Combustion Science*, *30*(2), 219–230. DOI: 10.1016/j.pecs.2003.10.004

Denz, P. R., Sauer, C., Waldhör, E. F., Schneider, M., & Vongsingha, P. (2021). Smart textile sun-shading: Development of functional adaptex prototypes. *Journal of Facade Design and Engineering*, *9*(1), 101–116.

Di Bitonto, M. G., & Torres, N. M. R. (2025). Energy Efficiency through Fog. *Sustainable Built Environment for People and Society, 129.*

Dias, I. S., Flores-Colen, I., & Silva, A. (2021). Critical analysis about emerging technologies for building's facade inspection. *Buildings (Basel, Switzerland)*, *11*(2), 53. DOI: 10.3390/buildings11020053

Ding, S., & Liu, D. (2006). Accelerated aging on sealants and a quantificational way to evaluate the durability. *Journal of Testing and Evaluation*, *34*(2), 79–82. DOI: 10.1520/JTE14067

Dixit, M. K., Fernández-Solís, J. L., Lavy, S., & Culp, C. H. (2010). Identification of parameters for embodied energy measurement: A literature review. *Energy and Building*, *42*(8), 1238–1247. DOI: 10.1016/j.enbuild.2010.02.016

Diz-Mellado, E., Mascort-Albea, E., Romero-Hernández, R., Galán-Marín, C., Rivera-Gómez, C., Ruiz-Jaramillo, J., & Jaramillo-Morilla, A. (2021). Non-destructive testing and finite element method integrated procedure for heritage diagnosis: The Seville Cathedral case study. *Journal of Building Engineering*, *37*, 102134. DOI: 10.1016/j.jobe.2020.102134

Djuric-Mijovic, D., Savić, J., Milanović, D., & Cilic, A. (2018). Seismic damage mitigation of the glazed building façade. Facta Universitatis, Series. *Architecture and Civil Engineering*, *16*(3), 343–353. DOI: 10.2298/FUACE171115013D

Dong, C. (2007). Failure mode and effects analysis based on fuzzy utility cost estimation. *International Journal of Quality & Reliability Management*, *24*(9), 958–971. DOI: 10.1108/02656710710826216

Douglas, J., & Ransom, B. (2013). *Understanding building failures* (4th ed.). Routledge. DOI: 10.4324/9780203125175

Duffie, J. A., & Beckman, W. A. (2013). *Solar Engineering of Thermal Processes*. John Wiley & Sons. DOI: 10.1002/9781118671603

Eapen, J., Li, J., & Yip, S. (2007). Beyond the Maxwell limit: Thermal conduction in nanofluids with percolating fluid structures. *Physical Review. E*, *76*(6), 062501. DOI: 10.1103/PhysRevE.76.062501 PMID: 18233882

Eastman, J. A., Choi, S. U. S., Li, S., Yu, W., & Thompson, L. J. (2001). Anomalously increased effective thermal conductivities of ethylene glycol-based nanofluids containing copper nanoparticles. *Applied Physics Letters*, *78*(6), 718–720. DOI: 10.1063/1.1341218

Efthymiou, E., Cöcen, Ö. N., & Russo Ermolli, S. (2010). Sustainable aluminium systems. *Sustainability (Basel)*, *2*(9), 3100–3109. DOI: 10.3390/su2093100

Elangovan, M., Srimanickam, B., Čep, R., Saranya, A., & Ramachandran, M. (2022). Experimental study of a hybrid solar collector using TiO_2/water nanofluids. *Energies*, *15*(12), 4425. DOI: 10.3390/en15124425

Elgaaly, M., & Seshadri, A. (1998). Steel built-up girders with trapezoidally corrugated webs. *Engineering Journal (New York)*, *35*(1), 1–11. DOI: 10.62913/engj.v35i1.694

Elghaish, F., Matarneh, S. T., Talebi, S., Abu-Samra, S., Salimi, G., & Rausch, C. (2021). Deep learning for detecting distresses in buildings and pavements: A critical gap analysis. *Construction Innovation*, *22*(3), 554–579. DOI: 10.1108/CI-09-2021-0171

Elguezabal, P., & Garay, R. (2015). Experiences when employing different alternatives for envelope upgrading. *Journal of Facade Design and Engineering*, *3*(1), 81–89. DOI: 10.3233/FDE-150032

Elkington, J. (1997). *Cannibals with forks: The triple bottom line of 21st century business*. Capstone Publishing.

Elmnefi, M., & Al-Khazraji, W. (2023). Thermal performance enhancement of parabolic trough collectors using CuO nanofluid. *International Journal of Numerical Methods for Heat & Fluid Flow*, *33*(9), 3124–3163. DOI: 10.1108/HFF-11-2022-0659

El-Mowafy, B. N., Elmokadem, A., & Waseef, A. A. (2022). Evaluating adaptive facade performance in early building design stage: An integrated daylighting simulation and machine learning. In *Recent Advances in Advanced Machine Learning Approaches for Data Science* (pp. 211–223). Springer., DOI: 10.1007/978-3-031-03918-8_20

European Commission. (2020). A new circular economy action plan: For a cleaner and more competitive Europe. https://eur-lex.europa.eu/legal-content/EN/TXT/?uri=CELEX:52020DC0098

European Committee for Standardization. (2011). *EN 15978:2011 Sustainability of construction works: Assessment of environmental performance of buildings: Calculation method*. CEN.

Fakhabi, M. M., Hamidian, S. M., & Aliehyaei, M. (2024). Exploring the role of the Internet of Things in green buildings. *Energy Science & Engineering*, *12*(9), 3779–3822. DOI: 10.1002/ese3.1840

Falorca, J. F., & Lanzinha, J. (2020). Facade inspections with drones: Theoretical analysis and exploratory tests. *International Journal of Building Pathology and Adaptation*, *39*(2), 235–258. DOI: 10.1108/IJBPA-07-2019-0063

Faltynova, M., Matouskova, E., Sedina, J., & Pavelka, K. (2016). Building facade documentation using laser scanning and photogrammetry and data implementation into BIM. *The International Archives of the Photogrammetry, Remote Sensing and Spatial Information Sciences*, *B3*(B3), 215–220. DOI: 10.5194/isprs-archives-XLI-B3-215-2016

Faragalla, A. M. A., & Asadi, S. (2022). Biomimetic design for adaptive building façades: A paradigm shift towards environmentally conscious architecture. *Energies*, *15*(15), 5390. DOI: 10.3390/en15155390

Farhan, M., Hassan, F., Zubair, M., Jamil, F., Shahid, I., Khiadani, M., & Ali, H. M. (2025). Experimental performance of auto-tracking parabolic trough solar collector with nanofluids. *Journal of Thermal Analysis and Calorimetry*. Advance online publication. DOI: 10.1007/s10973-025-14001-5

Farid, M., Khudhair, A. M., Razack, S. A. K., & Al-Hallaj, S. (2004). A review on phase change energy storage: Materials and applications. *Energy Conversion and Management*, *45*(9-10), 1597–1615. DOI: 10.1016/j.enconman.2003.09.015

Fassman-Beck, E. A., Simcock, R., Wang, S., & Gruner, D. (2013). 4.6 Vegetated stormwater biofilters: Green roofs and living walls. In *Handbook of Water Sensitive Planning and Design.*

Faton, C., & Hermans, J. (2025). Designing under Constraints: How Entrepreneurs Make the Future in Belgium's Construction Industry. Journal of Business Venturing Design, 100033.

Feizbahr, M., & Pourzanjani, P. (2024). Nanotechnology in Construction: Innovations, Applications, and Impacts. *environment, 1, 7.*

Fernando, D., Navaratnam, S., Rajeev, P., & Sanjayan, J. (2023). Study of technological advancement and challenges of façade system for sustainable building: Current design practice. *Sustainability (Basel)*, *15*(19), 14319. DOI: 10.3390/su151914319

Ficapal, A., & Mutis, I. (2019). Framework for the detection, diagnosis, and evaluation of thermal bridges using infrared thermography and unmanned aerial vehicles. *Buildings (Basel, Switzerland)*, *9*(8), 179. DOI: 10.3390/buildings9080179

Fiore, P., Donnarumma, G., Falce, C., & Sicignano, G. (2021). A tool for the evaluation of energy-environmental retrofit interventions on opaque walls using ETICS. *Sustainability (Basel)*, *13*(7), 4063. DOI: 10.3390/su13074063

Flansbjer, M., Williams Portal, N., Vennetti, D., & Mueller, U. (2018). Composite behaviour of textile reinforced reactive powder concrete sandwich façade elements. *International Journal of Concrete Structures and Materials*, *12*(1), 71. DOI: 10.1186/s40069-018-0301-4

Fotopoulou, E., Zafeiropoulos, A., Terroso-Saenz, F., Şimşek, U., González-Vidal, A., Tsiolis, G., Gouvas, P., Liapis, P., Fensel, A., & Skarmeta, A. (2017). Providing personalized energy management and awareness services for energy efficiency in smart buildings. *Sensors (Basel)*, *17*(9), 2054. DOI: 10.3390/s17092054 PMID: 28880227

Fu, H., & Gray, K. A. (2021). The key to maximizing the benefits of antimicrobial and self-cleaning coatings is to fully determine their risks. *Current Opinion in Chemical Engineering*, *33*, 100715. DOI: 10.1016/j.coche.2021.100761 PMID: 36569284

Fylan, F., & Glew, D. (2021). Barriers to domestic retrofit quality: Are failures in retrofit standards a failure of retrofit standards? *Indoor and Built Environment*, *31*(3), 710–718. DOI: 10.1177/1420326X211027197

Garg, R., Garg, R., & Eddy, N. O. (2024). Applications of nanocoatings: Smart and sustainable protection. *Sustainable Approach to Protective Nanocoatings, IGI Global.* https://www.igi-global.com/chapter/applications-of-nanocoatings/330071

Gasparri, E., Brambilla, A., Lobaccaro, G., Goia, F., Andaloro, A., & Sangiorgio, A. (2022). Façade innovation: between 'product' and 'process. In *Rethinking Building Skins* (pp. 1–13). Woodhead Publishing. DOI: 10.1016/B978-0-12-822477-9.00025-5

Ghorbanalavi, F. (2018). Performance characteristics and practical applications of common building polymeric materials for building envelope. *Journal of Energy and Economic Development*, *3*(2), 13–33.

Ghosh, J., Rupanty, N. S., Noor, T., Asif, T. R., & Islam, T. (2025). Functional coatings for textiles: Advancements in flame resistance, antimicrobial defense, and self-cleaning performance. *RSC Advances*, *15*(12), 9876–9891. DOI: 10.1039/D4RA08567B

Gibilisco, G., Monteleone, A., Rodonò, G., & Sapienza, V. (2024, October). Self-Sufficient and Responsive Textile. In Proceedings of the 11th International Conference of Ar. Tec.(Scientific Society of Architectural Engineering): Colloqui. AT. e 2024-Volume 2 (Vol. 611, p. 345). Springer Nature.

Giro-Paloma, J., Martínez, M., Cabeza, L. F., & Fernández, A. I. (2015). Types, methods, techniques, and applications for microencapsulated phase change materials (MPCM): A review. *Renewable & Sustainable Energy Reviews*, *53*, 1059–1075. DOI: 10.1016/j.rser.2015.09.040

Goh, C. S., Goh, H. H., Omar, M. F., & Toh, T. C. (2013). Leadership styles and project success in the construction industry. *International Journal of Project Management*, *31*(7), 998–1012.

Gonçalves, M., Figueiredo, A., Almeida, R., & Vicente, R. (2024). Dynamic façades in buildings: A systematic review across thermal comfort, energy efficiency and daylight performance. *Renewable & Sustainable Energy Reviews*, *182*, 114934. DOI: 10.1016/j.rser.2024.114474

González-Lezcano, R. A. (2021). Advancements in sustainable architecture and energy efficiency. *Google Books.* https://books.google.com

Gopalsamy, V., Rajasekaran, K., & Baccoli, R. Experimental and analytical study of nanofluid-based parabolic trough collectors.

Grant, M. J., & Booth, A. (2009). A typology of reviews: An analysis of 14 review types and associated methodologies. *Health Information and Libraries Journal*, *26*(2), 91–108. DOI: 10.1111/j.1471-1842.2009.00848.x PMID: 19490148

Grillo, E., Milardi, M., & Olivieri, F. (2022). A review of innovative materials for the design of adaptive biomimetic façades. *Advances in Urban Planning and Architectural Design*, *67*, 231–245. DOI: 10.1007/978-3-031-16261-8_19

Gupta, M., Arora, N., Kumar, R., Kumar, S., & Dilbaghi, N. (2014). A comprehensive review of experimental investigations of forced convective heat transfer characteristics for various nanofluids. *International Journal of Mechanical and Materials Engineering*, *9*(1), 11. DOI: 10.1186/s40712-014-0011-x

Gupta, V., MacMillan, I. C., & Surie, G. (2004). Entrepreneurial leadership: Developing and measuring a cross-cultural construct. *Journal of Business Venturing*, *19*(2), 241–260. DOI: 10.1016/S0883-9026(03)00040-5

Hamann, R., Smith, J., Tashman, P., & Marshall, R. S. (2017). Why do SMEs go green? An analysis of wine firms in South Africa. *Business & Society*, *56*(1), 23–56. DOI: 10.1177/0007650315575106

Han, B., Wang, R., Liang, Y., Dai, Y., Zhang, L., & Shen, J. (2015). Life cycle assessment of ceramic façade material and its comparative analysis with three other common façade materials. *Journal of Cleaner Production*, *99*, 86–93. DOI: 10.1016/j.jclepro.2015.03.032

Hannoura, A. A.. (2025). Thermal properties of heat transfer fluids in solar collectors. *Journal of Thermal Analysis and Calorimetry*. Advance online publication. DOI: 10.1007/s10973-025-14912-3

Hargreaves, T. (2011). Practice-ing behaviour change: Applying social practice theory to pro-environmental behaviour change. *Journal of Consumer Culture*, *11*(1), 79–99. DOI: 10.1177/1469540510390500

He, G., Wu, Q., Li, Z., Ge, W., Lv, D., & Cong, L. (2021). Ventilation performance of solar chimney in a test house: Field measurement and validation of plume model. *Building and Environment*, *193*, 107648. DOI: 10.1016/j.buildenv.2021.107648

Heigermoser, D., De Soto, B. G., Abbott, E. L. S., & Chua, D. K. H. (2019). BIM-based Last Planner System tool for improving construction project management. *Automation in Construction*, *104*, 246–254. DOI: 10.1016/j.autcon.2019.03.019

Heinzel, T. (2018). Reactive architecture, augmented textiles, domotics and soft architecture fabrication: On electronic and reactive textiles in domestic contexts. *Textile (Oxford)*, *16*(1), 34–61. DOI: 10.1080/14759756.2017.1332907

He, L., Standen, P., & Coetzer, A. (2017). The perceived personal characteristics of entrepreneurial leaders. *Small Enterprise Research*, *24*(2), 97–119. DOI: 10.1080/13215906.2017.1334225

He, Y., Yu, H., Ozaki, A., Dong, N., & Zheng, S. (2017). Long-term thermal performance evaluation of green roof system based on two new indexes: A case study in Shanghai area. *Building and Environment*, *120*, 13–28. DOI: 10.1016/j.buildenv.2017.04.001

Hong, S., Choi, A., & Sung, M. (2018). Impact of bi-directional PV blind control method on lighting, heating and cooling energy consumption in mock-up rooms. *Energy and Building*, *176*, 1–16. DOI: 10.1016/j.enbuild.2018.07.022

Hooi, H. C., Ahmad, N. H., Amran, A., & Rahman, S. A. (2016). The functional role of entrepreneurial orientation and entrepreneurial bricolage in ensuring sustainable entrepreneurship. *Management Research Review*, *39*(12), 1616–1638. DOI: 10.1108/MRR-06-2015-0144

Hosseinalipour, S. M., Al-Taee, A. A. H., & Asiaei, S. (2024). Improving energy consumption for dynamic intake and exhaust façades: A new setup design: Part B. *Building Services Engineering Research and Technology*, *45*(5), 613–639. DOI: 10.1177/01436244241264273

Houtman, R. (2015). Materials used for architectural fabric structures. In *Fabric Structures in Architecture* (pp. 101–121). Woodhead Publishing. DOI: 10.1016/B978-1-78242-233-4.00003-6

Htet, A., Gupta, S. K., & Oruganti, S. K. (2025). Modular adaptive facades: Designing a scalable framework for retrofitting urban buildings. *SGS Engineering & Sciences, 1*(2), NA.

Htet, A., Liana, S. R., Aung, T., & Bhaumik, A. (2023). Economic and environmental impact analysis of green building materials in facade engineering. *International Journal of Membrane Science and Technology, 10*(2), 2805–2814. DOI: 10.15379/ijmst.v10i2.2973

Htet, A., Liana, S. R., Aung, T., & Bhaumik, A. (2023). The Intersection of Facade Engineering and Building Information Modeling: Opportunities and Challenges. *Journal of Technology Innovations and Energy, 2*(3), 94–110. DOI: 10.56556/jtie.v2i3.643

Htet, A., Liana, S. R., Aung, T., Bhaumik, A., & Giri, O. P. (2025). The Role of Silicone in Enhancing the Weatherability of Facade Materials. *Journal of Innovations in Business and Industry, 3*(4), 201–214. DOI: 10.61552/JIBI.2025.04.00

Huedo, P., Mulet, E., & López-Mesa, B. (2016). A model for the sustainable selection of building envelope assemblies. *Environmental Impact Assessment Review, 57*, 63–77. DOI: 10.1016/j.eiar.2015.11.005

Hussain, M.. (2024). Direct absorption solar collectors working with nanofluids: A review. *Journal of Thermal Analysis and Calorimetry, 149*(15), 7919–7946. DOI: 10.1007/s10973-024-13304-3

Hussein, A. M., Awad, A. T., & Alı, H. H. M. (2024). Evaluation of the thermal efficiency of nanofluid flows in flat plate solar collector. *Journal of Thermal Engineering, 10*(2), 299–307. DOI: 10.18186/thermal.1448578

Hussein, H. S. (2023). The state of the art of nanomaterials and its applications in energy saving. *Bulletin of the National Research Center, 47*(1), 76. DOI: 10.1186/s42269-023-00984-4

Ibrahim, A., Al-refaei, A. A. A., Alshuhumi, S., Al-Hidabi, D., & Ateeq, A. (2024). The effect of entrepreneurial leadership on employee's creativity and sustainable innovation performance in education sector: a literature review. *Business Development via AI and Digitalization: Volume 2*, 143-154.

Iftikhar, H. B., & Aslam, S. (2025). Textile Architecture in Transition: Global Trends and Emerging Applications in Pakistan. [DSSR]. *Dialogue Social Science Review, 3*(6), 612–628.

Inoue, T., Matsumura, S., & Seike, T. (2003). Analysis of façade engineering in Japan. *Journal of Asian Architecture and Building Engineering, 2*(2), b77–b82. DOI: 10.3130/jaabe.2.b77

International Energy Agency (IEA). (2021a). *Net Zero by 2050: A Roadmap for the Global Energy Sector.* https://www.iea.org/reports/net-zero-by-2050

International Energy Agency (IEA). (2021b). *Tracking Buildings 2021.* https://www.iea.org/reports/tracking-buildings-2021

International Organization for Standardization. (2006a). *ISO 14040:2006 Environmental management: Life cycle assessment: Principles and framework.* ISO.

International Organization for Standardization. (2006b). *ISO 14044:2006 Environmental management: Life cycle assessment: Requirements and guidelines.* ISO.

International Organization for Standardization. (2011). *ISO 15686-1:2011 Buildings and constructed assets: Service life planning: Part 1: General principles and framework.* ISO.

Isaia, F., Fiorentini, M., Serra, V., & Capozzoli, A. (2021). Enhancing energy efficiency and comfort in buildings through model predictive control for dynamic facades with electrochromic glazing. *Journal of Building Engineering, 43*, 102535. DOI: 10.1016/j.jobe.2021.102535

ISO. (2006). *ISO 14040:2006 Environmental management Life cycle assessment Principles and framework.* International Organization for Standardization.

Iturralde Carrera, L. A., Díaz-Tato, L., Constantino-Robles, C. D., Garcia-Barajas, M. G., Zapatero-Gutiérrez, A., Álvarez-Alvarado, J. M., & Rodríguez-Reséndiz, J. (2025). Advances in Mounting Structures for Photovoltaic Systems: Sustainable Materials and Efficient Design. *Technologies, 13*(5), 204. DOI: 10.3390/technologies13050204

Jaafar, A. D., Al-Timimi, D. A. H., Alatabe, M. J. A., Shorbaz, M., & Kadhim Abed, B. (2024). Thermal conductivity of nanofluid: a mini review.

Jalil, W. A. (2020, February). Smart textiles for the architectural façade. []. IOP Publishing]. *IOP Conference Series. Materials Science and Engineering, 737*(1), 012078. DOI: 10.1088/1757-899X/737/1/012078

Jamilu, G., Abdou, A., & Asif, M. (2024). Dynamic facades for sustainable buildings: A review of classification, applications, prospects and challenges. *Energy Reports*, *10*, 4123–4140. DOI: 10.1016/j.egyr.2024.05.047

Jelle, B. P. (2011). Traditional, state-of-the-art and future thermal building insulation materials and solutions: Properties, requirements and possibilities. *Energy and Building*, *43*(10), 2549–2563. DOI: 10.1016/j.enbuild.2011.05.015

Jelle, B. P., Breivik, C., & Drolsum Rokenes, H. (2012). Building integrated photovoltaic products: A state-of-the-art review and future research opportunities. *Solar Energy Materials and Solar Cells*, *100*, 69–96. DOI: 10.1016/j.solmat.2011.12.016

Jha, P., Das, B., Gupta, R., Mondol, J. D., & Ehyaei, M. A. (2023). Review of recent research on photovoltaic thermal solar collectors. *Solar Energy*, *257*, 164–195. DOI: 10.1016/j.solener.2023.04.004

Jha, S. K., Sankar, A., Zhou, Y., & Ghosh, A. (2024). Incorporation of phase change materials in buildings. *Construction Materials*, *4*(4), 676–703. DOI: 10.3390/constrmater4040037

Jiao, Y., Chu, L., Zhu, M., Wei, Y., Wang, W., Wang, H., & Wang, H. (2021). UV-curable hyperbranched polyester sealant with tunable mechanical properties. *Journal of Biomaterials Applications*, *36*(4), 731–739. DOI: 10.1177/08853282211000527 PMID: 33940972

Jim, C., & Peng, L. L. (2012). Substrate moisture effect on water balance and thermal regime of a tropical extensive green roof. *Ecological Engineering*, *47*, 9–23. DOI: 10.1016/j.ecoleng.2012.06.020

Jiménez-Pulido, C., Jiménez-Rivero, A., & García-Navarro, J. (2020). Sustainable management of the building stock: A Delphi study as a decision-support tool for improved inspections. *Sustainable Cities and Society*, *61*, 102184. DOI: 10.1016/j.scs.2020.102184

Ji, X., Luo, R., Zhuang, Y., Gao, X., Lim, W., & Qu, Z. (2024). Seismic behavior of a stone curtain wall system with undercut bolt anchorage under various loading protocols. *Earthquake Engineering & Structural Dynamics*, *54*(2), 706–723. DOI: 10.1002/eqe.4277

Joe, J., Min, S., Oh, S.-H., Jung, B., Kim, Y. M., Kim, D. W., Lee, S. E., & Yi, D. H. (2022). Development of simplified building energy prediction model to support policymaking in South Korea: Case study for office buildings. *Sustainability (Basel)*, *14*(10), 6000. DOI: 10.3390/su14106000

Jones, M., Huynh, T., Dekiwadia, C., Daver, F., & John, S. (2020). Mycelium composites: A review of engineering characteristics and growth kinetics. *Journal of Bionic Engineering*, *17*(5), 843–872.

Junhao, X., Yingying, Z., & Jigang, X. (2017). Off-axial failure analysis of polytetrafluoroethylene-coated woven glass fibers under different loading rates. *Journal of Industrial Textiles*, *47*(3), 310–330. DOI: 10.1177/1528083716647198

Kaggwa, A., & Carson, J. K. (2019). Developments and future insights of using nanofluids for heat transfer enhancements in thermal systems: A review of recent literature. *International Nano Letters*, *9*(4), 277–288. DOI: 10.1007/s40089-019-00281-x

Kalsi, S., Kumar, S., Kumar, A., Alam, T., & Dobrotă, D. (2023). Thermophysical properties of nanofluids and applications in heat transfer. *Arabian Journal of Chemistry*, *16*(11), 105272. DOI: 10.1016/j.arabjc.2023.105272

Kalsi, S., Kumar, S., Kumar, A., Alam, T., Sharma, A., & Yadav, A. S. (2025). A review on hybrid nanofluids for heat transfer: Advancements, synthesis, challenges and applications. *Discover Applied Sciences*, *7*(7), 698. DOI: 10.1007/s42452-025-07141-8

Kamel, E., & Memari, A. M. (2022). Residential building envelope energy retrofit methods, simulation tools, and example projects: A review of the literature. *Buildings (Basel, Switzerland)*, *12*(7), 954. DOI: 10.3390/buildings12070954

Kamel, M. S., Syeal, R. A., & Abdulhussein, A. A. (2016). Heat transfer enhancement using nanofluids: A review of the recent literature. *American Journal of Nano Research and Applications*, *4*(1), 1–5.

Kaneki, Y., Takeuchi, T., Miyazaki, K., & Iwata, M. (2008). Studies on integrated facade engineering Structural performance of integrated facades. *AIJ Journal of Technology and Design*, *14*(27), 137–142. DOI: 10.3130/aijt.14.137

Kassem, M., & Mitchell, D. M. (2015). Bridging the gap between selection decisions of facade systems at the early design phase: Issues, challenges and solutions. *Journal of Facade Design and Engineering*, *3*(2), 165–183. DOI: 10.3233/FDE-150037

Keblinski, P., Eastman, J. A., & Cahill, D. G. (2005). Nanofluids for thermal transport. *Materials Today*, *8*(6), 36–44. DOI: 10.1016/S1369-7021(05)70936-6

Keblinski, P., Phillpot, S. R., Choi, S. U. S., & Eastman, J. A. (2002). Mechanisms of heat flow in nanofluids. *International Journal of Heat and Mass Transfer*, *45*(4), 855–863. DOI: 10.1016/S0017-9310(01)00175-2

Keles, Ç., & Yazıcıoğlu, F. (2023). Analyzing the environmental sustainability of primary schools' facades within the scope of life cycle assessment in Turkey and recommendations for improvement. *Smart and Sustainable Built Environment*, *12*(2), 298–324. DOI: 10.1108/SASBE-04-2021-0072

Khan, A. A., Lin, S., Huang, X., & Usmani, A. (2021). Facade fire hazards of bench-scale aluminum composite panel with flame-retardant core. *Fire Technology*, *59*(1), 5–28. DOI: 10.1007/s10694-020-01089-4

Khan, I., Zedadra, O., Guerrieri, A., & Spezzano, G. (2024). Occupancy prediction in IoT-enabled smart buildings: Technologies, methods, and future directions. *Sensors (Basel)*, *24*(11), 3276. DOI: 10.3390/s24113276 PMID: 38894069

Kibert, C. J. (2016). *Sustainable Construction: Green Building Design and Delivery* (4th ed.). Wiley.

Kim, S. D., & Koo, J. H. (2024). Development and material characterization of quartz-reinforced UHTR/aerogel composites: morphology, thermal, and flammability properties. *In AIAA SCITECH 2024 Forum (p.* 0364). Göppert, K., & Paech, C. (2015). Retractable roof solutions for sports stadiums and arenas. In P. Culley & J. Pascoe (Eds.), Stadium and arena design.

Kim, H., Lamichhane, N., Kim, C.-S., & Shrestha, R. (2023). Innovations in building diagnostics and condition monitoring: A comprehensive review of infrared thermography applications. *Buildings (Basel, Switzerland)*, *13*(11), 2829. DOI: 10.3390/buildings13112829

Klein, T. (2013). *Integral Facade Construction*. TU Delft OPEN Publishing., DOI: 10.59490/ABE.2013.3.611

Klimkowska, A., Cavazzi, S., Leach, R., & Grebby, S. (2022). Detailed three-dimensional building facade reconstruction: A review on applications, data and technologies. *Remote Sensing (Basel)*, *14*(11), 2579. DOI: 10.3390/rs14112579

Kohler, N., & Hassler, U. (2002). The building stock as a research object. *Building Research and Information*, *30*(4), 226–236. DOI: 10.1080/09613210110102238

Konbr, U., & Mamdouh, H. (2022). A proposed strategy to evaluate nanomaterials in construction to boost sustainable architecture. *Civil Engineering and Architecture*, *10*(6), 2329-2339. https://www.academia.edu/81868293

Kouris, L. A. S., & Triantafillou, T. C. (2018). State-of-the-art on strengthening of masonry structures with textile reinforced mortar (TRM). *Construction & Building Materials*, *188*, 1221–1233. DOI: 10.1016/j.conbuildmat.2018.08.039

Kozlovska, M., Petkanic, S., Vranay, F., & Simko, M. (2023). Enhancing energy efficiency and building performance through BEMS-BIM integration. *Energies*, *16*(17), 6327. DOI: 10.3390/en16176327

Kubba, S. (2012). *Handbook of green building design and construction: LEED, BREEAM, and Green Globes* (2nd ed.). Butterworth-Heinemann.

Kuda, A., Yadav, M., & Ali, S. M. (2025). Performance Evaluation of Etfe and Auto Heal for Responsive Kinetic Facades: Improving Building Efficiency and Urban Built Environment.

Kuda, A., & Yadav, M. (2022). Opportunities and challenges of using nanomaterials and nanotechnology in architecture: An overview. *Materials Today: Proceedings*, *62*(3), 10094–10099. DOI: 10.1016/j.matpr.2022.07.052

Kültür, S., Türkeri, N., & Knaack, U. (2019). A holistic decision support tool for facade design. *Buildings (Basel, Switzerland)*, *9*(8), 186. DOI: 10.3390/buildings9080186

Kumar, A., Patel, C., Maurya, S. K., & Patel, D. (2013). *Characteristics of incompatible and strong oxidizing agents in e-glass random fibre reinforced composites*. MANFEX.

Kumar, P., & Imam, B. (2013). Footprints of air pollution and changing environment on the sustainability of built infrastructure. *The Science of the Total Environment*, *444*, 85–101. DOI: 10.1016/j.scitotenv.2012.11.056 PMID: 23262326

Kumar, S., & Kumar, A. (2023). *Review of nanofluids in parabolic trough solar collectors*. Energy Sources Part A.

Kusure, B. D., Warkhedkar, R. M., Harde, P. R., & Shirke, P. K. (2014). Heat transfer enhancement by using nanofluids in heat pipe—A review. *Applied Mechanics and Materials*, *592*, 932–938. DOI: 10.4028/www.scientific.net/AMM.592-594.932

Lahjiri, M., & Benhida, Z. (2025). Bridging the Sustainability Gap: A Behavioral Science Approach for Entrepreneurs and Managers. In Intersecting Corporate Governance, Financial Performance, and Long-Term Value Creation (pp. 293-320). IGI Global.

Lameira, G., Ramos, R. J. G., Valentim, N., & Cruz, C. O. (2023). Good architecture matters: The architect's perspective on design for ageing and energy efficiency. *Buildings (Basel, Switzerland)*, *13*(4), 979. DOI: 10.3390/buildings13041067

Latifah, N. L., Zulwaqar, K. H., Andini, K., & Nisa, A. S. (2022). Building envelope design with glass curtain wall to reduce OTTV, study case: WU Tower Building at Bandung, Indonesia. *International Journal of Built Environment and Scientific Research*, *6*(2), 97–110. DOI: 10.24853/ijbesr.6.2.97-110

Laufs, W., & Verboon, E. (2013). Innovative façade design and products. In Façade Tectonics Institute / ASCE proceedings (pp. 154-193). American Society of Civil Engineers. DOI: 10.1061/9780784412701.ch10

Lazar, N., & Chithra, K. (2022). Role of culture in sustainable development and sustainable built environment: A review. *Environment, Development and Sustainability*, *24*(5), 11544–11565. DOI: 10.1007/s10668-021-01691-8

Lee, E., Selkowitz, S., Bazjanac, V., Inkarojrit, V., & Kohler, C. (2002). High-performance commercial building façades.

Lee, A. D., Shepherd, P., Evernden, M., & Metcalfe, D. (2018). Optimizing the architectural layouts and technical specifications of curtain walls to minimize use of aluminium. *Structures (Oxford)*, *13*, 8–25. DOI: 10.1016/j.istruc.2017.10.004

Lee, J. H., Ostwald, M. J., & Kim, M. J. (2021). Characterizing smart environments as interactive and collective platforms: A review of the key behaviors of responsive architecture. *Sensors (Basel)*, *21*(10), 3417. DOI: 10.3390/s21103417 PMID: 34069004

Lee, Y., & Ng, E. (2024). Critical review of current understanding of passive façade design in residential buildings. *Environmental Research: Infrastructure and Sustainability*, *4*(3), 032001. DOI: 10.1088/2634-4505/ad78fc

Linden, S., & Bossche, N. V. (2025). Experimental and numerical assessment of rainwater intrusion in ventilated facades with open joints. *Journal of Building Physics*, *49*(2), 173–206. DOI: 10.1177/17442591251324122

Li, R., Dai, Y., & Wang, R. Z. (2015). Experimental and theoretical analysis on thermal performance of solar thermal curtain wall in building envelope. *Energy and Building*, *87*, 324–334. DOI: 10.1016/j.enbuild.2014.11.029

Liu, H.-C., Liu, L., & Liu, N. (2013). Risk evaluation approaches in failure mode and effects analysis: A literature review. *Expert Systems with Applications*, *40*(2), 828–838. DOI: 10.1016/j.eswa.2012.08.010

Liu, M., Liu, C., Xie, H., Zhao, Z., Zhu, C., Lu, Y., & Bu, C. (2023). Analysis of the Impact of Photovoltaic Curtain Walls Replacing Glass Curtain Walls on the Whole Life Cycle Carbon Emission of Public Buildings Based on BIM Modeling Study. *Energies*, *16*(20), 7030. DOI: 10.3390/en16207030

Loonen, R. C., Trčka, M., Cóstola, D., & Hensen, J. L. (2015). Climate adaptive building shells: State-of-the-art and future challenges. *Renewable & Sustainable Energy Reviews*, *25*, 483–493. DOI: 10.1016/j.rser.2013.04.016

Luna-Navarro, A., Hunt, G. R., & Overend, M. (2022). Dynamic facades: An exploratory campaign to assess occupant multi-domain environmental satisfaction and facade interaction. *Building and Environment*, *211*, 108703. DOI: 10.1016/j.buildenv.2021.108703

Luo, D., Zhang, T., Zheng, W., & Nie, Q. (2026). Intelligent detection and energy-driven repair of building envelope defects for improved thermal and energy performance. *Energies*, *19*(2), 351. DOI: 10.3390/en19020351

Luo, R., & Edlund, B. (1996). Shear capacity of plate girders with trapezoidally corrugated webs. *Thin-walled Structures*, *26*(1), 19–44. DOI: 10.1016/0263-8231(96)00006-7

Lu, Y., & Weng, L. (2018). Smart construction: A new era of digital transformation. *Engineering, Construction, and Architectural Management*, *25*(4), 495–513.

Lyons, A. (2013). Polytetrafluoroethylene (PTFE) and Ethylene Tetrafluoroethylene Copolymer (ETFE). *In Construction Materials Reference Book (pp.* 269-278). Routledge.

Mahamude, A. S. F., Kamarulzaman, M., Harun, W., Kadirgama, K., Ramasamy, D., Farhana, K., Bakar, R., Yusaf, T., Subramanion, S., & Yousif, B. (2022). A comprehensive review on efficiency enhancement of solar collectors using hybrid nanofluids. *Energies*, *15*(4), 1391. DOI: 10.3390/en15041391

Maheshwari, A., Bell, F., Gallo, E., Ibanescu, M., Robertson, E., Fairless, C., Logan, F., Schmuhl, M. J., Cheung, H., Rein, M., Danielescu, A., & Alistar, M. (2025). Self-Cleaning Textiles for Publicly Shared Soft Surfaces: An Accessible and Scalable Photocatalytic Polyester System with Embedded Light-Emitting Diodes. *ACS Omega*, *10*(4), 3978–3987. DOI: 10.1021/acsomega.4c09837 PMID: 39926538

Mahmoud, A. H. (2023). Studying the impact of the role of nanomaterials technology in developing the performance of buildings and their effect on the architectural formation. *Journal of Al-Azhar University Engineering Sector*, *18*(68), 1371–1380. https://journals.ekb.eg/article_304211.html. DOI: 10.21608/auej.2023.283051

Maity, S., Singha, K., & Singha, M. (2012). Textiles in earthquake-resistant constructions. *Journal of Safety Engineering*, *1*(2), 17–25. DOI: 10.5923/j.safety.20120102.01

Malik, N., Nirupama, E., Selvakumar, P., Sankpal, U., Nathsharma, S. K., Sharma, M., & Manjunath, T. C. (2025). 2D Materials for Energy Storage and Conversion. In Mhadhbi, M. (Ed.), *Innovations in Next-Generation Energy Storage Solutions* (pp. 1–24). IGI Global Scientific Publishing., DOI: 10.4018/979-8-3693-9316-1.ch001

Mangherini, G., Diolaiti, V., Bernardoni, P., Andreoli, A., & Vincenzi, D. (2023). Review of Façade Photovoltaic Solutions for Less Energy-Hungry Buildings. *Energies*, *16*(19), 6901. DOI: 10.3390/en16196901

Manimaran, M., Norizan, M. N., Kassim, M. H. M., Adam, M. R., Abdullah, N., & Norrahim, M. N. F. (2025). Stability and thermal conductivity of water-based hybrid nanofluids. *RSC Advances*, *15*(18), 14088–14125. DOI: 10.1039/D5RA00844A PMID: 40313320

Manso, M., & Castro-Gomes, J. (2015). Green wall systems: A review of their characteristics. *Renewable & Sustainable Energy Reviews*, *41*, 863–871. DOI: 10.1016/j.rser.2014.07.203

Mansy, A. A. Y. (2019). Can Ethylene Tetra Fluoro Ethylene cushions double skin façade applications improve buildings envelope performance, reduce energy consumption and enhance visual and thermal comfort in UAE buildings? (Master's thesis, The British University in Dubai).

Massafra, A., Costantino, C., Predari, G., & Gulli, R. (2023). Building information modeling and building performance simulation-based decision support systems for improved built heritage operation. *Sustainability (Basel)*, *15*(14), 11240. DOI: 10.3390/su151411240

Masubuchi, M. (2013). Conceptual and structural design of adaptive membrane structures with spoked wheel principle-folding to the perimeter.

Mathur, P., Sharma, C., & Azeemuddin, S. (2024). Autonomous inspection of high-rise buildings for facade detection and 3D modeling using UAVs. *IEEE Access : Practical Innovations, Open Solutions*, *12*, 18251–18258. DOI: 10.1109/ACCESS.2024.3360209

Matin, N. H., & Eydgahi, A. (2019). Factors affecting the design and development of responsive facades: A historical evolution. *Intelligent Buildings International*, *12*(4), 257–270. DOI: 10.1080/17508975.2018.1562414

Mazzucchelli, E. S., Angelo, L., & Tattoni, S. (2017). Analysis and control of façade claddings structural issues. *Rivista Tema*, *3*(1). Advance online publication. DOI: 10.30682/tema0301i

Mendes, M. V. A. da S., Pavón, E., & Melo, M. B. (2022). Damages mapping of façade using aerophotogrammetry and thermography inspection: Zoroastro Artiaga Museum - Art Deco in Brazil. *Matéria (Rio de Janeiro)*, *27*(3), e20220031. Advance online publication. DOI: 10.1590/1517-7076-rmat-2022-0031

Meza, F. J., Becque, J., & Hajirasouliha, I. (2020). Experimental study of cold-formed steel built-up beams. *Journal of Structural Engineering*, *146*(7), 04020126. Advance online publication. DOI: 10.1061/(ASCE)ST.1943-541X.0002677

Mirzabeigi, S., Razkenari, R., & Crovella, P. (2025). Building envelope thermal anomaly detection using an integrated vision-based technique and semantic segmentation. *Buildings (Basel, Switzerland)*, *15*(15), 2672. DOI: 10.3390/buildings15152672

Mishra, S., Unnikrishnan, L., Nayak, S. K., & Mohanty, S. (2019). Advances in piezoelectric polymer composites for energy harvesting applications: A systematic review. *Macromolecular Materials and Engineering*, *304*(1), 1800463. DOI: 10.1002/mame.201800463

Monticelli, C., & Zanelli, A. (2022). *9. Design for Re-manufacturing (DfRem) of short chains from design-to-construction: the case of textile-based tertiary architecture.* ORKS FOR TERTIARY ARCHITECTURES.

Mottahedeh, R. (2023). Circularity and design: material reuse and sustainable practices in Architecture.

Mousavi, S. M.. (2025). Surface engineered nanofluids in geothermal systems: Experimental and numerical investigation. *Journal of Molecular Liquids*.

Mydin, M. A. O., Sarpin, N., Zainol, R. M., Odeh, R. A., & Nawi, M. N. M. (2024). The impact of climatological factors on the multifaceted and multisystemic deficiencies of building anatomy. *Journal of Advanced Research in Applied Sciences and Engineering Technology*, *50*(1), 308–329. DOI: 10.37934/araset.50.1.308329

Nabwey, H. A., Armaghani, T., Azizimehr, B., Rashad, A. M., & Chamkha, A. J. (2023). Review of nanofluid heat transfer in porous media. *Nanomaterials (Basel, Switzerland)*, *13*(5), 937. DOI: 10.3390/nano13050937 PMID: 36903815

Naeem, N., Abdin, A., & Saleh, A. (2024). An approach to using shape memory alloys in kinetic façades to improve the thermal performance of office building spaces. *Civil Engineering and Architecture*, *12*(3), 1067-1079. https://www.researchgate.net/publication/375012719

Nagar, S., & Sreenivasa, S. (2025). Thermal performance investigations using nanofluids and PCM in parabolic trough collectors. *Proceedings of the IMechE Part A.*

Nardi, I., Lucchi, E., de Rubeis, T., & Ambrosini, D. (2018). Quantification of heat energy losses through the building envelope: A state-of-the-art analysis with critical and comprehensive review on infrared thermography. *Building and Environment*, *146*, 190–205. DOI: 10.1016/j.buildenv.2018.09.050

Narola, D., Meena, A., & Meena, K. (2019). Effect of corrugation angle in strength of trapezoidal corrugated web steel built-up beam. *Proceedings of the International Conference on Structural Engineering and Construction*. DOI: 10.11159/icsect19.143

Naseer, A. N., Al-Gahtani, K. S., Altuwaim, A., Alsanabani, N. M., & Almohsen, A. S. (2024). Selecting building façade materials by integrating stepwise weight assessment ratio analysis and weighted aggregated sum product assessment into value engineering. *Sustainability (Basel)*, *16*(11), 4611. DOI: 10.3390/su16114611

Nassan, T. H., Heris, S. Z., & Bagheban, S. H. N. (2023). Experimental investigation of nanofluid heat transfer in a square cross-sectional duct. *arXiv preprint arXiv:2311.07447*.

Naveed, A. (2021). Thermal Performance Analysis of Flat-Plate Solar Collector Using Nanofluids (Doctoral dissertation, University of Malaya).

Nazari, M., & Matusiak, B. (2024). Daylighting simulation and visualisation: Navigating challenges in accuracy and validation. *Energy and Building*, *312*, 114188. DOI: 10.1016/j.enbuild.2024.114188

Nečasová, B., Liška, P., Šlanhof, J., Sedlák, P., & Motyčka, V. (2020). Long adhesive joints in façade applications exposed to wind suction. *Proceedings of the Institution of Mechanical Engineers. Part E, Journal of Process Mechanical Engineering*, *234*(5), 488–498. DOI: 10.1177/0954408920945065

Niazy, D., Metwally, E. A., Rifat, M., Awad, M. I., & Elsabbagh, A. (2023). A conceptual design of circular adaptive façade module for reuse. *Scientific Reports*, *13*(1), 20552. DOI: 10.1038/s41598-023-47593-9 PMID: 37996556

Niedostatkiewicz, M. (2022). Emergency condition of the ceramic facade of a residential building. *Inżynieria Bezpieczeństwa Obiektów Antropogenicznych*, *2*(2), 32–38. DOI: 10.37105/iboa.140

Nishanth, L., Patil, N. N., & Ganesh, N. A. (2023). Mechanical, Durability and Embodied Energy Analysis of Geopolymer Concrete with Fly Ash, GGBFS and Glass Fiber. In *E3S Web of Conferences* (Vol. 455, p. 03006). EDP Sciences.

Nor-Aishah, H., Ahmad, N. H., & Thurasamy, R. (2020). Entrepreneurial leadership and sustainable performance of manufacturing SMEs in Malaysia: The contingent role of entrepreneurial bricolage. *Sustainability (Basel), 12*(8), 3100. DOI: 10.3390/su12083100

Nowak, K., & Byrdy, A. (2018). Effect of mounting brackets on thermal performance of buildings with ventilated facades. *Journal of Building Physics, 43*(1), 46–56. DOI: 10.1177/1744259118790759

NPSP BV. (2019, October 19). Nabasco biocomposite with high fire safety class [News item]. Retrieved November 4, 2025, from https://www.npsp.nl/en/news-item/nabasco-biocomposite-high-fire-safety-class

Oesterle, E., Lieb, R.-D., Lutz, M., & Heusler, W. (2014). *Double-skin façades: Integrated planning*. Prestel.

Oestreich, C. R., Lazzari, P. M., Dutra, V. F. P., & Silva Filho, L. C. P. D. (2024). Bending reinforcement of reinforced concrete beams with alkali-resistant fiberglass textile-reinforced concrete with different surface treatments. *Revista IBRACON de Estruturas e Materiais, 18*(1), e18102. DOI: 10.1590/s1983-41952025000800002

Okeke, G., Witharana, S., Antony, S. J., & Ding, Y. (2011). Computational analysis of factors influencing thermal conductivity of nanofluids. *Journal of Nanoparticle Research, 13*(12), 6365–6375. DOI: 10.1007/s11051-011-0389-9

Orlowski, K., Shanaka, K., & Mendis, P. (2018). Design and development of weatherproof seals for prefabricated construction: A methodological approach. *Buildings (Basel, Switzerland), 8*(9), 117. DOI: 10.3390/buildings8090117

Otafy, S. S., Nassar, K., Abou-Taleb, E. M., & Aly, N. M. (2025). Sustainable Textile-Based Water Harvesting Systems: An Overview. *Egyptian Journal of Chemistry, 68*(6), 147–164. DOI: 10.21608/ejchem.2024.304309.10014

Özen, G. (2022). Design and analysis of deployable reciprocal frames (Doctoral dissertation, Izmir Institute of Technology).

Paech, C., Göppert, K., & Wilbrenninck, S. (2014). Fast-track design and build of a FIFA compliant stadium: The Hazza Bin Zayed Stadium in Al Ain/UAE. *Stahlbau, 83*(6), 394–399. DOI: 10.1002/stab.201410166

Pakdel, E., Fang, J., Fang, J., Sun, L., Wang, X., & Wang, X. (2018). Nanocoatings for smart textiles. *Smart textiles: wearable nanotechnology, 247-300.*

Panduro, E. A. C., Finotti, F., Largiller, G., & Lervåg, K. Y. (2022). A review of the use of nanofluids as heat-transfer fluids in parabolic-trough collectors. *Applied Thermal Engineering*, *211*, 118346. DOI: 10.1016/j.applthermaleng.2022.118346

Paoletti, I., & Nastri, M. (2024). *Construction of the Façade Systems*. Springer Nature Switzerland., DOI: 10.1007/978-3-031-49608-0

Park, J., Hengevoss, D., & Wittkopf, S. (2018). Industrial data-based life cycle assessment of architecturally integrated glass-glass photovoltaics. *Buildings (Basel, Switzerland)*, *9*(1), 8. DOI: 10.3390/buildings9010008

Pasternak, H., & Kubieniec, G. (2010). Plate girders with corrugated webs. *Journal of Civil Engineering and Management*, *16*(2), 166–171. DOI: 10.3846/jcem.2010.17

Patel, D. (2023). Exploring the frontiers of microfluidics: challenges and future prospects. In *Advances in MEMS and Microfluidic Systems* (pp. 11–31). DOI: 10.4018/978-1-6684-6952-1.ch002

Patel, D., Singh, R. P., Rajput, R. S., & Tiwari, P. (2022, December). Thermophysical properties and applications of nanofluids—A review. In 2022 5th International Conference on Contemporary Computing and Informatics (IC3I) (pp. 1324–1328). IEEE.

Patel, D. (2024). Sustainable renewable energy sources and emerging technologies. In *Optimization Techniques for Hybrid Power Systems* (pp. 343–361). Renewable Energy, Electric Vehicles, and Smart Grid. DOI: 10.4018/979-8-3693-0492-1.ch015

Patel, D. (2025). Emerging sustainable nanomaterials and their applications and future scope. In *Advances in Sustainable Materials* (pp. 107–135). Elsevier. DOI: 10.1016/B978-0-443-13849-2.00005-3

Patel, D., Pandey, G., Mourya, V. K., & Kumar, R. (2024). Sustainable base isolation: A review. *Sadhana*, *49*(2), 173. DOI: 10.1007/s12046-024-02511-1

Patel, D., Sharma, A., & Mishra, A. (2021, November). Study of convective heat transfer characteristics of nanofluids in a circular tube. In *2021 International Conference on Technological Advancements and Innovations (ICTAI)* (pp. 264–267). IEEE. DOI: 10.1109/ICTAI53825.2021.9673432

Patel, D., Varma, I. P., & Khan, F. A. (2022). A review advanced vehicle with automatic pneumatic bumper system using two cylinder. *International Research Journal of Engineering and Technology*, *9*, 264–267.

Patel, H. R.. (2022). Review on design aspects of parabolic trough collectors. *International Journal of Renewable Energy Research*, *12*(3), 1566–1588.

Peng, J., Lu, L., & Yang, H. (2011). Review on life cycle assessment of energy payback and greenhouse gas emission of solar photovoltaic systems. *Renewable & Sustainable Energy Reviews*, *19*, 255–274. DOI: 10.1016/j.rser.2012.11.035

Pereira, C., de Brito, J., & Silvestre, J. D. (2020). Harmonised classification of the causes of defects in a global inspection system: Proposed methodology and analysis of fieldwork data. *Sustainability (Basel)*, *12*(14), 5564. DOI: 10.3390/su12145564

Pereira, J., Souza, R., Moreira, A., & Moita, A. (2023). A review on the nanofluids–PCMs integrated solutions for solar thermal heat transfer enhancement purposes. *Technologies*, *11*(6), 166. DOI: 10.3390/technologies11060166

Perez, G., Sirvent, P., Sanchez-Garcia, J. A., & Guerrero, A. (2021). Improved methodology for the characterization of thermochromic coatings for adaptive façades. *Solar Energy*, *221*, 310–319. DOI: 10.1016/j.solener.2021.10.062

Perini, K., & Rosasco, P. (2013). Cost–benefit analysis for green facades and living wall systems. *Building and Environment*, *70*, 110–121. DOI: 10.1016/j.buildenv.2013.08.012

Petresevics, F., & Nagy, B. (2022). FEM-based evaluation of the point thermal transmittance of various types of ventilated façade cladding fastening systems. *Buildings (Basel, Switzerland)*, *12*(8), 1153. DOI: 10.3390/buildings12081153

Pittau, F., Krause, F., Lumia, G., & Habert, G. (2018). Fast-growing bio-based materials as an opportunity for storing carbon in exterior walls. *Building and Environment*, *129*, 117–129. DOI: 10.1016/j.buildenv.2017.12.006

Poddar, S., & Guha, S. (2021). Daylight Illumination and Building Architecture-Effect at Workplace. [AMT]. *Advancement in Management and Technology*, *2*(1), 11–16. DOI: 10.46977/apjmt.2021v02i01.002

Poirazis, H. (2006). Double skin façades for office buildings. *Energy and Building*, *38*(3), 284–292.

Pomponi, F., & Moncaster, A. (2016). Embodied carbon mitigation and reduction in the built environment – What does the evidence say? *Journal of Environmental Management*, *181*, 687–700. DOI: 10.1016/j.jenvman.2016.08.036 PMID: 27558830

Pomponi, F., & Moncaster, A. (2017). Circular economy for the built environment: A research framework. *Journal of Cleaner Production*, *143*, 710–718. DOI: 10.1016/j.jclepro.2016.12.055

Pomponi, F., & Moncaster, A. (2017). Scrutinising embodied carbon in buildings: The next performance gap made manifest. *Renewable & Sustainable Energy Reviews*, *81*, 2431–2442. DOI: 10.1016/j.rser.2017.06.049

Popoola, O., & Aluko, O. O. (2025). Entrepreneurial Enhancement through Commercial Architecture in Ondo State, Nigeria. *Journal of the Nigerian Institute of Town Planners*, *30*(3), 43–63.

Popovic, P. L., & Vla, D. (2020). Assessment and repair of exterior facades. *Scientific Journal of Civil Engineering*, *9*(1), 43–51. DOI: 10.55302/SJCE2091043p

Porgar, S., Oztop, H. F., & Salehfekr, S. (2023). A comprehensive review on thermal conductivity and viscosity of nanofluids and their application in heat exchangers. *Journal of Molecular Liquids*, *386*, 122213. DOI: 10.1016/j.molliq.2023.122213

Pracucci, A., Magnani, S., & Casadei, O. (2020). The integration of vacuum insulated glass in unitized façade for the development of innovative lightweight and highly insulating energy efficient building envelope The results of Eensulate façade system design. *Designs*, *4*(4), 40. DOI: 10.3390/designs4040040

Pracucci, A., Vandi, L., Morganti, L., Fernández, A. G., Nunez Diaz, M., Navarro Muedra, A., Győri, V., Kouyoumji, J.-L., & Astudillo Larraz, J. (2024). Design and simulation for technological integration of bio-based components in façade system modules. *Buildings (Basel, Switzerland)*, *14*(4), 1114. DOI: 10.3390/buildings14041114

Prieto, A., Knaack, U., Auer, T., & Klein, T. (2019). COOLFACADE: State-of-the-art review and evaluation of solar cooling technologies on their potential for façade integration. *Renewable & Sustainable Energy Reviews*, *101*, 395–414. DOI: 10.1016/j.rser.2018.11.015

Procaccini, G., & Monticelli, C. (2024). The application and potentialities of textile facade retrofit strategies for energy-efficient and resilient buildings. *E3S Web of Conferences, 546*, 03005. https://doi.org/DOI: 10.1051/e3sconf/202454603005

Procaccini, G., & Monticelli, C. (2024). The application and potentialities of textile facade retrofit strategies for energy-efficient and resilient buildings. *In E3S Web of Conferences (Vol.* 3, pp. 1-5).

Procaccini, G. (2025). Textile materials and retrofit strategies. *New frontiers for energy efficiency. Techne : Journal of Technology for Architecture and Environment*, *29*(1), 299–308. DOI: 10.36253/techne-16600

Procaccini, G., Prieto, A., Knaack, U., Monticelli, C., & Konstantinou, T. (2023). Textile Membrane for Façade Retrofitting: Exploring Fabric Potentialities for the Development of Innovative Strategies. *Buildings (Basel, Switzerland)*, *14*(1), 86. DOI: 10.3390/buildings14010086

Rabajczyk, A., Zielecka, M., Klapsa, W., & Dziechciarz, A. (2021). Self-Cleaning coatings and surfaces of modern building materials for the removal of some air pollutants. *Materials (Basel)*, *14*(9), 2161. DOI: 10.3390/ma14092161 PMID: 33922766

Rae, D. (2017). Developing entrepreneurial leadership: The challenge for sustainable organisations. *International Journal of Work Innovation*, *2*(1), 76–100. DOI: 10.1504/IJWI.2017.080725

Raiden, A., & King, A. (2021). Social value, organisational learning, and the sustainable development goals in the built environment. *Resources, Conservation and Recycling*, *172*, 105663. DOI: 10.1016/j.resconrec.2021.105663

Rajput, R. S., Patel, D., & Singh, R. P. (2023). Applications and feasibility of large-scale solar-powered Peltier refrigeration systems.

Rashidi, M. M., & Zinatloo-Ajabshir, S. (Eds.). (2024). *Nanofluids: Preparation.* Applications and Simulation Methods.

Rashidzadeh, Z., & Matin, N. H. (2023). A comparative study on smart windows focusing on climate-based energy performance and users' comfort attributes. *Sustainability (Basel)*, *15*(3), 2294. DOI: 10.3390/su15032294

Rathore, P. K. S., Gupta, N. K., Yadav, D., Shukla, S. K., & Kaul, S. (2022). Thermal performance of the building envelope integrated with phase change material for thermal energy storage: An updated review. *Sustainable Cities and Society*, *76*, 103445. DOI: 10.1016/j.scs.2022.103690

Raudaschl, M., Levak, T., Riewe, R., Triantafyllidis, G., Drnda, E., Popek, S., Schlegl, D., Funke-Kaiser, D., & Lund, A. (2022, September). Piezoelectric Textile Facade for the energy supply of active sensor technology with regard to data management for circular economy in building construction. [). IOP Publishinging.]. *IOP Conference Series. Earth and Environmental Science*, *1078*(1), 012037. DOI: 10.1088/1755-1315/1078/1/012037

Reckien, D., Salvia, M., Pietrapertosa, F., Simoes, S., Olazabal, M., De Gregorio Hurtado, S., Geneletti, D., Lorencova, E. K., D'Alonzo, V., Krook-Riekkola, A., Fokaides, P., Ioannou, B., Foley, A., Orru, H., Orru, K., Wejs, A., Flacke, J., Church, J., Feliu, E., & Heidrich, O. (2019). Dedicated versus mainstreaming approaches in local climate plans in Europe. *Renewable & Sustainable Energy Reviews*, *112*, 948–959. DOI: 10.1016/j.rser.2019.05.014

Renko, M., El Tarabishy, A., Carsrud, A. L., & Brännback, M. (2015). Understanding and measuring entrepreneurial leadership style. *Journal of Small Business Management*, *53*(1), 54–74. DOI: 10.1111/jsbm.12086

Riahinezhad, M., Hallman, M., & Masson, J.-F. (2021). Critical review of polymeric building envelope materials: Degradation, durability and service life prediction. *Buildings (Basel, Switzerland)*, *11*(7), 299. DOI: 10.3390/buildings11070299

Robertson, K. (2012). *Felt Space: Responsive Textiles*. Fabric Dwellings and Precarious Housing.

Robinson, L. A. (2005). Structural Opportunities of ETFE (ethylene tetra fluoro ethylene) (Doctoral dissertation, Massachusetts Institute of Technology).

Roque, E., & Santos, P. (2017). The effectiveness of thermal insulation in lightweight steel-framed walls with respect to its position. *Buildings (Basel, Switzerland)*, *7*(1), 13. DOI: 10.3390/buildings7010013

Ruiz, F., Aguado, A., Serrat, C., & Casas, J. R. (2019). Condition assessment of building façades based on hazard to people. *Structure and Infrastructure Engineering*, *15*(10), 1346–1365. DOI: 10.1080/15732479.2019.1621907

Runge, D., Surholt, F., Uhlemann, J., & Stranghöner, N. (2025). ETFE-foils in pneumatically retractable civil engineering structures. *Architectural Engineering and Design Management*, •••, 1–26. DOI: 10.1080/17452007.2025.2548549

Sadegh, S. O., Gasparri, E., Brambilla, A., & Globa, A. (2022). Kinetic facades: An evolutionary-based performance evaluation framework. *Journal of Building Engineering*, *53*, 104408. DOI: 10.1016/j.jobe.2022.104408

Şahin, Y. D., & Okur, M. (2025). The types of textiles used in the façade and roofing systems of stadium facilities in Turkey. *Fibres and Textiles*, *32*(1), 96–101. DOI: 10.15240/tul/008/2025-1-019

Saidur, R., Leong, K. Y., & Mohammed, H. A. (2011). A review on applications and challenges of nanofluids. *Renewable & Sustainable Energy Reviews*, *15*(3), 1646–1668. DOI: 10.1016/j.rser.2010.11.035

Sainger, G. (2018). Leadership in digital age: A study on the role of leader in this era of digital transformation. *International Journal on Leadership*, *6*(1), 1.

Saleh, A. N., Attar, A. A., Ahmed, O. K., & Mustafa, S. S. (2021). Improving the thermal insulation and mechanical properties of concrete using Nano-SiO2. *Results in Engineering*, *12*, 100284. DOI: 10.1016/j.rineng.2021.100303

Sandak, A., Sandak, J., Brzezicki, M., & Kutnar, A. (2019). *Bio-based building skin*. Springer. DOI: 10.1007/978-981-13-3747-5

Santamaria-Fernandez, S., Eizaguirre-Iribar, A., & Olano-Azkune, X. (2020). Pathologies of a glass building envelope that affect durability and comfort. DBMC 2020 Proceedings. DOI: 10.23967/dbmc.2020.105

Saraf, A. K., Goyal, R., & Yusop, A. F. (2025). Application of nanofluids under laminar flow in heat exchangers. *Discover Applied Sciences*, *7*(10), 1154. DOI: 10.1007/s42452-025-07726-3

Saraswat, S., & Johari, R. Review of advancements in heat transfer enhancement through nanofluids.

Schulz, J., Kent, D., Crimi, T., Glockling, J. L. D., & Hull, T. R. (2020). A critical appraisal of the UK's regulatory regime for combustible façades. *Fire Technology*, *57*(1), 261–290. DOI: 10.1007/s10694-020-00993-z

Schwark, J., & Müller, J. (1996). High performance silicone-coated textiles: Developments and applications. *Journal of coated fabrics, 26(1), 65-77.*

Seghezzi, E., Locatelli, M., Pellegrini, L., Pattini, G., Di Giuda, G. M., Tagliabue, L. C., & Boella, G. (2021). Towards an occupancy-oriented digital twin for facility management: Test campaign and sensors assessment. *Applied Sciences (Basel, Switzerland)*, *11*(7), 3108. DOI: 10.3390/app11073108

Şeker, Ş. (2019). Analyzing occupational risks of pharmaceutical industry under uncertainty using a Bow-Tie analysis. *International Journal of Optimization and Control: Theories & Applications*, *9*(2), 113–124. DOI: 10.11121/ijocta.01.2019.00655

Sezegen, A., & Edis, E. (2020). Product innovation types: A discussion considering building facade products. *Engineering, Construction, and Architectural Management*, *27*(9), 2379–2408. DOI: 10.1108/ECAM-10-2018-0454

Shahruddin, S., & Husain, S. H. (2024). Navigating paradoxes of identity and leadership in the age of digital transformation of construction industry: Architects' experiences and perceptions. *Construction Management and Economics*, *42*(7), 591–609. DOI: 10.1080/01446193.2023.2293901

Shang, W. L., & Lv, Z. (2023). Low carbon technology for carbon neutrality in sustainable cities: A survey. *Sustainable Cities and Society*, *93*, 104512. DOI: 10.1016/j.scs.2023.104489

Shan, X., Liu, K., & Sun, P.-L. (2017). Risk analysis on leakage failure of natural gas pipelines by fuzzy Bayesian network with a Bow-Tie model. *Scientific Programming*, *3639524*, 1–11. Advance online publication. DOI: 10.1155/2017/3639524

Sharma, B., Gatóo, A., Bock, M., & Ramage, M. (2015). Engineered bamboo for structural applications. *Construction & Building Materials*, *81*, 66–73. DOI: 10.1016/j.conbuildmat.2015.01.077

Sheng, Y., Ward, W. O. C., Arbabi, H., Álvarez, M., & Mayfield, M. (2022). Deep multimodal learning for residential building energy prediction. *IOP Conference Series. Earth and Environmental Science*, *1078*(1), 012038. DOI: 10.1088/1755-1315/1078/1/012038

Silva, A., Gaspar, P. L., de Brito, J., & Neves, L. C. (2015). Probabilistic analysis of degradation of façade claddings using Markov chain models. *Materials and Structures*, *49*(7), 2871–2892. DOI: 10.1617/s11527-015-0692-5

Silva, M. M., Ferreira, F. A. F., & Santos, S. P. (2014). Methodologies for failure mode and effect analysis: A literature review. *Quality and Reliability Engineering International*, *30*(7), 989–1005.

Silver, P., Silver, P., & McLean, W. (2015). *Air Structures*.

Singh, S. (2021). ArchiTextile: A Review on Application of Textiles in Architecture. [JTATM]. *Journal of Textile & Apparel Technology & Management*, *12*(1).

Sivaji, C., Chidambaram, P., Ramachandran, M., & Murugan, A. (2022). Performance analysis of facade materials using VIKOR method. *REST Journal on Advances in Mechanical Engineering*, *1*(2), 41–49. DOI: 10.46632/jame/1/2/6

Sluyts, Y., Rychtarikova, M., & Glorieux, C. (2024). Acoustic characterisation of ETFE cushions and their impact on the acoustic comfort of building users.

Snyder, H. (2019). Literature review as a research methodology: An overview and guidelines. *Journal of Business Research*, *104*, 333–339. DOI: 10.1016/j.jbusres.2019.07.039

Solanki, K., Balster, T., Maywald, C., & Monticelli, C., & PADILLA MARCOS, M. Á. (2024). Evaluation of the wind load history of environmentally exposed ETFE foils. *In Proceedings of the IASS 2024 Symposium (pp.* 1-6).

Solis-Guzman, J. A. I. M. E., Martínez-Rocamora, A., & Marrero, M. (2014). *(pp* (Vol. 1). Methodology for determining the carbon footprint of the construction of residential buildings. *In Assessment of Carbon Footprint in Different Industrial Sectors*. Springer Singapore.

Soltan, Y. I., Nasser, M. S., Almomani, F., Mahmoud, K. A., & Onaizi, S. A. (2024). Thermal conductivity of different materials nanofluids including MXenes and metal-organic frameworks. *Journal of Materials Research and Technology*, *31*, 2723–2761. DOI: 10.1016/j.jmrt.2024.07.023

Souza, A. L. R., de, , & Romeiro, T. R. (2024). Influence of the exposure degree on the degradation of facades of buildings in Brasília, Brazil. *Buildings (Basel, Switzerland)*, *14*(1), 133. DOI: 10.3390/buildings14010133

Spence, S. M. J. (2006). *Database-Assisted Design for Wind*. National Institute of Standards and Technology., DOI: 10.6028/NIST.BSS.180

Srivastava, G., Nakrani, D., & Ghoroi, C. (2020). Performance of combustible facade systems with glass, ACP and firestops in full-scale, real fire experiments. *Fire Technology*, *56*(4), 1575–1598. DOI: 10.1007/s10694-019-00943-4

Stamatis, D. H. (2003). *Failure mode and effect analysis: FMEA from theory to execution* (2nd ed.). ASQ Quality Press.

Stazi, F., Tomassoni, E., & Di Perna, C. (2016). Super-insulated wooden envelopes in Mediterranean climate: Summer overheating, thermal comfort optimization, environmental impact on an Italian case study. *Energy and Building*, *138*, 716–732. DOI: 10.1016/j.enbuild.2016.12.042

Strasser, D., Reber, A., Biesinger, A., & Eicker, U. (2018). Daylight Performance of a Translucent Textile Membrane Roof with Thermal Insulation. *Buildings (Basel, Switzerland)*, *8*(9), 118. DOI: 10.3390/buildings8090118

Straub, A. (2009). Dutch standard for condition assessment of buildings. *Structural Survey*, *27*(1), 23–35. DOI: 10.1108/02630800910941665

Straube, J. F., & Burnett, E. F. P. (2005). *Building science for building enclosure design*. Building Science Press.

Straube, J. F., & Burnett, E. F. P. (2005). *Building science for building enclosures*. Building Science Press.

Struhala, K., Čekon, M., Slávik, R., & Kalousek, M. (2018). Life cycle assessment of solar façade concepts based on transparent insulation materials. *Sustainability (Basel)*, *10*(11), 4212. DOI: 10.3390/su10114212

Sujitha, R., Sunmathi, N., Manikandan, R. K., Arunprasad, J., Rajkumar, S., Sharma, S., Sharma, K., Li, C., & Tag Eldin, E. M. (2022). Analytical and experimental study on cold-formed steel built-up sections for bending. *Materials (Basel)*, *15*(20), 7140. DOI: 10.3390/ma15207140 PMID: 36295206

Sulakatko, V., & Vogdt, F. U. (2018). Construction process technical impact factors on degradation of the external thermal insulation composite system. *Sustainability (Basel)*, *10*(11), 3900. DOI: 10.3390/su10113900

Suresh, C., Hotta, T. K., & Saha, S. K. (2022). Phase change material incorporation techniques in building envelopes for enhancing the building thermal comfort: A review. *Energy and Building*, *264*, 112086. DOI: 10.1016/j.enbuild.2022.112225

Tabrizikahou, A., Hadzima-Nyarko, M., Kuczma, M., & Lozančić, S. (2021). Application of shape memory alloys in retrofitting of masonry and heritage structures based on their vulnerability revealed in the Bam 2003 earthquake. *Materials (Basel)*, *14*(24), 7700. DOI: 10.3390/ma14164480 PMID: 34443003

Takeuchi, T., Yasuda, K., & Iwata, M. (2006). Studies on integrated building facade engineering with high-performance structural elements. IABSE Reports, 33-40. DOI: 10.2749/222137806796185526

Takeuchi, T., Koyano, K., Yasuda, K., Yuasa, K., & Iwata, M. (2006). Studies on integrated facade engineering: Proposal for integrated facade and its evaluation. [Transactions of AIJ]. *Journal of Environmental Engineering*, *71*(601), 81–88. DOI: 10.3130/aije.71.81_1

Takhmasib, M., Lee, H. J., & Yi, H. (2023). Machine-learned kinetic facade: Construction and artificial intelligence enabled predictive control for visual comfort. *Automation in Construction*, *156*, 105093. DOI: 10.1016/j.autcon.2023.105093

Tanuja, T. N.. (2023). Flow and heat transfer analysis on micropolar fluid through porous media. *Frontiers in Materials*, *10*, 1216757. DOI: 10.3389/fmats.2023.1216757

Tarkhan, N., Szczesniak, J. T., & Reinhart, C. (2024). Facade feature extraction for urban performance assessments: Evaluating algorithm applicability across diverse building morphologies. *Sustainable Cities and Society*, *105*, 105280. DOI: 10.1016/j.scs.2024.105280

Tejedor, B., Casals, M., Gangolells, M., & Roca, X. (2017). Quantitative internal infrared thermography for determining in-situ thermal behaviour of façades. *Energy and Building*, *151*, 187–197. DOI: 10.1016/j.enbuild.2017.06.040

Temme // Obermeier GmbH. (n.d.). PVC/PES - PVC coated polyester fabrics - the all-rounder. Retrieved November 1, 2025, from https://www.to-experts.com/en/membranes/pvcpes/

Toader, T. P., Dico, C., & Mircea, A. C. (2024). Thermal insulation of buildings using innovative materials based on nanoparticles. *European Journal of Materials Science and Engineering*, *9*(1), 1–10. https://www.ejmse.ro. DOI: 10.36868/ejmse.2024.09.01.063

Tokbolat, S., Naizabekov, Y., & Mariani, S. (2020). The impacts of different façade types on energy use in residential buildings. E3S Web of Conferences, 172, 24010. DOI: 10.1051/e3sconf/202017224010

Trencher, G., Broto, V. C., Takagi, T., Sprigings, Z., Nishida, Y., & Yarime, M. (2016). Innovative policy practices to advance building energy efficiency and retrofitting: Approaches, impacts and challenges in ten C40 cities. *Environmental Science & Policy*, *66*, 353–365. DOI: 10.1016/j.envsci.2016.06.021

Trubina, N., Lupíšek, A., Pešta, J., & Vlasatá, B. (2024). Demountable structures: Evaluation of a circular design approach through life cycle assessment. *IOP Conference Series. Earth and Environmental Science*, *1402*(1), 012040. DOI: 10.1088/1755-1315/1402/1/012040

U.S. Green Building Council. (2020). *LEED v4.1 Building Design and Construction.* https://www.usgbc.org/leed/v41

U.S. Green Building Council. (2023). *LEED v4.1 building design and construction.* USGBC.

Uhlemann, J. (2016). Elastic constants of architectural fabrics for design purposes (Doctoral dissertation, Duisburg, Essen, Universität Duisburg-Essen, Diss., 2016).

Ujma, A., & Pomada, M. (2019). Analysis of the temperature distribution in the place of fixing the ventilated facade. *E3S Web of Conferences, 97*, 01041. https://doi.org/DOI: 10.1051/e3sconf/20199701041

Ulrich, R. S., Simons, R. F., Losito, B. D., Fiorito, E., Miles, M. A., & Zelson, M. (1991). Stress recovery during exposure to natural and urban environments. *Journal of Environmental Psychology*, *11*(3), 201–230. DOI: 10.1016/S0272-4944(05)80184-7

United Nations Framework Convention on Climate Change. (2015). *Paris Agreement.* https://unfccc.int/process-and-meetings/the-paris-agreement/the-paris-agreement

United Nations. (2018). *World Urbanization Prospects: The 2018 Revision.* https://population.un.org/wup

Valladares-Rendón, L. G., Schmid, G., & Lo, S. L. (2017). Review on energy savings by solar control techniques and optimal building orientation for the strategic placement of façade shading systems. *Energy and Building*, *140*, 458–479. DOI: 10.1016/j.enbuild.2016.12.073

Van Dongen, P., Britton, E., Wetzel, A., Houtman, R., Ahmed, A. M., & Ramos, S. (2022). Suntex: Weaving solar energy into building skin. *Journal of Facade Design and Engineering*, *10*(2), 141–160. DOI: 10.47982/jfde.2022.powerskin.9

Vandi, L., Calcagni, M. T., Belletti, F., Pandarese, G., Martarelli, M., Revel, G. M., Docter, V., & Pracucci, A. (2024). Structural health monitoring for prefabricated building envelope under stress tests. *Applied Sciences (Basel, Switzerland)*, *14*(8), 3260. DOI: 10.3390/app14083260

von Ahsen, A. (2008). Cost-oriented failure mode and effects analysis. *International Journal of Quality & Reliability Management*, *25*(5), 466–476. DOI: 10.1108/02656710810873871

Wang, M., Wang, C. C., Cai, H. Q., Li, Y. Y., Zhang, Q. W., & Yi, H. B. (2019). Molecular dynamics simulation study on distinctive hydration characteristics of highly coordinated calcium chloride complexes. *Journal of Molecular Liquids*, *274*, 261–269. DOI: 10.1016/j.molliq.2018.10.136

Wang, T. H., Huang, Y., & Park, J. (2022). Development of daylight glare analysis method using an integrated parametric modelling approach: A comparative study of glare evaluation standards. *Buildings (Basel, Switzerland)*, *12*(11), 1810. DOI: 10.3390/buildings12111810

Wang, T., Ward, G. J., & Lee, E. S. (2018). Efficient modeling of optically-complex, non-coplanar exterior shading: Validation of matrix algebraic methods. *Energy and Building*, *174*, 464–483. DOI: 10.1016/j.enbuild.2018.06.022

Wang, X. Q., & Mujumdar, A. S. (2007). Heat transfer characteristics of nanofluids: A review. *International Journal of Thermal Sciences*, *46*(1), 1–19. DOI: 10.1016/j.ijthermalsci.2006.06.010

Wang, X., Li, W., Luo, Z., Wang, K., & Shah, S. P. (2022). A critical review on phase change materials (PCM) for sustainable and energy efficient building: Design, characteristic, performance and application. *Energy and Building*, *256*, 111783. DOI: 10.1016/j.enbuild.2022.111923

Wei, G., Wan, F., Zhou, W., Xu, C., Ye, Z., Liu, W., Lei, G., & Xu, L. (2023). BFD-YOLO: A YOLOv7-based detection method for building façade defects. *Electronics (Basel)*, *12*(17), 3612. DOI: 10.3390/electronics12173612

Widomski, P., Maksymowicz, P., Trzaska, O., Mayer, P., Kaczyński, P., Berbesz-Wyrodek, A., & Kogut, M. S. (2025). The impact of destructive climatic factors on the mechanical and performance properties of structural materials. *Materials (Basel), 18*(13), 2970. DOI: 10.3390/ma18132970 PMID: 40649459

Winter, F. (2022). 3.3 Glass fibers. *From Construction Materials to Technical Gases, 55, 316.*

Wójcik, R., & Bomberg, M. (2016). On interior rehabilitation of buildings with historic facades. *Journal of Building Physics, 40*(2), 144–161. DOI: 10.1177/1744259116662280

Wole-Osho, I., Okonkwo, E. C., Abbasoglu, S., & Kavaz, D. (2020). Nanofluids in solar thermal collectors: Review and limitations. *International Journal of Thermophysics, 41*(11), 157. DOI: 10.1007/s10765-020-02737-1

Wu, C., Hassanein, M. F., Deng, H., Zhang, Y.-M., & Shao, Y.-B. (2020). Shear buckling response of S690 steel plate girders with corrugated webs. *Thin-walled Structures, 157*, 107015. DOI: 10.1016/j.tws.2020.107015

Wu, P., Wang, X., & Wang, J. (2016). A critical review of the use of 3D printing in construction. *Automation in Construction, 68*, 21–31. DOI: 10.1016/j.autcon.2016.04.005

Wu, Y., & Flemmer, C. L. (2020). Glass curtain wall technology and sustainability in commercial buildings in Auckland, New Zealand. *International Journal of Built Environment and Sustainability, 7*(2), 57–65. DOI: 10.11113/ijbes.v7.n2.495

Xie, J., & Sawyer, A. O. (2021). Simulation-assisted data-driven method for glare control with automated shading systems in office buildings. *Building and Environment, 196*, 107801. DOI: 10.1016/j.buildenv.2021.107801

Xu, C., Xi, Q., & Ma, Q. (2021). Technical characteristics and energy performance of double skin faces: A review. *E3S Web of Conferences, 248*, 01029. https://doi.org/DOI: 10.1051/e3sconf/202124801029

Xuan, Y., & Li, Q. (2000). Heat transfer enhancement of nanofluids. *International Journal of Heat and Fluid Flow, 21*(1), 58–64. DOI: 10.1016/S0142-727X(99)00067-3

Yadav, D., & Sanserwal, M. (2023). Effects of factors on thermal conductivity and rheology of CNT nanofluids. *Journal of Thermal Analysis and Calorimetry, 148*(5), 1723–1763. DOI: 10.1007/s10973-022-11821-7

Yang, J., Yang, X., Wang, J., Chin, H. H., & Sundén, B. (2022). Review on thermal performance of nanofluids with/without magnetic fields. *Frontiers in Energy Research, 10*, 822776. DOI: 10.3389/fenrg.2022.822776

Yang, L., Liu, K., Ou, R., Qian, P., Wu, Y., Tian, Z., Zhu, C., Feng, S., & Yang, F. (2024). Surface defect-extended BIM generation leveraging UAV images and deep learning. *Sensors (Basel), 24*(13), 4151. DOI: 10.3390/s24134151 PMID: 39000929

Yang, M., Zhang, Z., Yuan, J., Wu, L., Li, P., Zhao, X., & Men, X. (2020). Enhanced mechanical and tribological properties of Kevlar/PTFE-phenolic composites by improving interfacial properties by aramid nanofibers. *Polymer Composites, 41*(10), 4192–4201. DOI: 10.1002/pc.25704

Yang, X., Liu, S., Zou, Y., Ji, W., Zhang, Q., Ahmed, A., Han, X., Shen, Y., & Zhang, S. (2022). Energy-saving potential prediction models for large-scale building: A state-of-the-art review. *Renewable & Sustainable Energy Reviews, 156*, 111992. DOI: 10.1016/j.rser.2021.111992

Yan, Y., Chen, W., Hu, J., & Tariq, I. (2025). Long-term testing study of tensile ETFE, PTFE and PVDF membrane structure integrated thin Si-PV. *Solar Energy, 289*, 113307. DOI: 10.1016/j.solener.2025.113307

Yasmin, H., Giwa, S. O., Noor, S., & Sharifpur, M. (2023a). Thermal conductivity enhancement of metal oxide nanofluids: A critical review. *Nanomaterials (Basel, Switzerland), 13*(3), 597. DOI: 10.3390/nano13030597 PMID: 36770558

Yasmin, H., Giwa, S. O., Noor, S., & Sharifpur, M. (2023b). Experimental exploration of hybrid nanofluids for solar and thermal energy storage. *Nanomaterials (Basel, Switzerland), 13*(2), 278. DOI: 10.3390/nano13020278 PMID: 36678031

Yazdi, M., Mohammadpour, J., Li, H., Huang, H.-Z., Zarei, E., Pirbalouti, R. G., & Adumene, S. (2023). Fault tree analysis improvements: A bibliometric analysis and literature review. *Quality and Reliability Engineering International, 39*(5), 1639–1659. DOI: 10.1002/qre.3271

Yuan, Z., & Jia, Y. (2021). Mechanical properties and microstructure of glass fiber and polypropylene fiber reinforced concrete: An experimental study. *Construction & Building Materials, 266*, 121048. DOI: 10.1016/j.conbuildmat.2020.121048

Yucer, C. T., & Hepbasli, A. (2010). Thermodynamic analysis of a building using energy analysis method. *Energy and Building, 43*(2–3), 536–542. DOI: 10.1016/j.enbuild.2010.10.019

Yu, J., Chang, W.-S., & Dong, Y. (2022). Building energy prediction models and related uncertainties: A review. *Buildings (Basel, Switzerland), 12*(8), 1284. DOI: 10.3390/buildings12081284

Zagubień, A., Wolniewicz, K., & Wesolowski, M. (2025). Overview of textile tensile surface structures with the largest uniform area. *Textile Research Journal, 95*(7-8), 879–892. DOI: 10.1177/00405175241268836

Zemitis, J., Borodinecs, A., Geikins, A., Kalamees, T., & Kuusk, K. (2016). Ventilation system design in three European geo cluster. *Energy Procedia, 96*, 285–293. DOI: 10.1016/j.egypro.2016.09.151

Zhang, C., Wang, F., Zou, Y., Dimyadi, J., Guo, B. H. W., & Hou, L. (2023). Automated UAV image-to-BIM registration for building facade inspection using improved generalised Hough transform. *Automation in Construction, 153*, 104957. DOI: 10.1016/j.autcon.2023.104957

Zhang, S., Zhao, Y., Ding, H., Qiu, J., & Guo, Z. (2021). Recycling flue gas desulfurization gypsum and phosphogypsum for cemented paste backfill and its acid resistance. *Construction & Building Materials, 275*, 122170. DOI: 10.1016/j.conbuildmat.2020.122170

Zhang, W., Wu, Y., & Calautit, J. K. (2022). A review on occupancy prediction through machine learning for enhancing energy efficiency, air quality and thermal comfort in the built environment. *Renewable & Sustainable Energy Reviews, 167*, 112716. DOI: 10.1016/j.rser.2022.112704

Zhang, X., Gu, H., & Fujii, M. (2006). Effective thermal conductivity and diffusivity of nanofluids. *Journal of Applied Physics, 100*(4). Advance online publication. DOI: 10.1063/1.2259789

Zhang, Y., Xu, S., Xue, J., & Zhang, Q. (2017). Anisotropic mechanical properties and constitutive relations of PTFE coated glass fibers. *Composite Structures, 179*, 601–616. DOI: 10.1016/j.compstruct.2017.07.076

Zhao, J., & Ma, P. (2022). Comparative Study on the tribological properties of PTFE coated Kevlar fabric and hybrid PTFE/Kevlar fabric. *Fibers and Polymers, 23*(4), 1111–1118. DOI: 10.1007/s12221-022-4421-8

Zheng, L., Mueller, M., Luo, C., & Yan, X. (2021). A review of data-driven approaches for occupant's behaviour in building energy conservation. *Energy Procedia*. Advance online publication. DOI: 10.46855/energy-proceedings-7460

Zheng, L., Ye, Y., Zhuang, J., & Zheng, Y. (2020). Impact tensile behaviors of PVDF building coated fabrics. *Advances in Civil Engineering*, *2020*(1), 1620760. DOI: 10.1155/2020/1620760

Zhou, D., Zhao, C. Y., & Tian, Y. (2012). Review on thermal energy storage with phase change materials (PCMs) in building applications. *Applied Energy*, *92*, 593–605. DOI: 10.1016/j.apenergy.2011.08.025

Zwanzig, S. D., Lian, Y., & Brehob, E. G. (2013). Numerical simulation of phase change material composite wallboard in a multi-layered building envelope. *Energy Conversion and Management*, *69*, 27–40. DOI: 10.1016/j.enconman.2013.02.003

About the Contributors

Arkar Htet holds a Ph.D. in Management from Lincoln University College, Malaysia, with a dissertation focused on the relationship between leadership styles and project success in Myanmar's construction industry. He also holds a Master of Science in Civil Engineering from the same university, where his research addressed "Optimizing Glass Façade Design for Aesthetics and Energy Efficiency in Myanmar's Building Industry." His academic foundation also includes a Bachelor of Computer Technology from the University of Computer Studies, Yangon, and an MBA from SMC University, Switzerland. Dr. Htet is currently the Managing Director of Htet Myat Arkar Co., Ltd., Academic Dean of Mullion Façade Engineering School, and Program Director of R&T Academy. He is also a Postdoctoral Fellow at Eudoxia Research University, Newcastle, where his research focuses on adaptive façade systems for climate resilience in urban buildings. With over a decade of experience in façade engineering, project management, and sustainable building systems, Dr. Htet has made notable contributions to both academic and industry practice across Southeast Asia. He holds multiple professional qualifications, including a Diploma in Construction Project Management, Executive Diploma in Data Analytics, and certification in Façade Engineering from ASTI Dubai. Dr. Htet has authored and co-authored over 40 peer-reviewed articles and book chapters published in international journals and edited volumes. His key research areas include green façade innovation, building information modeling (BIM) integration, AI-driven risk prediction, and lifecycle performance of façade materials. He is committed to bridging engineering principles with digital innovation to drive resilient and energy-efficient building envelopes in the era of climate change.

Theingi Aung holds a Ph.D. in Management from Lincoln University College, Malaysia, with a doctoral thesis titled "The Effect of Risk Management Practices on Project Success of the Construction Industry in Myanmar." She currently serves as the Director of Htet Myat Arkar Co., Ltd., Research and Development Director at

Mullion Facade Engineering School, and Founder and CEO of R&T Academy. Her academic background includes a Master of Business Administration from UCSI University, Malaysia, with a dissertation on consumer behavior in online shopping, and a Bachelor of Arts from the University of Distance Education, Yangon. In addition, she holds multiple professional diplomas in accounting, financial analysis, risk management, and business management. Dr. Aung is an active researcher and has authored more than 30 scholarly articles and book chapters across peer-reviewed journals and international conferences. Her work spans areas such as façade engineering, risk management, consumer behavior, green building materials, AI applications in construction, and sustainability practices. She continues to contribute to interdisciplinary research and is currently pursuing a Post Doctoral Fellow program at Eudoxia Research University, Newcastle.

***

Om Prakash Giri is an Assistant Professor at Pokhara University, Pokhara, Gandaki, Nepal. He holds a Ph.D. in Engineering from Lincoln University College. His academic and research interests include engineering, construction management, sustainable development, and applied research in the built environment. He has authored and co-authored more than 20 scholarly articles and book chapters, contributing to academic literature in engineering and related interdisciplinary fields. Dr. Giri has been actively involved in teaching, research supervision, curriculum development, and scholarly publication. Through his academic work, he supports the advancement of engineering education and research practice in Nepal and beyond. His contributions reflect a strong commitment to academic excellence, research collaboration, and the practical application of engineering knowledge.

Sriparna Guha is an Assistant Professor in the Department of Management at Amity University (Kolkata). She has completed her PhD in Management from Brainware University, Kolkata, India. She has done her post graduation in the field of economics from Rabindra Bharati University and has completed Post Graduate Diploma in Applied Economics and Information Management from Globsyn Business School, Kolkata, and PGDM in Business Analytics from AIMA, India. She has authored many research papers in national and international journal of repute. She has also attended different national and international workshops and conferences.(ORCID: 0000-0003-3092-3722).

Mukesh Gulani is associated with Datta Meghe Institute of Management Studies, where he contributes to the academic environment through his involvement in management education. He is committed to fostering a strong foundation in business and management principles among students, emphasizing both theoretical understanding and practical application. Through his teaching and academic engagement, he supports the development of critical thinking, leadership skills, and professional growth. Mr. Gulani is known for his dedication to student learning and his active participation in institutional and scholarly activities within the field of management studies.

Imran Zahoor Khan is an accomplished academician currently serving as an Assistant Professor at Datta Meghe Institute of Management Studies. He is dedicated to the field of management education, contributing to both teaching and academic development. With a strong commitment to fostering analytical thinking and professional skills among students, he actively engages in delivering lectures, mentoring learners, and supporting institutional initiatives. His academic interests are centered on contemporary management practices, and he strives to integrate theoretical knowledge with practical applications. Known for his student-centric approach and collaborative mindset, Mr. Khan continues to contribute to the academic community through his teaching excellence and involvement in scholarly activities.

Leevesh Kumar is an Associate Professor Civil Engineering Dr. Leevesh Kumar is a distinguished academician and researcher with over 13 years of progressive experience in the field of civil and construction engineering across India, Ethiopia, the United Kingdom, and Bangladesh. Currently serving as an associate professor at the World University of Bangladesh and holding a concurrent part-time role with Liverpool John Moores University and the University of East London (UK), Dr. Kumar has built a diverse and global portfolio in higher education and construction technology research. He holds a Ph.D. in Civil Engineering from Career Point University (India), preceded by both M.Tech and B.Tech degrees in Civil Engineering from Bhagwant University. His educational foundation is complemented by a postgraduate diploma and several certifications from prestigious institutions like IIT Madras and NITTTR Chandigarh. Throughout his academic journey, Dr. Kumar has remained committed to advancing sustainable practices in the construction industry. Dr. Kumar's research focuses on sustainable construction materials, with specific interest in the mechanical and durability

Sui Reng Liana is an Associate Professor at Lincoln University College. He holds an MBA and a Doctor of Business Administration from Chungnam National University, Daejeon, Korea. His academic and professional work focuses on management, business administration, research development, and higher education leadership. He is the Founder of International Research and Management College,

Myanmar, where he contributes to academic capacity building, professional education, and research-oriented learning. He is also one of the co-founders of Into-Vertex Research and Solutions Co., Ltd., supporting research, advisory, and management-related professional services. With his experience in both academic and institutional development, Dr. Liana continues to contribute to management education, research collaboration, and the advancement of professional learning in Myanmar and the wider region.

Anirban Mandal is currently working as a faculty member in ICFAI Business School, Kolkata. He has worked as an Associate Professor and Head of the Department in the School of Management and Commerce at Brainware University. He has done his PhD from KIIT University, Bhubaneswar, India. He has more than 16 years of teaching experiences in different renowned colleges and institutions of West Bengal and also a visiting faculty in Pelita Harapan University, Jakarta, Indonesia. He has authored many research papers in national and international journals of repute. He acted as resource person in different national and international seminars (ORCID: 0000-0002-5130-5424).

TC Manjunath was born on 6th February 1967. He did his B.E. from RV College of Engg. in 1989 (B'lore Univ), M.E. in Robotics & Automation from Govt.'s LD College of Engg., (Gujarat Univ., Ahmedabad) in 1992 with FCD-First Rank & PhD in Systems & Control Engg. from IIT Bombay in the year 2007, got best research scholar award for the entire IITB. He is currently working as the Dean Research (R & D) and as Professor in Dept. of Computer Science & Engg. with specialization in IoT, Cyber Security, Blockchain Technology in Rajarajeswari College of Engineering, Bengaluru, Karnataka. He has got a teaching (academic), research & administrative experience of more than 35+ years (10 years as Principal & 8 years as HoD, 1 year as Dean Research & rest as faculty) and has published more than 1500+ research papers in various national & international journals & conferences, with 175+ patents, 100+ book-chapters, 25 awards, 100s of Keynote talks and also published 25 text books in various field of science, engineering & technology, completed projects worth Rs. 9 Crores till date & guided 10 PhD scholars in VTU. He is a member of 30 Professional Society Chapters and is a fellow of IE, IETE, AMIEE, Sr. Member of IEEE and a chartered engineer.

Moe Min Min is a Co-founder of Mullion Façade Engineering School, Myanmar. He contributes to professional training in façade engineering, with a focus on the applications and methods of façade work. His teaching and professional practice emphasize practical installation methods, site procedures, façade system applications, and workmanship requirements in building façade projects. Through

Mullion Façade Engineering School, he supports the development of industry-relevant skills for learners, technicians, and construction professionals. Mr. Moe Min Min plays an important role in connecting classroom-based learning with real construction practice, particularly in the areas of façade implementation, material handling, and project-site application. His contribution supports the advancement of practical façade education and the strengthening of technical capacity within Myanmar's façade construction industry.

Yin Hlaing Min holds a doctoral degree from Colorado Technical University. He is the Founder of the National Association of Sales Professionals, Myanmar, and Co-founder of Into-Vertex Research and Solutions Co., Ltd. His professional interests include sales management, business development, professional training, organizational growth, and applied management practice. Through his leadership roles, he has contributed to the development of professional networks and capacity-building initiatives for sales and business professionals in Myanmar. His work bridges academic knowledge and practical business application, particularly in the areas of sales professionalism, market development, and management training. Dr. Yin Hlaing Min continues to support professional education, research-based advisory services, and the development of business competencies for individuals and organizations.

Santosh Kumar Nathsharma is an accomplished academician serving in the Department of Chemistry at Stewart Science College, located in Cuttack, Odisha (753001), India. He is dedicated to teaching and research in the field of chemistry, contributing to the academic growth of undergraduate and postgraduate students. With a strong foundation in chemical sciences, Dr. Nathsharma has been actively involved in delivering quality education, guiding student projects, and fostering scientific curiosity. His academic interests may include core and applied areas of chemistry, and he plays a vital role in promoting research-oriented learning within the institution. Through his commitment to education and scholarly engagement, Dr. Nathsharma continues to contribute to the development of the academic community in Odisha.

Selvakumar P is currently an Associate Professor in the Department of Science and Humanities at Nehru Institute of Technology, Coimbatore, Tamil Nadu, India. Prior to joining Nehru Institute of Technology, he held positions as an Assistant Professor in the Department of Science and Humanities various engineering colleges in Coimbatore, Tamil Nadu, India, spanning from 2011 to 2023. Dr.P. Selvakumar's academic and research interests lie primarily in the field of phytochemistry. He has made significant contributions to his field, with an impressive record that includes

30 journal publications, 12 patents, and 20 conference papers. Additionally, he has organized 5 seminars and conferences, further demonstrating his commitment to academic and scientific exchange. Throughout his career, Dr. P.Selvakumar has been recognized for his dedication and excellence in teaching, having been honored with several awards, including the Best Teacher Awards. He holds a prominent position in the academic community, contributing actively to research.

Sandeep Poddar is the Deputy Vice Chancellor for Research and Innovation at Lincoln University College, Malaysia. He graduated with Honours in Zoology from the University of Calcutta and obtained a Postgraduate Diploma in Dietetics from the All India Institute of Hygiene and Public Health. He earned an MSc in Zoology, specializing in Biochemical Genetics, from Dayalbagh Educational Institute with distinction, and an MBA from Lincoln University College. He completed his Ph.D. in Zoology at Vivekananda Institute of Medical Sciences, focusing on cytotoxicity, and later pursued postdoctoral research on hemoglobinopathies and oral cancer mutation. Dr. Poddar has published more than 140 research papers in reputed indexed journals and books, filed more than 15 patents, organized international conferences, and edited academic books in Malaysia, Australia, and India. He also serves as guest editor for reputed Scopus-indexed journals.

Rukmathan R is an Assistant Professor of Management Studies at Dr. N.G.P. Arts and Science College, Coimbatore. With a career spanning over 12.5 years in academia, he specializes in the dual domains of Human Resource Management and Marketing. He holds an MBA and a Ph.D., and his scholarly work frequently examines professional performance, notably his study on emotional intelligence within the nursing sector. He is committed to bridging the gap between theoretical management frameworks and real-world industrial applications.

Ninad Rithe is a dedicated management student currently pursuing his Master of Business Administration (MBA) in the Department of MBA at Datta Meghe Institute of Management Studies, Nagpur, Maharashtra, India. He demonstrates a strong interest in Human Resource, leadership development and collaborative teamwork, along with a keen focus on effective communication, analytical thinking, and problem-solving. With a commitment to continuous learning and professional excellence, Ninad aims to apply academic knowledge to real-world business challenges while contributing meaningfully to organizational growth and team success.

Ansuman Samal is currently working as a Professor in the Faculty of Hospitality & Tourism Management at Siksha O Anusandhan Deemed to be University, Bhubaneswar, Odisha. He has more than 18 years of experience both in the field of marketing research and in academics. He has completed his graduation in Information Technology from Manipal Academy of Higher Education, MBA from KIIT Deemed to be University and PhD from Siksha O Anusandhan Deemed to be University. As an acclaimed scholar, he has received the Chancellor's Gold Medal from KIIT Deemed to be University for securing highest mark during the MBA programme. Before joining Siksha O Anusandhan, he was working as a senior researcher at IMRS Advisory Pvt. Ltd. In Bengaluru. As an academician, he has published more than 80 numbers of research articles in various National and International journals of repute. He has also authored 5 Books and has participated in more than 100 numbers of National and International Conferences, workshops, seminars / webinars and presented his research works. He has successfully guided 2 numbers of PhD students and another 4 students are currently continuing their research work under his guidance. His research interests include services marketing, hospitality & tourism management, consumer behavior, quality management, and strategic management.

Zakir Shaikh is an Assistant Professor, Department of Accounting and Finance, School of Business Administration, Kingdom University, Bahrain. Prior to this, he worked more than 28 years in commercial and academic industry. He has published numerous articles in referred journals and presented many papers in various conferences, both local and abroad in the different area of Accounting, management and Islamic Banking and Finance. He has also participated in a variety of seminars, forums, workshops and international conferences. He obtained his Ph.D. (Islamic Banking and Finance) from India in 2018. His doctoral thesis explored the entrepreneurial phenomenon from an Islamic perspective and argued for profit and loss sharing (PLS) contracts as viable alternatives to conventional interest-based financing instruments. He holds Master degree in Commerce, Business Administration & Finance and Control. He is a member of AAA, Chartered Institute of Islamic Finance Professionals (CIIF), Malaysia; Indian Accounting Association; Orissa Commerce Association, India and associated with Accounting & Auditing Organization for Islamic Financial Institution (AAOIFI), Bahrain

Vijay Uprikar is a passionate Faculty of Marketing Management at Datta Meghe Institute of Management Studies, Nagpur. With over 20 years of blended experience in academics and pharmaceutical industry, he demonstrates a unique blend of corporate insights and academic expertise to the world of marketing education. A firm believer in innovative pedagogy, have developed study materials that hold six copyrights, designed to enhance conceptual clarity in marketing management. His interdisciplinary research explores Mahatma Jotirao Phule's contributions to Entrepreneurial Social Responsibility (ESR), integrating historical and social perspectives into contemporary management frameworks. With a strong foundation in management education, Dr. Uprikar has contributed significantly to the academic and professional development of students through his expertise and dedication. His work focuses on fostering innovation, leadership, and excellence in the field of management. A respected figure in the academic community, Dr. Uprikar continues to play a vital role in shaping future business leaders.

Vandana V.C. is a distinguished academician and commerce educator serving in the Department of Commerce and Management at Government First Grade College, Malur, located behind the Railway Station, Kolar, Karnataka, India. With strong academic credentials including M.Com. and Ph.D., she has built a respected career marked by dedication to teaching, research, and academic authorship. With over 15 years of teaching and academic experience, Dr. Vandana V.C. has developed deep expertise in Human Resource Management, Business Management, and Finance. She is widely appreciated for her learner-centered teaching approach, where she focuses on making complex commerce and management concepts clear, practical, and application-oriented. Her classes emphasize conceptual clarity, real-world business examples, and interactive learning methods that enhance student understanding and engagement. Dr. Vandana V.C. has made significant scholarly contributions through the authorship of multiple textbooks in commerce and management subjects, supporting both undergraduate and postgraduate learners. In addition, she has actively contributed to the academic community by presenting papers at conferences and publishing research articles in peer-reviewed journals, reflecting her ongoing commitment to research and knowledge dissemination. Beyond classroom teaching, she contributes to curriculum development, student mentoring, and institutional academic initiatives. Her work consistently supports academic excellence, professional readiness, and ethical understanding in commerce education. Through her sustained contributions in teaching, writing, and research, Dr. Vandana V.C. continues to play an important role in strengthening commerce and management education at the collegiate level.

Nay Win is the Program Director of Mullion Façade Engineering School, Myanmar. He specializes in façade engineering, with particular expertise in structural calculation and the design of building façade systems. Through his role at Mullion Façade Engineering School, he provides professional training for learners and practitioners seeking technical knowledge in façade design, structural performance, and practical engineering applications. His teaching focuses on helping students understand the calculation principles, load considerations, and design requirements involved in modern façade systems. Mr. Nay Win contributes to professional education by linking technical theory with real project conditions, supporting the development of skilled façade professionals in Myanmar and southeast Asia. His work reflects a strong commitment to capacity building, practical training, and technical advancement in the façade industry.

Index

A

B

C

D

E

F

M

N

O

P

R

S

T

www.ingramcontent.com/pod-product-compliance
Lightning Source LLC
LaVergne TN
LVHW081536300926
849161LV00006B/1453
9798337360232